Theoretical Principles
in Astrophysics and Relativity

Theoretical Principles
in Astrophysics and Relativity

Edited by Norman R. Lebovitz
William H. Reid
Peter O. Vandervoort

The University of Chicago Press
Chicago and London

The University of Chicago Press, Chicago 60637
The University of Chicago Press, Ltd., London

88 87 86 85 84 83 82 81   1 2 3 4 5

Library of Congress Cataloging in Publication Data

Main entry under title:

Theoretical principles in astrophysics and
  relativity.

  Papers presented at a symposium on theoretical
principles in astrophysics and relativity, held at
the University of Chicago, May 27–29, 1975,
in honor of S. Chandrasekhar.
  Includes index.
  1. Stars—Congresses. 2. Astrophysics—
Congresses. 3. Relativity—Congresses.
I. Lebovitz, Norman. II. Reid, William Hill,
1926–   III. Vandervoort, Peter O.
QB799.T48   523.01   76-25636
ISBN 0-226-46989-1 (cloth)
        0-226-46990-5 (paper)

Contents

Photograph by Luis Medina and Jose Lopez.

Preface

On May 27–29, 1975, a Symposium on Theoretical Principles in Astrophysics and Relativity was held at the University of Chicago in honor of S. Chandrasekhar, in his 65th year. This book, the proceedings of that symposium, is likewise dedicated to Chandra. We know that his many friends throughout the world join with us in this expression of our admiration and affection.

Chandra's research has ranged wide. It is not possible in a volume of this size to cover all the areas in which he has made important contributions. We have tried to select those areas that reflect the increasing influence that the directions of research in astrophysics and in relativity have on one another. We hope and believe that this book will be useful to those interested in research in either or both of these fields.

It is a pleasure to thank the speakers for participating in the symposium and for preparing their articles for publication. We also thank the National Science Foundation for its financial support of the symposium and the University of Chicago Press for its assistance with the publication of this volume.

1   The Study of   Martin
Stellar Structure   Schwarzschild

Introduction

In a normal scientific colloquium it would be my assignment, under the general topic given me, to present to you a review of selected significant recent achievements in the field of stellar structure and evolution. The special character of this symposium, however, gives me, I believe—and I hope you agree—the permission and rare opportunity to review with you not the scientific topic of stellar structure but rather that human endeavor we call the study of stellar structure. Since I have no qualifications as a science historian, I would not want to concentrate on the actual history of this scientific endeavor, with its attendant difficult task of proper references and correct assignment of credit. Instead I would like to concentrate on the physical principles and processes as they have been introduced one by one into the study of stellar structure during the past 100 years, leading up to the present fascinatingly unfinished state.

I will restrict myself to those processes dominating the simplest stars in their various evolution phases. Accordingly I shall omit topics like instabilities (such as pulsations or thermal flashes), rotation and the related topic of close binaries, and magnetic fields. I am truly aware that these difficult topics are vital and in many cases decisive parts of the theory of the stellar interior. I feel justified in omitting them only because nearly all of them will be covered separately by subsequent speakers.

Hydrostatic Equilibrium

The apparent unchangeability of the majority of stars convinced researchers early that in general stars must be in strict hydrostatic equilibrium, i.e., that the gravitational force must be balanced by a pressure force everywhere within a star. The study of stellar models in hydrostatic equilibrium was well under way 100 years ago, and its classical period may be considered as essentially concluded when Emden wrote his summarizing book on this topic in 1907.

This first phase of stellar model making seems to me remarkable and instructive by the following circumstance. All investigators active in this development were fully aware that the application of hydrostatic equilibrium was by itself not sufficient to achieve unique stellar models. A pressure-density relation had

Martin Schwarzschild is at the Princeton University Observatory, Princeton, New Jersey.

to be more or less arbitrarily assumed. Much as is now the habit for formally representing an as yet unknown relation, a simple power law was used to represent the missing pressure-density relation, and the researchers of that epoch judged it worthwhile to study the thus defined polytropic star models in greatest detail. The value of such extensive theoretical work, based in part on pure assumption, is frequently questioned. The case we are here discussing is an example where such work has proved thoroughly valuable: the next phase of the study of stellar structure could surely not have proceeded as remarkably rapidly as it did if it had not been for the availability of extensive tabulations and full analysis of polytropic stellar models.

The story of the study of stellar models with hydrostatic equilibrium as the only equilibrium condition has two additional chapters, which occurred historically far later than the classical first chapter. In 1930 Chandrasekhar derived the pressure-density relation which holds in that density range in which electron degeneracy dominates the equation of state, and then he immediately proceeded to compute the corresponding stellar models. By this development he laid the foundation for the theory of white dwarfs, including the Chandrasekhar limit, i.e., the maximum mass possible for such stars. In contrast to the classical development of polytropic models, Chandrasekhar's development of white dwarf models contains no arbitrariness in the choice of the pressure-density relation—notwithstanding Eddington's violent and amazingly persistent objection to this relation, an objection which by now appears to have been based more on faith than on physics.

The third chapter in this study was opened by Oppenheimer and Volkoff in 1939 when they, for the first time, constructed models for neutron stars with a pressure-density relation dominated by ion degeneracy. This relation had to be modified for the effects of nuclear forces, effects even now not definitively known. Accordingly, this chapter of stellar structure, including the astronomically important determination of the upper limit for the masses of neutron stars, is still unfinished. This last chapter differs from the preceding ones in one fascinating aspect. The first neutron star models were derived as purely theoretical constructs without any observed astronomical objects known at that time to which they might apply. It took some 30 years after this initial theoretical construct for it to find its proper companion in the real world by the discovery of pulsars.

Energy Transport

Even though the fundamental importance of the role which energy transport mechanisms must play in the stellar interior was realized very early, I am not aware of any major efforts to determine such processes during the phase of polytrope building, though it appears that very general considerations of convective transport played a role in the choice of the pressure-density relations used during that phase.

In 1906, one year before Emden published his definitive book on polytropes, Schwarzschild showed that radiative energy transport was likely to be an important process for stars. But it was Eddington who, starting in 1916, solidly implanted radiative transport into the theory of the stellar interior. Even with this transport mechanism clearly defined, the problem of deriving a stellar model still did not permit a unique solution as long as the nuclear sources were unknown. Nevertheless, Eddington achieved the insight that the structure of a stellar model would not be greatly affected by the precise distribution of the nuclear sources within the star over a reasonable range of such distributions, a range we now know to be adequate for main-sequence stars but not for later evolution phases. The most striking result of this development by Eddington was the derivation of the theoretical mass-luminosity law which could be brought into fine agreement with the observed mass-luminosity relation for main-sequence stars when sufficient hydrogen was admitted as a major constituent of the stellar interior. Eddington curiously delayed the acceptance of sufficient hydrogen; when he finally took this step, it was with reluctance since it spoiled the balance between gas and radiation pressure for a star of average mass, a balance which appears to have had a high philosophical attraction for him. His book on the stellar interior, written in 1926, gives a lucid account of this phase of the study of stellar structure. We now know that Eddington's "standard model" is an amazingly good representation of most main-sequence stars, though we also know that it seriously fails to represent the main features of red giants.

After this period in which radiative energy transport ruled supreme, the role of convective transport in the stellar interior was reconsidered by Biermann and Cowling in the early 1930s. These investigations were based on Prandtl's mixing-length approach, the only approach for approximating turbulent convection then available. They showed convincingly that convective cores as well as deep convective envelopes might be expected in stellar models, depending on the specific character of the nuclear energy sources. These early investigations in which the superadiabatic gradient was generally neglected were followed up in due course by investigations, particularly on deep convective envelopes, in which the superadiabatic gradient was found in many cases to be far from ignorable and was computed by a detailed application of the mixing-length approach. This procedure was clearly dangerous but, I believe, unavoidable.

To complete the list of presently known energy transport mechanisms within stars, I should add the discovery that ordinary conduction is highly effective wherever the density is sufficiently high for electron degeneracy to affect the equation of state.

One more development properly belongs at this point of the story of the study of stellar structure, though actually it was not until 1962 that Hayashi accomplished it in a decisive manner. For a star of a given mass and composition which is in hydrostatic equilibrium and exploits the energy transport mechanisms just discussed, one can derive a limit, the Hayashi limit, which can be represented in the Hertzsprung-Russell diagram by a more or less vertical line. The significance of this line is that no models for the star can exist—under the conditions stated—to the right of the Hayashi limit, ie.., at lower effective temperatures, whatever the character of the energy sources of the star might be. The existence of the Hayashi limit tells us that the life of a star, from its hydrostatic pre–main-sequence contraction to its dying phase, is restricted in terms of the Hertzsprung-Russell diagram to run its course between the main sequence and the Hayashi limit.

Nuclear Sources

For the study of stellar structure, 1938 turned out to be a red-letter year. It was in this year that Bethe and von Weizsäcker independently detailed the nuclear processes by which hydrogen transmutation into helium occurs within stars. It was not the concept of such transmutation that was new, since the idea that the main sources of stellar energy must be "subatomic" had been accepted throughout the preceding period. What was new was the detailed enumeration of the specific nuclear processes through which hydrogen fusion proceeds within the stellar interior and, equally importantly, the daring tentative derivation of the rates with which these processes proceed, depending on the prevailing temperature.

This fundamental addition turned the study of stellar structure from one containing a substantial degree of arbitrariness to one in which definitive models could be derived for any given star in any given state of evolution. The basic character of this great change is not diminished by the realization that the physical concepts and processes have time and again undergone—and are sure to continue to undergo in the future—alterations, both qualitatively and quantitatively.

In the same eventful year Chandrasekhar published his first book, *An Introduction to the Study of Stellar Structure*. This book is of unique character, in part because of its timing. It contains a complete exposition of the preceding periods, including a most rigorous discussion of polytropic and fully degenerate models, as well as radiative transport and models dominated by it. At the same time it contains in its last chapter, added just before publication, a first report on the new developments regarding the nuclear sources for stars, and thus it points acutely in the direction of the next period of this research field.

The introduction of explicit nuclear sources was the opening step for the study of stellar evolution. The first endeavor in this direction was the construction of inhomogeneous stars, the inhomogeneity being a direct consequence of the nuclear transmutations. This endeavor had a most peculiar early history. The first substantive step in this endeavor was taken by Öpik who early in 1938, prior to the appearance of the famous papers by Bethe and von Weizsäcker, published two papers on inhomogeneous stellar models. His insight into the probable consequences of nuclear processes—though the details of them were still unknown—was so high that he was capable of constructing models representing quite advanced phases. In fact, the second of these papers contains one model that has all the essential characteristics of modern advanced evolution models containing two burning shells. For reasons that I now believe rest less on objective considerations than on weaknesses in the human character which time and again interfere with effective cooperation, none of us then working in the field studied these papers carefully; the consequence was an unnecessary delay in the development of our field.

The second effort in the endeavor of constructing inhomogeneous models was made by Hoyle and his group, but you might say for the wrong reason. They estimated at that time that accretion of hydrogen-rich interstellar matter should play a major role in the life of a star and thus cause internal inhomogeneities. In spite of this "wrong" reason this work was widely studied and contributed much to the first insight in the structure of inhomogeneous stellar models.

On this side of the Atlantic the study of inhomogeneous stars was taken up soon after 1938, particularly by Gamow and Chandrasekhar and their colleagues. This development led promptly to a temporary impasse. The discovery of the Schönberg-Chandrasekhar limit proved that nondegenerate stars in thermal equilibrium could exhaust their hydrogen fuel only in a quite limited central fraction of their entire mass. This scientifically fascinating impasse was not resolved as fast as it might have been in consequence of a short but spirited controversy about the possible influence of degeneracy on this limit.

Thus the study of stellar evolution was launched by a set of scientific undertakings, all brilliant but also violently disconnected. When the dust from this great launching settled, the study of stellar evolution was well under way.

Gravitational Sources

The idea that gravitational contraction can provide an energy source for stars is as old as the theory of the stellar interior itself. What is left of this classical idea now that we recognize nuclear processes as the main energy sources for most stars during most of their life?

The answer to this question appears to be that gravitational energy sources are the means by which a star can transit from one long-duration evolution phase to the next one. If such a transition is violent and accordingly dynamical, the effects of gravitational energy release can be spectacular. For the moment, however, let us restrict ourselves to more calm hydrostatic transitions.

The most classical of such evolutionary transitions is the one from an interstellar cloud to a main-sequence star. The theory of this transition has had a checkered history. First, this transition was thought to consist of a rather uneventful, basically homologous contraction throughout. Next, after the discovery of the Hayashi limit the transition was believed to consist of a short and harmless collapse to the Hayashi limit, followed by a long slide down along this limit and ending with the last portion of the classical contraction path. Most recently a penetrating analysis has shown that the dynamical phase represents a much larger portion of this transition phase than had previously been anticipated, with a corresponding shortening of the slide down the Hayashi limit.

This first transition phase starting from an interstellar cloud has a special character for those stars which have such low mass as never to reach hydrogen-burning temperatures prior to becoming degenerate. Since the onset of degeneracy precludes any further heating, these featherweight stars never have the opportunity to tap any of their major nuclear fuels. Thus for featherweight stars, in stark contrast to heavier stars, the first transition phase leads directly from an interstellar cloud to an ever cooling degenerate dead state. Even though the general outline of the short life of a featherweight star seems fairly well understood, I have the impression that not as much attention has been given to this topic recently (particularly as regards cooling rates) as the potential importance of featherweight stars to the overall makeup of a typical galaxy might warrant.

It would seem irrelevant for today's purposes to enter here into a detailed account of the role of gravitational energy sources during subsequent evolutionary transition phases. The further a star progresses in its evolution, the more frequently it has to call on its gravitational resources in different portions of its interior to manage the transitions from one nuclear source to the next. This circumstance is reflected by the fact that in modern stellar evolution calculations normally the gravitational sources are taken account of right along with the nuclear sources, i.e., that thermal equilibrium in the classical sense for the stellar interior is not used as a tolerable approximation for advanced phases.

I would, however, like to come back at this point to the impasse caused by the discovery of the Schönberg-Chandrasekhar limit referred to earlier. For low-mass stars this impasse was found to be resolved along the lines suggested by Gamow, namely, the occurrence of degeneracy in the stellar core, which lifts the Schönberg-Chandrasekhar limit. This way out, however, is not available to the more massive stars since their relatively lower densities do not lead to degeneracy. Here the resolution of the dilemma consists exactly in permitting the star to use its gravitational resources, however stingily, to heat up its core by contraction to the temperatures necessary to initiate helium burning in the center and thus to start the star off on its next slow evolution phase.

Undramatic Mass Ejection

During the initial period of the study of stellar evolution there existed one school in Cambridge according to which on the average the star's mass increases during its evolution by the process of accretion, a process I have already referred to. At the same time there existed a school in Moscow who believed, on the basis of statistical observational data, that main-sequence stars, or at least upper main-sequence stars, steadily decrease in mass. Between these two diverging views it seems that we stellar model makers on this continent took the easy way out by assuming a star's mass to stay substantially constant during its life. It appears now that we have fared quite well with this prosaic assumption as far as the early evolution phases are concerned. On the other hand, observational evidence has been rapidly accumulating for substantial mass ejection from stars in advanced evolution phases and particularly in the final phases of their lives. Indeed, it seems now rather definite that we cannot follow any star to its death without properly taking account of mass loss—except presumably for the featherweight stars.

All the observational evidence for mass ejection from stars and the concurrent theoretical considerations may be divided roughly into three areas. The first area is centered around the solar wind which has recently been measured directly by space probes in fine detail. These measurements show that the rate of mass loss represented by the solar wind is so low as to be irrelevant for the Sun's evolution. The cause of the solar wind appears to be understood by a three-step mechanism. The first step consists of the emission of acoustical or hydromagnetic waves by subphotospheric turbulence. The second step consists of the deposition of the energy of these waves in the corona, thus keeping the corona hot, while the final step can be thought of as a simple evaporation of the hot corona causing a steady stream of matter away from the Sun. If we transplant this solar wind mechanism from the Sun to other stars with as proper scaling as we know how, the equivalent mass ejection rates appear to be always below the level which would have evolutionary consequences.

The second area of mass ejection refers to hot giants and supergiants. Early observational evidence in this area was found in the spectra of P Cygni stars. Much additional evidence has been obtained during the past decade by ultraviolet observations from rockets and satellites. Though the quantitative interpretation of these observations is still far from certain, present estimates based on these data strongly suggest that the indicated steady mass ejection rate is sufficiently high not to be ignorable in the theory of the late evolution phases of massive stars. Tentative theoretical investigations suggest that this substantial mass ejection is caused by radiation pressure acting in resonance lines. It is, however, still far from clear whether such radiation pressure can cause the observed mass loss unaided or requires the help of other mechanisms, such as the first two steps of the solar wind mechanism. Clearly in this second area, which is of distinct relevance to the theory of stellar evolution, we are not yet in a state to derive the expected mass loss rates directly from theoretical considerations.

The same statement can also be made for the third area, which involves mass ejection from red giants and supergiants. Early evidence of substantial mass ejection from such stars was derived from the spectra of double stars with one red supergiant component. But recent infrared investigations, in which the thermal emission from the grains embedded in the ejected matter has been measured, not only have shown the generality of substantial mass ejection from this class of stars but have also improved the estimates of the rate of ejection. Again the result is that mass ejection from red giants and supergiants, which presumably represent the late evolution phases of low- and medium-weight stars, is amply high enough to have evolutionary consequences. In this area present theoretical estimates suggest that the main driving mechanism for the ejection is radiation pressure on the grains contained in the ejected matter. But many of the detailed steps of this process—which in this case includes even the difficult problem of grain formation—seems still far from definite and at best in the very first phases of rough quantitative estimates.

Altogether then, as far as the unspectacular mass ejection is concerned we find ourselves in the following situation. No new evidence has appeared to suggest that our normal approximation of constant mass needs a revision as far as early evolution phases are concerned. In contrast, for late evolution phases the assumption of constant mass appears to be at least dangerous if not downright invalid in most cases. I feel sure that the causes as well as the consequences of substantial mass ejection from highly evolved stars will be a major subject in the study of stellar evolution for a good number of years.

Active Dynamics and Nucleosynthesis

All the developments in the study of stellar structure I have recounted this far are based on hydrostatic equilibrium. Even the processes causing quiet, steady mass ejection require only the mild modification of hydrostatic equilibrium represented by stationary flow. The overwhelming dominance of the concept of hydrostatic equilibrium over the past studies of stellar structure clearly reflects that the overwhelming majority of observed stars are in a very steady state. But observations have long told us that there also are some very significant exceptions, ranging from pulsating stars to supernovae. In spite of the rarity of these dynamical phases, they were early suspected of being of extraordinary importance, a suspicion which by now has developed into a well based conviction. It is accordingly not surprising that active dynamical evolution phases and the intimately connected fundamental topic of nucleosynthesis have become a leading part of the study of stellar structure in the most recent epoch.

Stellar pulsation has been the first of these dynamical topics to be studied in depth. It was one of Eddington's major achievements to lay the foundation for this field, largely on the basis of linear perturbation theory. But he also succeeded in delineating the possible physical processes which might energize pulsating stars and which should act dissipatively for all other stars. One of the processes discussed by Eddington was subsequently found to be indeed the driving mechanism for all major classes of pulsating stars. In 1963 Christie initiated the study of nonlinear finite-amplitude pulsations by a daring endeavor employing modern computers. This research field has blossomed in recent years into an extensive activity. The skeptics among you might well wonder: What is the justification for researchers spending effort and computer funds on trying to find out finicky details, such as the exact characteristics of a pulsating star which has a hump on its light curve one hour after its maximum, or the exact position of the demarcation line in the Hertzsprung-Russell diagram which separates stars pulsating in the fundamental mode from those pulsating in the first overtone? My answer to such skeptical questions is that today's studies of stellar pulsations are a clear example of exploiting tricky perturbed cases with the aim of finding the solutions to fundamental problems, solutions that appear not to be obtainable by the study of unperturbed cases. This procedure, after all, is a long established tool in all physical sciences. Thus modern pulsation research, though heavily involved in the theoretical reproduction of observed details, is aimed not at the understanding of these details for their own sake. Rather it is aimed at using the understanding of such details to test uncertain points in the fundamental assumptions underlying all stellar interior theory, as well as possibly deriving fundamental data such as, for example, the helium content of the oldest stellar population in our Galaxy.

Let us turn to wilder dynamic phenomena. While pulsational phases appear to leave stars essentially unaffected as far as their general evolution is concerned—with the possible exception that pulsations might enhance the quiet mass ejection rate—other dynamical phases now under study do represent strong transition phases in the evolution of stars. Starting at the birth of a star, we have earlier considered the second portion of the transition from an interstellar cloud to a main-sequence star. In contrast to this second portion which proceeds in hydrostatic equilibrium, the first portion of this transition proceeds truly dynamically because during this portion the dissociation of molecular hydrogen and the ionization of hydrogen and helium depress the ratio of specific heats below the classical critical value of $4/3$. Hence a star's life starts with a hydrodynamic collapse—however slow this collapse may be in consequence of the low densities involved. Decisive computations of this collapse phase have been carried out and have given us much improved starting points for the subsequent hydrostatic contraction, as I have mentioned earlier. However, presently available investigations are quite naturally mainly concerned with the simplest case, that of spherical symmetry, though some computations already take account of the first complication, moderate rotation. But dense interstellar clouds are obviously not likely to be orderly at all. Again it cannot be our assignment to follow through all actually occurring classes of disorderliness. But surely it will be an aim in the future to follow sufficiently disorderly cases through their initial evolution to see how close double, or multiple, stars might be formed, or even how small bound clusters of stars are formed. It will be fascinating to see whether we will in fact need the brute force tool of three-dimensional dynamic calculations to gain reliable initial insight leading toward the eventual solutions of these problems. Finally, I feel duty bound to at least just state that the formation of a planetary system clearly is part of the general topic of the early phases of a star's formation. I feel hesitant to bring this topic up at all. Even if I were young now, I would—at least I hope I would—consider this particular problem not yet ready for a deductive attack. Shouldn't we at least wait until the planetologists have solved the ever so basic question as to the major chemical composition of the planets? Indeed, negatively, the immense complexity of the solar system gives me (on occasion) nightmares regarding the specific point of our standard assumption that stars, once they arrive on the main sequence, are homogeneous in chemical composition.

The other cases of dynamical evolution phases which are under active investigation at present refer to the very final phases of a star's life. The most moderate of these cases appears to be the one in which a lightweight star ejects a planetary nebula during the last throes of nuclear burning within the star and just prior to its cooling down to a white dwarf. The present working hypothesis for the cause of this dynamical ejection can be described as an extreme case of pulsational instability. This mechanism appears to be most effective for low-mass stars. In a number of calculations this instability has been followed to substantial amplitudes, under the assumption of spherical symmetry. These calculations each cover a sufficient time span to support strongly the working hypothesis that the process considered will lead in fact to mass ejection, but they do not cover long enough time spans to answer the question as to the total mass ejected by this process. This question is important, as we would surely like to know whether the ejection is limited to the envelope layers with a composition much like the original composition inherited by the star, or whether the ejection involves layers deep enough to be seriously affected by the nuclear transmutations during the past history of the star. If the latter were the case—which at present does not seem likely except for minor components—it could have a substantial influence on the progressive change of the chemical composition of the interstellar matter in the Galaxy. Even these one-dimensional dynamical computations have turned out to present more technical difficulties than we might have expected, largely owing to the complicated structure of the extremely extended envelopes of the red supergiants in question. In fact, it has recently been shown that in these cases the relevant differential equations combined with the standard local approximation of the mixing-length theory for convection are ill-posed, in the mathematical sense of this unfriendly adjective. Thus if one wants to represent the envelope with a sufficiently fine grid to represent securely the many physically different zones of the envelope, one finds oneself forced to represent the actual nonlocal character of convection in some approximation. In spite of these technical difficulties, I feel that it will be worthwhile in the near future to try to compute this entire ejection phase through to its termination, even if for no other reason than to establish whether estimates based on linear perturbation theory give us an adequate insight into the problem of the total amount of mass ejected and the character of the residual star.

For some more massive stars, life seems to end in a much more spectacular dynamical manner, namely, in the form of supernovae. The present main working hypothesis for the cause of such a sparkling explosion is not an envelope instability—like that presumed to be the cause of the ejection of planetary nebulae—but rather a core instability. Two candidates for such instabilities are presently in the running. Ignition of a nuclear fuel within a degenerate core might directly cause an explosion. Alternatively, an internal collapse caused by the core exceeding the Chandrasekhar limit might cause a sequence of dynamical events ending in the explosive ejection of the envelope. In both cases it appears plausible that, whatever happens in the core, the dynamics occurring in the envelope may, by and large, have the character of a single, very strong shock wave running outward with increasing peak temperatures. Thus while the collapsing core may—however hypothetically—provide us with neutron stars and black holes, the shock wave in the envelope may provide us with a perfect place for nucleosynthesis.

The idea of producing heavy elements within stars followed fast upon Baade's classification of stars in populations. However, the theory of stellar nucleosynthesis started in earnest only with the work which culminated in the spectacular paper of Burbidge, Burbidge, Fowler, and Hoyle in 1957. This paper contains the overall classification of the nuclear processes producing the heavy elements. This classification still seems to form the basis of the present-day theory of nucleosynthesis, though many modifications and additions have secured and enriched this topic vital to modern astrophysics. My subordinating nucleosynthesis to dynamical phases in stars in this presentation has the explicit intent of giving emphasis to one change in the overall picture of stellar nucleosynthesis that has been quite recently developed and that seems to me of particular importance. While in the early work in this field the physical conditions required for the production of the heaviest elements were more or less correctly identified, it was estimated that such conditions would be likely to be found only in the inner portions of collapsing stars and not in the envelopes. While the early estimates regarding the cores of the stars continue to seem basically correct, it now appears that the envelopes of highly evolved stars might also provide a seat for the production of heavy nuclei, albeit for the very short duration of an extremely strong shock wave. This turn in the theory would appear of particular significance since it provides, without any further hypotheses, for the ejection of the very heavy elements thus produced.

Flagrant Deficiencies

In the preceding two sections I have sketched two subject areas, quiet mass ejection and violent dynamical phases, which are relative newcomers to the study of stellar structure when viewed from the long-range view of the last 75 years, though both have strong old roots. From the tone of my discussion it should be clear that I feel very optimistic regarding substantive and exciting progress in these subjects in the near future. It might therefore be pleasant to stop my discussion here. However, this would have the danger of implying either that everything important in the study of stellar structure is done or that its progress is well under control. Clearly such assessment would in no way represent the actual situation. Flagrant deficiencies abound—even if we count only those that we are already aware of—and something needs saying about them.

Everybody in this field will have his personal list of the most outstanding deficiencies. May I be bold enough to describe my two top candidates for such a list of today's flagrant deficiencies in our field, namely, the missing theory for turbulent convection and the missing solar neutrinos? These two deficiencies, besides being flagrant indeed, also happen to be instructive examples, one of a very old unsolved problem—indeed, as old as the study of stellar structure itself—and the other of a basic discrepancy sprung on us suddenly and relatively recently.

The fundamental equations of hydrodynamics which presumably should form the basis of an exact statistical theory of turbulence have long been known. Very general methods have been developed with which the solution of these equations can be found for any particular case as long as such case falls into the laminar domain. Even the methods of deriving the limits of this domain, often expressible in terms of critical Reynolds or Rayleigh numbers, have been successfully developed. In contrast, for the turbulent domain at high Reynolds or Rayleigh numbers the deductive development of a methodology permitting the quantitative derivation of the relevant statistical characteristics of the fluid flow in any particular case has still not been achieved, in spite of the long time that this problem has been outstanding. It continues to be a source of puzzlement to me that this clearly difficult and, I would think, correspondingly challenging problem has still not provoked an effort of that magnitude and intellectual power and persistence to cause a decisive break away from our present impotence.

Under these circumstances it is only natural that those faced with problems involving turbulent flows—such as engineers, meteorologists, and astrophysicists—have taken refuge in various plausible but unproved and generally only semiquantitative approaches. The majority of us astronomers faced with such problems have applied Prandtl's mixing-length approach, which I believe has served us remarkably well as long as we required only the crudest statistical characteristics of turbulent convection. Now, however, we find ourselves more and more frequently requiring more precise and detailed characteristics such as, for example, the extent of convective overshooting into neighboring stable layers or the precise convective energy transport under circumstances varying rapidly with time. It seems to me that without too much exaggeration we may describe the procedure we are presently applying to the solution of these problems as a gradual extrapolation of the original mixing-length concept, a procedure which seems to me to be fast reaching the limit of even qualitative trustworthiness. Thus, until a major secure advance is made in the understanding of the statistical characteristics of turbulent flows, we will be forced to the study of stellar structure, particularly as far as advanced evolution phases and specifically dynamical phases are concerned, to more and more test the sensitivity of our results to the assumptions made regarding turbulent convection wherever it is involved.

While my first candidate for the list of flagrant deficiencies in our field is somewhat depressing, largely because of its old age, my second candidate for the list is in contrast very young, and all those actively involved in it seem emotionally confident, though without any secure logical reason, that this deficiency can be canceled from our list in the reasonably near future. As you all know, the missing solar neutrinos were caused by an act of Davis, who engaged in the daring and fundamentally important endeavor to measure the neutrinos predicted to be emitted by the Sun—and promptly did not find them. The resolution of this discrepancy might be searched for in three areas. First, there might be something wrong in the solar neutrino experiment. Second, there might be something wrong in the nuclear physics on which the chain of nuclear reactions presumed to occur in the Sun is based. Third, there might be something wrong with our solar model. Under these circumstances I was initially a little afraid that every specialist in one of these three areas would understandably assume that the flaw could not be in his area and that, in consequence, nobody would do anything to resolve this discrepancy. Thus this discrepancy would stay unresolved and would cast quite some shadow of distrust on any results of the study of stellar structure. My fears happily turned out quite unwarranted. On the contrary, a happy race seems to be on, with everyone in the relevant areas trying to solve the puzzle of the missing solar neutrinos. The experimentalists are continuing the present experiment and are preparing for a new search at critical lower energies. The nuclear physicists are strengthening the weakest links in the relevant chain of nuclear processes. We model builders seem to be ready to drop all restraints in an attempt to modify our solar model so that it will not emit any energetic neutrinos but will nevertheless keep on shining at its observed rate—with even the latter condition being occasionally forgotten. Whatever the outcome, the missing solar neutrinos exemplify a deficiency that is clearly far more stimulating than depressing and that has the potential of leading to basic new insight.

May I conclude with a general comment to which I think one would be led by any review of our topic. The study of stellar structure is not an undertaking static in either its specific topic or its methods. On the contrary, the study of stellar structure has consisted of a remarkably steady and rapid change in topics and techniques. I believe it is this characteristic of our field which has enabled it to persistently attract bright independent minds. It does so now, it has done so in the past, as for example, when Chandrasekhar introduced his white dwarfs at the time Eddington's accomplishments on main-sequence stars reached their completion.

2 Stellar Stability

Paul Ledoux

Introduction

Studies in stability are mostly attempts at compensating partially for our inability to analyze the families of solutions of the general partial differential equations we have to deal with, to pick up those of physical interest, and to follow their development in time, starting usually from singular points or singular paths in a very complex phase space like that of a star. Perhaps this is a somewhat hazardous approach, but I don't know of any better, although more powerful methods will, no doubt, be brought to bear on the stability problem in the future (cf. Thom 1972, 1974; Thompson 1975a, b).

We shall limit our review here to the case of purely spherical hydrostatic stars most of the time in thermal balance, thus excluding the important effects of rotation, magnetic fields, or tidal fields. Although I shall try to emphasize recent developments, I certainly cannot avoid repetition (Ledoux and Walraven 1958; Ledoux 1958, 1963b, 1965, 1969, 1974).

In the stellar case, most results have been obtained by the small perturbation method and the linear equations derived from it, the solution of the latter being used to study the effects of changes in the initial conditions. In that case, the definition of stability usually adopted (for generalizations and refinements cf., for instance, Leipholz 1970; Perdang 1975a) implies that the perturbations of all positions, velocities, and state variables remain smaller than a given small quantity ϵ for all times (or, more realistically, for some limited but significant time) provided that the initial perturbations are smaller than another small quantity $\eta(\epsilon)$.

After separation of the time by a factor e^{st}, the remaining problem can usually be reduced to an eigenvalue problem defining the normal modes; and, if they form a complete set, a necessary and sufficient condition of stability is that all the real parts $\Re(s)$ of the eigenvalues s be negative while instability will occur if at least one of the $\Re(s)$ is positive.

In some cases, the unperturbed configuration may also depend on parameters α_j. If the eigenvalues are explicit functions of these α_j the influence of the latter on stability can then be discussed directly, surfaces $\Re[s(\alpha_j)] = 0$ (marginal state) separating alternate regions of stability and instability in the parameters space. Very often, in the stellar case, this can only be partially realized on the basis of numerous numerical integrations for discrete values of the parameters.

Paul Ledoux is at the Institut d'Astrophysique, Université de Liège, Belgium.

15

As long as the structure of the differential equations does not change, when the α_j vary, a direct study of the influence of the latter in a small enough range is possible through the so-called sensitivity equations (Leipholz 1970), but this approach has received little explicit attention in the stellar case.

On the other hand, Poincaré's theory of linear series associated with the variations of such parameters and the existence of the critical points (bifurcation, etc.) which limit them has recently found new applications and extensions (Thompson 1969; Thompson and Hunt 1973; Thompson 1975a, b) and it has even proved useful, for spherical stars, in the interpretation of secular instabilities and of stellar evolution around phases corresponding to multiple hydrostatic solutions.

As to modifications of the structure of the fundamental equations by the addition of a perturbation function or operator, the usual well known methods have often been used for instance to take account of the effects of nonconservative terms or of new factors such as rotation, magnetic or tidal fields.

While variational methods have been useful in the conservative (adiabatic) approximation, they become rather cumbersome in the general nonconservative, non–self-adjoint problem, although Chandrasekhar (1961, appendix IV; cf. also Kantorovich and Krylov 1958; Schecter 1967) has given examples of applications to problems of the type considered here. He has also shown that, in such problems depending on a parameter, more conventional variational methods may also be found for determining the value of this parameter characterizing the marginal state $[\Re(s) = 0]$ as, on the average, the latter corresponds to a balance between input of energy and dissipation. This is particularly straightforward if the principle of exchange of stability applies. Up to now, little use has been made, in the stellar case, of more general approaches such as the Galerkin method or the Glansdorff and Prigogine generalized variational principle (1971).

The energy method generalized to take into account the internal energy of the compressible matter composing the star was used early by Thomas (1931; cf. also Ledoux 1958) as the basis for a very general discussion. In the adiabatic case, its explicit and detailed application (Tolman 1939) reduces exactly to the small perturbation method (Ledoux 1958), and it seems that it has yielded few precise results that could not be obtained at least as easily by the latter. In fact, other global approaches such as those provided by the virial theorem (Ledoux, 1945) or the generalized virial theorem (Chandrasekhar 1969) may have been more useful.

At this stage, one should perhaps raise the question of the kind of physical information we expect to gather from these stability studies. Of course, what we are really most interested in is change; but the first things we learn to build in the case of stars, as in many other cases, are models, purely static at first even without any energy exchange with the rest of the world. Then we try to take account of their thermal field by notions such as local thermodynamic equilibrium, and we add fluxes and nuclear energy generation which balance each other exactly. Only recently have we begun to build sequences of evolution taking into account the continual readjustment of the energetic balance as nuclear fuel is depleted. One might claim that it is already, in a way, considerations of stability and the impossibility of ensuring it at each of these successive steps that led us to the next. But of course we expect more than this from a detailed stability analysis. Could there be, for instance, violent instabilities leading to explosion or collapse of the star or to its disruption into, separate parts? Or could some models become the seat of amplified perturbations, radial or nonradial, susceptible to explain some of the intrinsic variables if nonlinear effects stabilize their amplitudes at some finite values? If not, could such pulsations increase to the point of giving rise to continuous mass ejection? What is the role of progressive waves or shock waves in these phenomena? Could traveling waves be energized in some region and dissipated in another? Is the evolution itself subject to accelerations and decelerations? Does it present branching or turning points related to the possibility, at some phases, of multiple solutions to the hydrostatic equations?

Radial Perturbations

The General Problem

The linearized equations expressing the conservation of mass, momentum, and energy are familiar. If the unperturbed model is in hydrostatic equilibrium and in thermal balance, the time may be separated in a factor e^{st}, and these equations may be combined to yield the general equation

$$s^3 r\xi - \frac{s}{\rho}\frac{1}{r^3}\left\{\frac{d}{dr}\left(\Gamma_1 p r^4 \frac{d\xi}{dr}\right) + r^3\xi\frac{d}{dr}[(3\Gamma_1 - 4)p]\right\}$$

$$= -\frac{1}{\rho}\frac{d}{dr}\left\{(\Gamma_3 - 1)\rho\left[\delta\epsilon_N - \delta\epsilon_\nu - \frac{d\delta L}{dm}\right]\right\}, \qquad (1)$$

where ϵ_N is the rate of conversion of nuclear energy into thermal energy and ϵ_ν is the emission rate of neutrinos at the expense of the thermal energy; Γ_1, Γ_3 are the generalized adiabatic exponents; $L(r)$ is the luminosity across the sphere of radius r containing the mass $m(r)$; and $\xi = \delta r/r$, δ denoting a Lagrangian variation following the motion.

The solutions must obey boundary conditions $\delta r = 0$, $\delta L = 0$ in $r = 0$ and $\delta p = 0$ at the surface, which imply that ξ and $d\xi/dr$ must be regular everywhere and that $d\delta L/dr$ must tend to zero at the surface. If the model possesses a realistic radiative atmosphere, the surface boundary conditions may have to be transferred to some optical level in that atmosphere (Baker and Kippenhahn 1962, 1965; Unno 1965; Castor 1971; Cox 1974b; Scuflaire 1975).

In a large fraction of the star, the nonconservative terms on the right are very small. If one neglects them, one obtains the so-called adiabatic approximation

$$\frac{d}{dr}\left(\Gamma_1 pr^4 \frac{d\xi}{dr}\right) + \xi\left\{r^3 \frac{d}{dr}[(3\Gamma_1 - 4)p] - s^2\rho r^4\right\} = 0 \qquad (2)$$

with the boundary conditions

$$\delta r = r\xi = 0 \quad \text{in} \quad r = 0 \,,$$

$$\delta p = -\Gamma_1 p\left(3\xi + r\frac{d\xi}{dr}\right) = 0 \quad \text{in} \quad r = R \,.$$

The problem is self-adjoint and admits a complete set of regular solutions ξ_0, ξ_1, ξ_2, ..., with zero, one, two, ... nodes corresponding to real eigenvalues $-(s_a{}^0)^2$, $-(s_a{}^1)^2$, $-(s_a{}^2)^2$ ... ordered by increasing values more often denoted by $(\sigma_a{}^0)^2$, $(\sigma_a{}^1)^2$, $(\sigma_a{}^2)^2$, ...

If all the $\sigma_a{}^2$ are positive, the roots $\pm\sigma_a{}^k = \pm i s_a{}^k$ define the angular frequencies of the various modes of oscillation and the star is said to be dynamically stable. Dynamical instability occurs if one or more of the $\sigma_a{}^2$ are negative. Of course, it enters always through the smallest eigenvalue $(\sigma_a{}^0)^2$, and for realistic models this is the only one that ever becomes negative. This, as is well known, happens when some appropriate average of Γ_1 taken over the whole star becomes smaller than $4/3$. The time scales associated with the $\sigma_a{}^k$, except very close to marginal stability, are relatively short, of the order of the time taken by an acoustic wave to travel some fraction of the radius depending on the mode considered.

At some distance below the surface depending on the model, the adiabatic approximation breaks down, but in many cases the heat capacity of the layers above this level (say M_a) is so small that $d\,\delta L/dr$ is already practically zero through it and ϵ_N and ϵ_ν are of course negligible there. In that case, the effect of the nonconservative terms can be evaluated by a perturbation method yielding a quasi-adiabatic correction $-\sigma_{\text{q.a.}}'$ to s:

$$\sigma_{\text{q a}}' = -\frac{1}{2(\sigma_a{}^k)^2 J_a{}^k} \int_0^{M_a} \left(\frac{\delta T}{T}\right)_a \left(\delta\epsilon_N - \delta\epsilon_\nu - \frac{d\delta L}{dm}\right)_a^k dm \,, \qquad (3)$$

with

$$J_a{}^k = \int_0^M (\xi_a{}^k)^2 r^2 dm \,,$$

where the subscript a means that the second member must be evaluated by means of the previously obtained adiabatic solution for the k mode considered. As defined, $\sigma_{\text{q.a.}}'$ is the damping constant of the oscillation whose amplitude will increase (vibrational instability or overstability) or decrease (vibrational stability) depending on whether $\sigma_{\text{q.a.}}'$ is negative or positive. Thus, in this case, which may be called *quasi-adiabatic*, the complete dynamical spectrum is represented by the two sets of eigenvalues

$$(s_{1,2}{}^k)_{q\,a} = \pm i\sigma_a{}^k - \sigma_{q\,a}{}'^k . \tag{4}$$

Dynamical instability (related to the conservative terms) occurs when σ_a is pure imaginary while vibrational instability (related to the nonconservative terms) occurs when $\sigma_{q.a.}'$ is negative.

This procedure, however, fails when the external nonadiabatic layer despite its small mass possesses sufficient heat capacity to affect significantly σ'. This occurs, in particular, for important classes of intrinsic variable stars (Cepheids, RR Lyrae; perhaps δ Scuti, long period variables) in which the incipient vibrational instability responsible for the observed pulsations finds its origin in the lower part of this external nonadiabatic (transition) region. It so happens that, in the appropriate stellar models, this transition region is the seat of the mid-stages of ionization of He^+ and possibly of He and H. This increases considerably its heat capacity resulting, through the so-called Γ and κ-mechanisms, in a rapid decrease outward of δL at compression implying a deposition of heat and the reverse at expansion, which of course constitutes a strong driving mechanism. As one moves higher up toward regions of lower heat capacity, the nonadiabatic effects become predominant and prevent δL from increasing again with r at compression $(d\delta L/dr \rightarrow 0)$ which otherwise could partly destroy the driving of the lower layers. Thus to be efficient the important ionization zones must fall in the critical transition region.

One should add that the appropriate stellar models for Cepheids and RR Lyrae stars are very highly centrally condensed so that the amplitudes are relatively very large in the external layers, giving a great weight to the destabilizing mechanism described which, in a period, deposit a positive amount of mechanical energy, say $(\Delta K)_P$, in the pulsation. On the other hand, as the total mass of the external layers with large amplitudes is very small, the maximum total kinetic energy $K_M = (\sigma_a{}^2/2)\int_0^M \delta r^2 dm$ is relatively small so that the amplification constant

$$-\sigma' = \frac{1}{2}\frac{\sigma_a}{2\pi}\frac{(\Delta K)_p}{K_M} \tag{5}$$

can be relatively large ($1/\sigma'$ from a few times to a few hundred times the period $P = 2\pi/\sigma_a$) for these variables. Furthermore, when the destabilizing factors occur in the external layers, there is always a chance for some of the higher modes (ξ^1, ξ^2, ξ^3, . . .) to become vibrationally unstable as well since in that region their amplitudes increase even more rapidly than that of the fundamental ξ_0 (cf. Cox 1974b for a comprehensive discussion of the applications to variable stars including nonlinear work).

Although some of these results have been established first by somewhat hybrid methods (Zhevakin 1953, 1963; Baker and Kippenhahn 1962, 1965; Cox 1963), the need for a direct numerical treatment of the complete nonadiabatic non–self-adjoint problem (eq. [1], + B.C.) was strongly felt as soon as the possible importance of the transition and the nonadiabatic external layers was realized. It took some time, however, to develop sufficiently powerful numerical methods (Baker 1968; Castor 1971; Iben 1971; Scuflaire 1975) capable of yielding directly the nonadiabatic complex eigenvalues

$$(s_{1,2}{}^k)_{n \; a} = \pm i\sigma_{n.a.}{}^k - \sigma_{n \; a.}{}'^k , \tag{6}$$

and the corresponding complex eigensolutions. In (6), $\sigma_{n.a.}{}^k$ is nearly always very close to $\sigma_a{}^k$, but $\sigma_{n.a.}{}'^k$ can be, as we have seen above, quite different (even in sign) from $\sigma_{q.a.}{}'^k$.

The secular (or thermal) stability problem corresponds to the third group, say s_3, of eigenvalues of (1) which are proportional to the nonconservative terms in the right-hand member. Because of this, $|s_3|$ is usually much smaller than $|\sigma_a|$ and, most often, one can simplify the secular problem by neglecting the dynamical term in s^3 in (1). This reduced problem is also highly non-Hermitian and can have real or complex eigenvalues $(s_3)_{ap.}$. However, it is obvious that this procedure fails if the dynamical $(1/|\sigma_a|)$ and secular time scales $(1/|s_3|)$ become of the same order as may happen close to dynamical marginal stability or in cases of extremely strong secular instability.

Dynamical Stability

Provided the star is not close to marginal stability, the adiabatic approximation yields very good values for the eigenfrequencies and the variational interpretation (Ledoux and Pekeris 1941) leads in particular to the following expression for the fundamental eigenvalue:

$$(\sigma_a{}^0)^2 = \min_u \frac{\int_0^R \Gamma_1 p r^4 \left(\dfrac{du}{dr}\right)^2 dr - \int_0^R r^3 u^2 \dfrac{d}{dr}[(3\Gamma_1 - 4)p]dr}{\int_0^R \rho r^4 u^2 dr} , \tag{7}$$

the minimum obtaining for $u = \xi^0$. However, the existence of a critical value $\bar{\Gamma}_1 = 4/3$ below which the star is dynamically unstable appears even more clearly on the integral expression of $(\sigma_a{}^0)^2$ derived from the virial (Ledoux 1945)

$$(\sigma_a{}^0)^2 = - \frac{\int_0^R r^3 \xi^0 \dfrac{d}{dr}[(3\Gamma_1 - 4)p]dr}{\int_0^R \rho r^4 \xi^0 dr} , \tag{8}$$

where ξ^0 keeps the same sign all through the interval.

If the star is dynamically stable ($\bar{\Gamma}_1 > 4/3$), a convenient approximation can be derived for the fundamental period from (7) or (8) using $u = \xi^0 = $ constant, namely,

$$(P_a{}^0)_d = \frac{Q}{[(\bar{\rho}/\bar{\rho}_\odot)(3\bar{\Gamma}_1 - 4)]^{1/2}} ,$$

where the "pulsation constant" Q varies from about 0.1 to 0.03 along a series of realistic models of increasing central condensation. This corresponds, in going from white-dwarfs to red supergiants, to periods in the range

$$2 \text{ seconds} \leq P_a{}^0 \leq 1000 \text{ days}.$$

For neutron stars the minimum fundamental period is of the order of a milli-second and corresponds to central densities of the order of 10^{15} g cm^{-3}. For higher densities the period increases again due to the general-relativistic effects which lead to dynamical instabilities past the maximum mass around 2–3 $M_\odot$ for central densities $\sim$2–4 $\times$ 10^{15} g cm^{-3}.

As we have recalled, dynamical instability always manifests itself through the fundamental solution for $\bar{\Gamma}_1 < 4/3$. For ordinary stars, practically the only possibilities for such instabilities occur either very early in the formation phases during the first isothermal contraction ($\Gamma_1 = 1$) or a little later during the collapse associated to the dissociation of H_2 and the ionization of H and He which reduce $\bar{\Gamma}_1$ below $4/3$ (Cameron 1962; cf. also Hayashi 1966, Narita *et al.* 1970; Larson 1973) or in the late stages of stellar evolution at very high temperatures and densities when nuclear dissociation equilibrium, especially past the iron peak (Hoyle 1946; Itoh 1969), has the same effect on $\bar{\Gamma}_1$.

It is interesting to note that while the fundamental stable mode ξ^0 increases toward the surface the more strongly the greater the central condensation and $(3\bar{\Gamma}_1 - 4)$, the unstable solution ξ^0, on the contrary, increases toward the center (Van der Borght 1968) in agreement with the results of direct nonlinear computation of the ensuing collapses which lead to very peaked central condensation.

Related cases of possible dynamical instabilities occur also in evolved red supergiants when they rise again along the Hayashi track, which might perhaps explain the formation of planetary nebulae (Abell and Goldreich 1966; Rose 1967; Roxburgh 1967; Paczynski and Ziolkowski 1968; Smith and Rose 1972; cf. also Wood, 1974) or in evolved massive stars where pair annihilation and production ($e^+ + e^- \rightleftharpoons h\nu$) occur on a sufficiently large scale (Rakavy and Shaviv 1967; cf. also Itoh 1969).

The value of $\bar{\Gamma}_1$ may also be reduced to a value very close to $4/3$ either by radiation in supermassive stars or by degeneracy in very condensed white dwarfs. In those circumstances, a small destabilizing factor such as general relativity corrections of the order GM/Rc^2 active in both types of stars (Fowler 1964; Chandrasekhar 1964b; Chandrasekhar and Tooper 1964) or the establishment of an equilibrium between electron captures and β-decay in the case of the condensed configurations (Schatzman 1958; Baglin 1968; Wheeler *et al.* 1968) may lower the effective $\bar{\Gamma}_1$ below $4/3$ and cause dynamical instability and collapse.

In the case of neutron stars, there may still be some difficulties with the equation of state (cf. C. Hansen, ed. 1974; Canuto 1974; Ferrini 1975) and complex problems associated with rapid rotation, strong magnetic fields, or crystalline mantles or cores, but roughly they are dynamically stable only in those ranges where the central density ρ_c and the total mass M increase together ($0.05 \leq M/M_\odot \leq 1.5$ to 3; $10^{14} \leq \rho$ (g cm^{-3}) $\leq 3 \times 10^{15}$) (cf. Wheeler 1966; Cameron 1970; Zel'dovich and Novikov 1971). However, recently Heintzman and Hillebrandt (1975) have shown that an anisotropic equation of state might allow higher masses to be stable toward radial perturbations, although their configurations are rather likely to be unstable toward nonradial modes.

In relation to the theory of supernovae (for a short review cf. Imshennik and Nadyazhin 1974), other violent collapses of the central regions (Colgate and White, 1966; Howard et al. 1972; Iben 1972; Buchler et al. 1974; Barkat et al. 1972; cf. also Chiu 1961, 1964) or reexplosions (Arnett 1968, 1969) have been advocated which, however, are not directly or mainly related to dynamical instability but are due either to a catastrophic emission of neutrinos or the sudden release of a very large amount of thermonuclear energy—processes which are really related to aspects of the secular stability problem. However, in such extreme conditions, where dynamical and secular time scales approach each other, our previous separation of the problem into a dynamical and a secular one becomes precarious. Although, in such cases, the nonlinear problem has often been tackled directly, a discussion of the complete *linear* problem (1) may still help in the interpretation of the results.

A somewhat similar case occurs when the configuration approaches very closely a state of marginal dynamical stability, the dynamical time scale ($1/\sigma_a$) increasing to values comparable to the vibrational or secular time scales. Recently Scuflaire (1975) has used his program for the complete linear nonadiabatic problem to study the behavior of the dynamical and secular eigenvalues of supermassive stars on the main sequence through their marginal state, due to the general relativity corrections, at a critical mass $M_c \approx 3.7 \times 10^5 \, M_\odot$.

The results are illustrated on Figure 1, where the eigenvalues for the fundamental modes are represented as function of mass across the critical mass, $\sim 3.7 \times 10^5 \, M_\odot$, the real parts by full lines, the imaginary parts by dashed lines. The thin lines correspond to the usual approximation: dynamical spectrum $(s_{1,2}{}^0)_{\text{q.a.}}$ with the imaginary part $\sigma_a{}^0$ computed from the adiabatic equation (2) and the real part $\sigma_{\text{q.a.}}{}'^0$ from (3); secular spectrum $(s_3)_{\text{ap.}}$ (always real here) computed from (1) with $s^3 = 0$. The thick lines correspond to the complete nonadiabatic problem. At relatively small distances on both sides of the critical mass, the approximate and exact solutions converge very nicely. However, close to the critical mass, which is slightly different in the exact solution, the approximate solutions are meaningless. Furthermore, if all eigenvalues become real on the unstable side, in both the approximate and the exact solutions, the transition, in the exact solution, occurs through all finite, nonzero eigenvalues. The filiation is also quite different since, in the complete problem, it is the real part $\sigma_{n.a.}'^{0}$ of the dynamical mode that gives rise, beyond the critical mass M_c, to the dynamical $(s_1^0)_{n.a.}$ and the secular s_3^0 modes which become simultaneously unstable (Demaret and Ledoux 1973) while the dynamically damped mode $(s_2^0)_{n.a.}$ beyond M_c results from the initially stable secular mode s_3^0. Perhaps this kind of discussion, if repeated for a few supermassive stars $(M > M_c)$ as they evolve at constant mass across the critical radii R_c, might

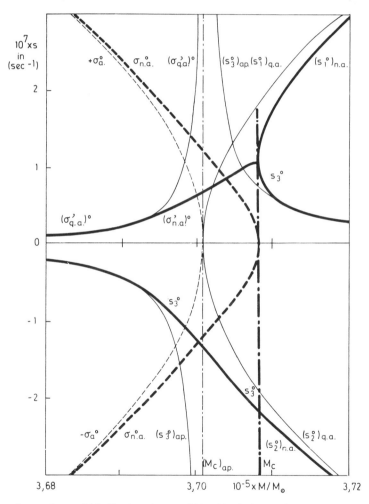

Figure 1. Approximate (*thin lines*) and exact (*thick lines*) eigenvalues around the critical mass of main sequence supermassive stars. *Full lines*, real parts; *dashed lines*, imaginary parts.

help interpret numerical results obtained directly in the nonlinear problem (Appenzeller and Fricke 1972a, b).

Vibrational Stability (Static Models)

Despite their obvious interest, we shall not come back to applications to variable stars which we have mentioned briefly before since the subject is reviewed thoroughly by Cox (1974b) for many types of variables.

Although in practically all these variable stars vibrational instability is related to the ionizations of H and He in the external layers, in other groups of stars on the main sequence it is due to the thermonuclear reactions. For the CNO cycle, all stars of usual chemical composition with masses greater than 60 to 120 $M_\odot$ are vibrationally unstable (Ledoux 1941; Schwarzschild and Härm 1959; Stothers and Simon 1970; Ziebarth 1970). This is easily understood since, as the mass increases, the role of radiation becomes more important and lowers the value of Γ_1 which, coupled to the rather low central condensation of main sequence stars, reduces considerably the increase of ξ_a, $(\delta\rho/\rho)_a$, etc., from the center to the surface. This means that in (3) the weight of the nuclear energy term $\delta\epsilon_N$ which is always destabilizing and important in the central region increases compared to the usually dissipative term in δL and finally overcomes its stabilizing effects above a certain mass.

Of course, this is not true of the higher modes which continue to increase sufficiently fast toward the surface to remain vibrationally stable in practically all cases. Thus when the destabilizing factor is deep down toward the center, as is nearly always the case with nuclear reactions, the vibrational instability induced concerns essentially the fundamental mode ξ_0 and the comments below refer mainly to this case.

The exact value of the upper limit to the stable masses found above depends of course on $\epsilon_{N,\rho} = (\partial \log \epsilon_N / \partial \log \rho)_T$, $\epsilon_{N,T}$, κ_ρ, and κ_T which characterize the sensitivity of ϵ and κ to variations in ρ and T and on the mean molecular weight $\bar{\mu}$ since the critical parameter determined by $(\sigma_{\text{q.a.}}')^0 = 0$ is really $(M/M_\odot)\bar{\mu}^2$.[1]

Thus for initially pure H stars ($\bar{\mu} \approx \frac{1}{2}$ and rather low $\epsilon_{N,T}$), the critical mass is much larger, of the order of 300 to 400 $M_\odot$ (Boury 1963, 1964; Stothers and

---

[1] *Note added in proof.* In a paper which has come to the attention of the author only recently, Sastri and Stothers (1974) have shown that the critical mass may be very sensitive to the opacity in the fairly external layers. Apart from the direct effects of changes in κ_T and κ_ρ, an increase in the opacity there may, for instance, result in creating an external convection zone and affect the overall structure in the star, considerably increasing its central condensation $\rho_0/\bar{\rho}$. In that case the amplitude of the oscillation will increase much more rapidly from the center to the surface, reducing the weight of the nuclear energizing region at the center. In some cases, according to the authors, the notion of a pulsational upper limit to the stellar masses may even vanish completely, but many of the factors involved deserve very careful analysis (cf. Stothers 1976, where new sources of instability related to the increased opacity are described).

Simon 1970), while for pure He stars ($\bar{\mu} \approx 4/3$, $\epsilon_{N,T}$ high) the critical mass is much smaller ($\sim$10 $M_\odot$) (Noels-Grötsch 1967). However, in the latter case, the addition of a rather small hydrogen-rich envelope is sufficient to increase the stability greatly (Van der Borght 1969; Hansen and Spangenberg 1971a, b).

One might then expect pure carbon stars to have even a smaller critical mass; however, the sensitivity of κ to T and ρ is stronger in this case, and if neutrino losses are neglected these stars remain vibrationally stable up to very large values of the mass. But if neutrino losses are taken into account, the corresponding stellar models are practically unstable at all masses (Noels-Grötsch 1967; Mariska and Hansen 1972), although thermal imbalance terms (see later) could affect this conclusion (Marshall and Van Horn 1973).

If we come back to the hydrogen main sequence, stars below the critical mass (60–120 $M_\odot$) soon become very stable toward radial perturbations and remain so down to some value somewhat smaller than 1 $M_\odot$. However, current stellar models for masses smaller than, say 0.3 $M_\odot$, are almost wholly convective (polytrope $n = 3/2$), and this implies a very low central condensation ($\rho_c/\bar{\rho} \approx 7$) and a low ratio ξ_R/ξ_0 ($\sim$1.3) which again strongly favors the thermonuclear term in (3) and leads, as Gabriel (1964, 1967) has shown, to vibrational instability with an e-folding time short compared with the evolutionary time scale. Boury and Noels (1973) find that even a 0.5 $M_\odot$ is still vibrationally unstable due mainly here to the comparatively high sensitivity ($\epsilon_{N,T}$) of ^{3}He burning. Opoien and Grossman (1974) have confirmed and refined these results on the basis of models computed with an improved treatment of gas thermodynamics and Coulomb interactions. There is, however, little observational evidence up to now of the effects of this vibrational instability in small mass stars, and one may still wonder if the models are really adequate or if our approximate treatment of the effects of convection is sufficient. Pre–main-sequence stars of small masses are also unstable during the temporary phases of deuterium burning (Toma 1972; Opoien and Grossman 1974).

Stars in the lower part of the helium-burning main sequence ($M \leq 0.3 M_\odot$) are also vibrationally unstable (Hansen and Spangenberg 1971a, b) but in this case mainly because of the lowering of the central condensation through the increasing role of nonrelativistic degeneracy which makes the configuration approach a polytropic structure ($n = 3/2$).

This brings us to the problem of the vibrational stability of white dwarfs where these effects of degeneracy should be maximum and should increase as relativistic degeneracy sets in, progressively lowering Γ_1 which tends to 4/3 for complete relativistic degeneracy. This more than compensates for the increase of the central condensation, and it was found early (Ledoux and Sauvenier-Goffin 1950; cf. also Baglin 1967; Vauclair 1971; Vila, 1971) that any nuclear reactions accounting for an appreciable part of the luminosity either at the center or in the external layers could lead to vibrational instability. The presence of nuclear reactions, at least in the degenerate part of the star, would also cause secular instability (Mestel 1952), but the two types of instability could occur simultaneously or a relatively small amount of fuel in the external partially degenerate layers might preferentially cause vibrational instability. However, for radial modes, the e-folding time is rather long (Van Horn et $al.$ 1972) since with ξ_0 nearly constant the maximum kinetic energy of the pulsation K_M is large and consequently (cf. [5]) the amplification factor is small.

It is always difficult to estimate the real effects of such vibrational instabilities without actual nonlinear computations. In the cases of various classes of variable stars (classical Cepheids, RR Lyrae stars, W Virginis stars, long period variables) amplifications and extensions of the pioneering nonlinear work of Christy (1962) by a series of authors (cf. Cox 1974b) has brought convincing evidence that, in those cases, the destabilizing factors in the external layers can lead to finite pulsations, the star reaching some kind of global limit cycle in a rather short time because the e-folding time is only a relatively small multiple (10 to 100) of the period. However, even in these cases, there might subsist some real difficulties concerning, for instance, the actual stability of the numerically found nonlinear cycle (Baker and von Sengbusch 1969; von Sengbusch 1975; Stellingwerf 1974) or the actual finite nonlinear behavior of models which are linearly unstable in more than one mode and the influence of the initial conditions on the resulting motion.

But when, as in many of the other cases mentioned above, the e-folding time is really a very large multiple of the period, the problem appears much more untractable. Recourse must be had then to some artifice, like the introduction of different momentum and energy time scales or the arbitrary multiplication of the velocities by a factor slightly larger than unity at regular time intervals, as used in the investigations of Appenzeller (1970a, b), Ziebarth (1970), Talbot (1971a, b) and Papaloizou (1973a, b) in the case of the effects of the vibrational instability above the critical mass M_c on the main sequence. One word should perhaps be added first concerning the stabilizing effect of the evolution which increases rapidly the central condensation of the model and the ratio ξ_R/ξ_0 so that, as shown first by Schwarzschild and Härm (1959), the effective critical mass M_c^* should always be taken somewhat larger than M_c since in the interval M_c to M_c^* the instability would not have the time to manifest itself.

But even with this reservation, the nonlinear results were rather different from the general expectations based on approximative arguments (cf., for instance, Osaki 1966; Simon and Stothers 1970) that for $M > M_c^*$, escape velocity would be reached in the external layers for $\Delta R/R$ of the order of unity and that rather large mass flow would result capable of reducing fairly quickly the mass below M_c^*. In fact, for masses in a fairly large range above M_c^* (say M_c^*– $2\,M_c^*$) the calculations reported above either led to no or to a small mass loss. In the latter case, Appenzeller (1970a) estimated that it was nevertheless sufficient in the case of a $130\,M_\odot$ (up to $4 \times 10^{-5}\,M_\odot$ per year) to result in a

P Cygni phenomenon with continuously expanding outer layers. In an extension to higher masses, Appenzeller (1970*b*) concluded that above 300 $M_\odot$, mass loss could become catastrophic. However, Talbot (1971*b*) found that even for masses as large as 10^3 $M_\odot$, the star still reached some kind of finite-amplitude regime more similar to the result of Ziebarth for a 100 $M_\odot$ star, without any serious indication of mass loss.

In all cases, the strong damping in the external layers which, after some time, balances the energizing in the core by the nuclear reactions is due to the development of strong shock waves and it may be that the somewhat contradictory results are related to the way of treating these shock waves and the local dissipation and redistribution of energy between thermal or mass motions and radiation. On the other hand, Papaloizou emphasizes the damping by nonlinear coupling with higher stable modes, a possibility already pointed out by Murphy (1967).

Also (cf. especially Appenzeller and Ziebarth), the total radius of the star increases considerably (a factor of 3 to 4) before a balance is reached between the stabilizing and destabilizing factors, this being due essentially to the lifting of the very external layers by the upward flux of momentum associated with outgoing compression waves or shocks, as pointed out by Whitney (1956) and by Osaki (1966). This in itself modifies the model toward one of greater effective central condensation and favors *a priori* the damping. On the other hand, stellar winds and mass loss occur in many types of stars with much smaller masses than M_c^*, and in the present state of our knowledge it is impossible to guess how much these phenomena might be enhanced by the vibrational instability above the critical mass.

If the latter cannot reduce quickly M to some value around M_c^* and since observations seem to confirm the absence of stars with masses greater than some upper limit which seems to have been recently brought up to some 120 $M_\odot$ (Conti and Burnichon 1975), some other limiting mechanism must be at work. For instance, Larson and Starrfield (1971; cf. also Appenzeller and Tscharnuter 1974; Larson 1973) have pointed out that, in a contracting cloud, the very condensed core forming early may generate enough radiation to stop the contraction of the external layers and finally push them off. However, if this process is as efficient as described, it may then be difficult to form a 120 $M_\odot$ at all, at least for a Population I chemical composition.

Vibrational Stability (Evolving Models)

The progress of stellar evolution computations brought to the fore the existence of phases of fast evolution with lifetimes t_s which might be comparable to the *e*-folding times of oscillations. This raises the question of the possible influence of the evolutionary motion on the oscillations and especially on the associated coefficients of vibrational stability, which should be affected to the first order in the small parameter $\epsilon = P/t_s$ whereas the period P of the oscillation should suffer only higher order corrections in ϵ.

In principle, this was already covered by the analysis of Thomas (1931) based on the energy method, and Ledoux (1958) elaborated it somewhat to make it more explicit. The first actual applications came much later and, except for an early attempt by Meurice (1964), they rested essentially on that same approach (Kato and Unno 1967; Okamoto 1967; Simon, 1970, 1971; Simon and Sastri 1972; Sastri and Simon 1973). Even the discussion by Axel and Perkins (1971) while starting from the detailed linearized pulsation equations still relied on the evaluation of the rate of change of energy to evaluate the vibrational stability. On the other hand, the general variational principle of Prigogine and Glansdorff (1965) was used by Unno (1968) to discuss the homologous pulsations of a star undergoing homologous contraction, but its extension to more realistic cases seems very complicated.

In view of a direct attack, the general perturbation equations of a slowly evolving (quasi-static) model were discussed both in the Eulerian and the Lagrangian formalisms in the Saas-Fee lectures (Ledoux 1969), the Lagrangian perturbation δf of any variable f denoting here the difference between the value $f(m, t)$ taken by f in the perturbed motion and the value $f_0(m, t)$ at the same instant and for the same mass element in the unperturbed motion.

While the Lagrangian linearized equations of conservation of mass and momentum keep the usual form but with coefficients slowly variable in time, a few extra terms appear in the conservation of thermal energy and namely a term in $T_0(dS/dt)_0$, S representing the entropy per unit mass, which characterizes the lack of balance between nuclear energy generation and the total energy flux $L(r)$ (thermal imbalance).

Combining these equations as before leads to a general partial differential equation, third order in time, which may be written (Ledoux 1969)

$$\frac{\partial^3 \delta r}{\partial t^3} - \frac{2}{r}\frac{\partial r}{\partial t}\frac{\partial^2 \delta r}{\partial t^2} - 4\pi r^2 \frac{\partial}{\partial m}\left[4\pi\Gamma_1 p\rho\,\frac{\partial}{\partial m}\left(r^2\,\frac{\partial \delta r}{\partial t}\right)\right]$$

$$+ 4\pi r^2 \frac{\partial}{\partial m}\left[\frac{\Gamma_1 p}{\rho}\left(\frac{\delta\Gamma_1}{\Gamma_1} + \frac{\delta p}{p} + \frac{\delta\rho}{\rho}\right)\frac{\partial\rho}{\partial t}\right]$$

$$+ 16\pi r^2 \frac{\partial}{\partial t}\left(\frac{\delta r}{r}\frac{\partial p}{\partial m}\right) - 4\pi r^2 \frac{\partial}{\partial m}\left[8\pi\Gamma_1 p\rho\,\frac{\partial}{\partial m}\left(r\delta r\,\frac{\partial r}{\partial t}\right)\right]$$

$$= -4\pi r^2 \frac{\partial}{\partial m}\left[(\Gamma_3 - 1)\rho\delta\left(\epsilon_N - \epsilon_\nu - \frac{dL}{dm}\right)\right]$$

$$-4\pi r^2 \frac{\partial}{\partial m}\left[\left(\frac{\delta\Gamma_3}{\Gamma_3 - 1} + \frac{\delta\rho}{\rho}\right)(\Gamma_3 - 1)\rho T\,\frac{dS'}{dt}\right], \qquad (9)$$

in which all coefficients are slowly varying functions of time.

Of course, for a closer analogy with (1), it can be transformed in an equation in $\zeta = \delta r/r$ which was favored by Cox *et al.* (1973) who adopted a time dependence of the form

$$\zeta(m, t) = \xi(m, t) \exp\left[\int_0^t s(t')dt'\right], \tag{10}$$

where $\xi(m, t)$ and $s(t)$ are now *slowly varying* functions of time. They showed then that if, following a procedure used by Ledoux (1963b), their general equation in ξ is multiplied by $r\xi^* dm$ and integrated over the whole mass, this yields a cubic equation in s which may be used to derive an integral expression for the dynamical roots $s_{1,2}$, at least up to the first order in the usual nonadiabatic term (first term in the right-hand member of [9]) and in the small quantity $\epsilon = P/t_s$ which characterizes the variations of the coefficients of (9) as well as the last term in the right-hand member.

The resulting expression for the real part of s (i.e., in our previous notations [4], $-\sigma_{q.a.}'$) is really quite complicated and contains no less than nine integrals (Cox *et al.* 1973; Cox *et al.* 1974), some of which have no clear physical meanings. Furthermore, at least in the first few papers of Cox and his associates, the eigenfunctions appearing in their various integral terms were not defined, although one could guess that they should be close to the usual time-independent *adiabatic* eigenfunctions of the model considered as static at some given time.

Demaret, who was interested in the possible influence of these evolutionary terms on the stability of supermassive stars, also tackled the complete small perturbations problem, which can also be written

$$\frac{\partial^2 \delta r}{\partial t^2} - A(r_0, t)\frac{\partial^2 \delta r}{\partial r_0^2} - B(r_0, t)\frac{\partial \delta r}{\partial r_0} - C(r_0, t)\delta r$$

$$= \frac{r^2}{\rho_0 r_0^2}\frac{\partial}{\partial r_0}[T\rho(\Gamma_3 - 1)\delta S], \tag{11}$$

where

$$A(r_0, t) = \frac{\Gamma_1 p \rho r^4}{\rho_0 r_0^4}, \quad B(r_0, t) = \frac{r^2}{\rho_0 r_0^2}\frac{\partial}{\partial r_0}\left(\Gamma_1 p \frac{\rho r^2}{\rho_0 r_0^2}\right) + \frac{2r}{\rho_0 r_0^2}\Gamma_1 p,$$

$$C(r_0, t) = \frac{4Gm}{r^3} + \frac{2r^2}{\rho_0 r_0^2}\frac{\partial}{\partial r_0}\left(\frac{\Gamma_1 p}{r}\right),$$

and

$$T\frac{d\delta S}{dt} = \left(\delta\epsilon_N - \delta\epsilon_\nu - \frac{\partial \delta L}{\partial m}\right) - \delta T\frac{dS}{dt}, \tag{12}$$

r_0 being an invariant Lagrangian coordinate related to m by $dm = 4\pi\rho_0 r_0^2 dr_0$. The solutions must of course satisfy the usual boundary conditions.

Generalizing the resolution in two steps of the classical problem, Demaret

(1974a, b, 1975a, b) considered first the case of so-called "isentropic" oscillations such that, at every instant, the entropy of any mass element in the perturbed motion is equal to that of the same mass element in the unperturbed configuration in slow evolution.

In that case, $\delta S = 0$ or $\delta p/p = \Gamma_1 \delta\rho/\rho$; and equation (11), whose right-hand member vanishes, can be written

$$\frac{\partial^2 \delta r_{Is}}{\partial t^2} = A(r_0, t) \frac{\partial^2 \delta r_{Is}}{\partial r_0^2} + B(r_0, t) \frac{\partial \delta r_{Is}}{\partial r_0} + C(r_0, t)\delta r_{Is}. \tag{13}$$

For any of the successive models in an evolutionary sequence, considered as "frozen in" at the corresponding time say t^*, equation (13) with $t = t^*$ admits solutions of the form $\eta(r_0)e^{st}$ and reduces to the adiabatic equation (2) where ξ is replaced by η/r_0. Assuming dynamical stability, its resolution at successive times t^* yields, for each mode, sets of real eigenfunctions $\eta_{Is}{}^k(r_0, \tau)$ and frequencies $\sigma_{Is}{}^k(\tau)$ functions of a time parameter $\tau = (P_0/t_s)t^*$ slowly variable. The eigensolutions are normalized, at each instant, by the condition

$$\int_0^M \eta^i(r_0, \tau)\eta^k(r_0, \tau)dm = \delta_{ik}. \tag{14}$$

With the help of these solutions, the partial differential equation (13) can be transformed into an infinite set of coupled *differential* equations of the second order in time with coefficients slowly variable which can be solved by asymptotic methods (cf. Feshchenko *et al.* 1967). To the first order in P_0/t_s, Demaret shows that the general solution may be written

$$\delta r_{Is}(r_0, t) = \sum_{k=0}^{\infty} \frac{\eta_{Is}{}^k(r_0, t)}{[\sigma_{Is}{}^k(t)]^{1/2}} \, \Re\left\{\alpha^k \exp\left[i \int_0^t \sigma_{Is}{}^k(t')dt' + \theta_{Is}{}^k(r_0, t)\right]\right\}, \tag{15}$$

where τ has been explicitly replaced by $P_0 t/t_s$.

The α^k are complex constants to be determined by the initial conditions and the $\theta_{Is}{}^k$ are complicated phase functions depending on the time derivatives of the $\eta_{Is}{}^i (i \neq k)$ of order P_0/t_s. While a time dependence of the form (10) appears automatically, the instantaneous frequencies $\sigma_{Is}{}^k = \sigma_a{}^k$ however are not affected. The appearance of the factor $[\sigma_{Is}{}^k(t)]^{-1/2}$ or $\exp\{-\frac{1}{2}\int_0^t[\dot\sigma_{Is}{}^k(t')/\sigma_{Is}{}^k(t')]dt'\}$ corresponds to a purely "isentropic" contribution to the coefficient of vibrational stability,

$$\sigma_{Is}{}'^k = \frac{1}{2}\frac{\dot\sigma_{Is}{}^k}{\sigma_{Is}{}^k} = -\frac{1}{2}\frac{\dot P_a{}^k}{P_a{}^k}, \tag{16}$$

which is typical of slowly varying systems as shown for instance by the theory of adiabatic invariants. Depending on whether the period increases or decreases due to the slow motion, the amplitude is amplified or damped. Because of the normalization (14), the dependence on t of the eigenfunctions $\eta_{Is}{}^k$ appearing in (15) does not contribute to the global coefficient of vibrational stability.

The same technique may be extended without difficulty to the general equation (11) at least if its right-hand member is small compared to the zero order isentropic terms. The general solution is again of the form (15) with an extra phase term $\theta_{N.\mathrm{Is}}{}^k(r_0, t)$ added and a global damping coefficient

$$\sigma_{\mathrm{q.Is}}{}'^k = \frac{1}{2}\frac{\dot{\sigma}_{\mathrm{Is}}{}^k}{\sigma_{\mathrm{Is}}{}^k} - \frac{1}{2(\sigma_{\mathrm{Is}}{}^k)^2}\int_0^M \left(\frac{\delta T}{T}\right)^k_{\mathrm{Is}} \left(\delta\epsilon_N - \delta\epsilon_\nu - \frac{d\delta L}{dm}\right)^k_{\mathrm{Is}} dm$$

$$+ \frac{1}{2(\sigma_{\mathrm{Is}}{}^k)^2}\int_0^M \left[\left(\frac{\delta T}{T}\right)^k_{\mathrm{Is}}\right]^2 \left(\epsilon_N - \epsilon_\nu - \frac{dL}{dm}\right) dm, \qquad (17)$$

where the subscript "Is" in the integrand means that the quantities have to be evaluated with the help of the space part of the isentropic solution.

Apart from the "isentropic" part (16) and taking into account the normalization (14), one recognizes the classical expression (3) with an extra term corresponding to the effects of thermal imbalance so that the physical meaning of this expression is straightforward.

Asymptotic methods have also been used later by Aizenman and Cox (1974a), and a subdivision of the treatment similar to that above has been adopted in another paper (Aizenman and Cox 1974b) where it is shown that expression (17) can be extended directly to the case of nonradial oscillations provided, in the varied expression, dL/dm be replaced by $(1/\rho)$ div $\mathbf{F}$.

It seems that some confusion has sometime arisen in these discussions leading for instance to different values of $\sigma_{\mathrm{q.Is}}{}'$ depending on whether it is computed with the eigenfunctions for the perturbations δf or for the relative perturbations $(\delta f/f)$. However, as Demaret (1975c) has noted later, this difficulty is lifted if the normalization of $(\delta r/r)$ is properly taken into account. On the other hand, since σ_{Is} is slowly variable in time, there are no reasons here why the vibrational stability coefficient $\sigma_{\mathrm{q.Is}}{}'$ derived above for δr should be the same as that derived from the energy, i.e.,

$$-\frac{1}{2}\frac{\left\langle\dfrac{dE}{dt}\right\rangle_p}{\langle E\rangle_p} = \sigma_E{}' = \sigma_{\mathrm{q.Is}}{}' - \frac{\dot{\sigma}_{\mathrm{Is}}}{\sigma_{\mathrm{Is}}},$$

where E is the total energy of the pulsation; nor is there any basis for attempts at forcing agreement (Davey and Cox 1974). Cases may indeed occur where δr is constant ($\sigma_{\mathrm{q.Is}}{}' = 0$) while the energy of the pulsation increases ($\dot{\sigma}_{\mathrm{Is}} > 0$) or decreases ($\dot{\sigma}_{\mathrm{Is}} < 0$); these cases are also well known in the adiabatic invariant theory of slowly varying systems.

Many illustrative applications to relatively simple models can be found in the work of Cox and his associates, and very generally the effects of evolution on vibrational stability are quite small compared with the classical effects (second term in eq. [17]). The corresponding e-folding time tends to be of the order of the evolutionary time scale t_s which in itself reduces the interest of the possible effects. However, while this is rigorously true of the "isentropic" effect (16), the direct thermal imbalance term (third term in eq. [17]) could perhaps become important if a region with a large value of TdS/dt (shell burning, for instance) could have a large weight in this integral. But even in such a case, treated by Cox (1974a), the correction to σ' was only of the order of 25 percent. Of course, when, classically, a star is close to a state of marginal vibrational stability, the evolutionary correction could change the sign of σ'; but it is doubtful whether such suppressed or induced slight instabilities are of real interest.

Secular Stability

In some sense all stars are secularly unstable since they evolve continually. However, some phases of this evolution are controlled by nuclear transmutations and have time scales of the order of $\tau_N = E_N/L$, where E_N is the total available thermonuclear energy at that phase and L is the corresponding luminosity. This time scale may be very long—for instance, during the hydrogen main-sequence phase where the evolutionary models are really very close to purely hydrostatic models in thermal balance ($L = \int_0^M \epsilon_N dm$).

For such models, perturbed at constant chemical composition, the secular stability problem, as already announced above, is concerned with the third group of eigenvalues s_3 of the general equation (1). Multiplying the latter, where the term in s^3 is neglected, by $r\xi^* dm$ and integrating over the whole mass yields a formal expression (Ledoux 1963b)

$$s_3 = -\frac{\int_0^M (\Gamma_3 - 1)\left(\frac{\delta\rho}{\rho}\right)_s^* \left(\delta\epsilon_N - \delta\epsilon_\nu - \frac{d\delta L}{dm}\right)_s dm}{\int_0^R 4\pi\Gamma_1 pr^4 \left|\frac{d\xi_s}{dr}\right|^2 dr - \int_0^R 4\pi r^3 |\xi_s|^2 \frac{d}{dr}[(3\Gamma_1 - 4)p]dr}, \quad (18)$$

where the subscript s means that the right-hand side must be evaluated with a solution of the differential system as simplified above. The denominator of equation (18), which is real, is exactly of the same form as the numerator in the variational expression (7) of $(\sigma_a^0)^2$ and will always be positive, whatever ξ_s, provided the star is dynamically stable.

The numerator shows that complex values of s_3 are possible since the products of $(\delta\rho/\rho)^*$ with some of the terms in the developments of $\delta\epsilon$ and $d\delta L/dm$ will not reduce automatically to real quantities. This is not surprising since the differential system, which reflects essentially the effects of the nonconservative terms, is non-Hermitian. This was actually confirmed for the first time numerically by Schwarzschild and Härm (1967b; cf. also Hansen et $al.$ 1970; Aizenman and Perdang 1971a, 1971b; Härm and Schwarzschild 1972; Gabriel and Noels 1972; Gabriel 1972).

If dynamical stability is ensured, the condition for secular stability reduces then to the condition that the real part of the numerator of (18) be positive. Or, if the secular solutions are real, to the condition

$$\int_0^M (\Gamma_3 - 1) \left(\frac{\delta \rho}{\rho} \right)_s \left(\delta \epsilon_N - \delta \epsilon_\nu - \frac{d\delta L}{dm} \right)_s dm > 0 , \qquad (19)$$

which may be compared formally with the numerator of (3) where $(\delta T/T)_a = (\Gamma_3 - 1)(\delta \rho/\rho)_a$ and which, on the contrary, should be negative for vibrational stability. Of course, in general, the solutions ξ_a or ξ_s with which these terms must be evaluated may be very different; nevertheless, this shows that qualitatively nonconservative factors favorable to vibrational instability would tend to reinforce secular stability and vice versa.

This problem was already considered by Russell (1925) and Jeans (1928) especially with a view to finding some necessary characteristics of the possible nuclear processes in stars.

Jeans used the homology transformation

$$\frac{\delta r}{r} = C , \quad \frac{\delta \rho}{\rho} = -3C , \quad \frac{\delta p}{p} = -4C , \quad \frac{\delta T}{T} = -C , \qquad (20)$$

as a solution, and obtained a criterion which can be recovered in a somewhat improved form by using (20) to evaluate the left-hand member of (19). If we neglect the neutrino production, we get

$$3\bar{\epsilon}_{N,\rho} + \bar{\epsilon}_{N,T} + (3\kappa_\rho + \kappa_T)_R > 0 , \qquad (21)$$

where the bars indicate appropriate space averages and R, the surface.

It is easy to generalize (21) to include the effects of neutrino emission (Ledoux 1969) and to show that this favors secular instability only if $\epsilon_{\nu,T}$ and/ or $\epsilon_{\nu,\rho}$ are larger than $\epsilon_{N,T}$ and/or $\epsilon_{N,\rho}$.

Once the high sensitivity of the thermonuclear reactions to the temperature ($\epsilon_{N,T} > 1$) was established, condition (21) was satisfied with such a safe margin for all ordinary gaseous stars with any realistic opacity as to make it look almost trivial. Nevertheless, it does correspond to a significant aspect of the problem which is the intrinsic built-in secular stability provided for ordinary stars by the physics of the thermonuclear reactions and of the opacity.

However, this is not the case for white dwarfs where, because of the degeneracy of the electron gas, the pressure is practically independent of temperature. As Lee noted (1950), this can lead, in the presence of nuclear reactions, to some indeterminacy in the model; and in 1952 Mestel showed that, in such a situation, secular instability must arise, leading to a continuous enhancement of thermonuclear energy generation. This conclusion was to find confirmation later in the helium flash along the sequence of red giants of small masses with degenerate helium cores.

Considering the importance of the problem of secular stability for stellar evolution with which it has narrow connections and the inadequacy of (20) which is an actual solution only in very artificial circumstances, a more general attack was necessary (Ledoux 1958, 1960, 1963b). But, in fact, active new interest in the problem of secular stability arose only later when Schwarzschild and Härm (1965), having encountered numerical difficulties in following the advanced stages of evolution of a 1 $M_\odot$ star with both hydrogen- and helium-burning shells, were able to show that these occurred exactly where the secular stability of the model failed.

Since then, the problem has been integrated numerically for a great variety of cases, specific effects of important factors have been analyzed, and various approximate criteria have been developed which may help in gaining a more intuitive view of the problem at least in special cases.

For instance, Henyey and l'Ecuyer (1969) and Henyey and Ulrich (1972) have emphasized the importance for this problem of the radiative diffusion times across the star or part of it and of the redistribution of entropy among the various regions of the star. This certainly plays a role in generating phase-shifts along the radius and complex eigenfunctions (cf. also Noels and Gabriel 1973; Hansen 1972).

Others like Kippenhahn and his co-workers (Kippenhahn et al. 1966; Kippenhahn 1970; Appenzeller and Kippenhahn 1971) and Rakavy and Shaviv (1968) have developed criteria in terms of a generalized specific heat covering all energy exchanges including the conversion into internal energy or vice versa of the gravitational energy released or absorbed by the hydrostatic readjustments of the configuration.

Demaret and Ledoux (1973) have shown that there is a straightforward correspondence between these criteria and the general expression (18) of s_3 and have discussed various cases like that of the supermassive stars in which secular instability occurs at the same time as dynamical instability (denominator becoming negative). Of course, in such cases, the secular instability that sets in cannot be very significant since the dynamical time scale soon becomes much shorter.

Another approach was opened by the analogy with the linear series of Poincaré (Ledoux 1960, 1963b, 1965) when the nuclear time scale τ_N is very long, so that one may think of the actual evolutionary models as a series of hydrostatic models built for successive values of the chemical composition. One may then expect that secular instability (perhaps together with dynamical instability) will occur at the singular points of the series, raising the question of the existence or of the multiplicity of the solutions of the hydrostatic equations at these points.

The existence of the upper Schönberg-Chandrasekhar (1942) limit to the relative mass $q_I = m_I/M$ of the isothermal helium core of a model having burned out its central hydrogen provided an early example of this kind of situation. If q_I, which can be taken as the parameter characterizing the chemical composition, is represented as a function for instance of the central density ρ_c, the diagram presents a series of maxima and minima (referred to also as "turning" points in the present context) and, in the case considered, for instance, by Gabriel and Ledoux (1967), there exist at least two possible models for $0.048 \leq q_I \leq 0.074$.

Since the original chemically homogeneous sphere ($q_I = 0$) is stable, one may expect that the branch rising from this initial configuration to the first maximum ($dq_I/d\rho_c > 0$) should also be stable. This was confirmed by the investigation of Gabriel and Ledoux (1967), who also found that the models became secularly unstable, beyond the first maximum, along the descending branch ($dq_I/d\rho_c < 0$). Thus, in this case, one could relate the qualitative changes in evolution and structure beyond the Schönberg-Chandrasekhar limit (Sandage and Schwarzschild 1952) to the existence of multiple solutions and the setting-in of secular instability.

Soon other examples were found for pure carbon stars (Gabriel and Noels-Grötsch 1968) or pure helium stars (Mariska and Hansen 1972; Hansen *et al.* 1972, where, however, some of the conclusions were at variance with the picture above) or even stars on the lower hydrogen main sequence (Stellingwerf and Cox 1972). In all these cases, stars on the lower end of the corresponding main sequences, with masses between the minimum mass which limits the latter and the Chandrasekhar limiting mass for degeneracy, admit at least two models: one of low and one of high degeneracy, the latter being secularly unstable.

Other examples concerned the central helium-burning phase where various methods often based on the continuity conditions at the junction between core and envelope but also sometimes involving sets of complete hydrostatic models were introduced to discuss the possibility of multiple hydrostatic solutions with a view to gain a better insight into the origin of the loops to the left in the Hertzsprung-Russell diagram (Refsdal and Weigert 1970; Kozlowski 1971; Lauterborn *et al.* 1971*a;* Lauterborn *et al.* 1971*b;* Dallaporta 1971; Lauterborn *et al.* 1972; Lauterborn 1972, 1973; Fricke and Strittmatter 1972; Murai 1974; cf. also Biermann and Kippenhahn 1971, for a model with carbon-oxygen cores).

In this context, the general problem of the existence and uniqueness of the solution of the hydrostatic equations has also received attention (Kähler 1972; Kähler and Weigert 1974; Perdang 1975*b*).

It is Paczynski (1972) who has carried furthest and in greatest detail the application of the concept of linear series and their implications for the multiplicity of hydrostatic models in thermal balance. To illustrate the problem, let us assume that the system of four linear homogeneous first-order differential equations (equivalent to eq. [1] with $s^3 = 0$) governing the secular stability of these models and the corresponding boundary conditions are written in terms of differences for some subdivision of the integration interval using, for instance, the same algorithm as in the Henyey method. The secular spectrum s_3 is found by setting the determinant $D(s)$ formed with the coefficient of these homogeneous algebraic equations equal to zero.

Let us suppose now that the Henyey method is applied to build the series of hydrostatic models in thermal balance considered here for given successive values of the chemical composition X or of a parameter q—for instance, the relative mass of a core of distinct chemical composition. To go from one model to the next corresponding to an increment ΔX or Δq, one has, in the Henyey linearized process of iteration, to solve a system of nonhomogeneous difference equations which can be compared with the homogeneous system of the secular problem above. If $s = 0$, the left-hand members of the two systems are identical while the right-hand members of the Henyey system represent the defect with which the previous model satisfies these equations when the coefficients are evaluated for the new composition. Thus these right hand members are functions of ΔX or Δq and could indeed be written explicitly, in a first approximation, by developing $\bar{\mu}$, κ, ϵ, β to the first order in ΔX or Δq when writing the linearized equations.

As long as $s_3 = 0$ is not an eigenvalue of $D(s)$, the nonhomogeneous equations in the Henyey method can be solved and a linear series of static models will ensue. However, if at one point, $D(s)$ admits a solution $s_3 = 0$, the linear non-homogeneous system, in general, has no solution unless at the same point the parameter X or q goes through an extremum ($\Delta q = 0$), i.e., q as a function of some other characteristic of the model, for instance ρ_c, has a "turning" point in the sense of Paczynski (1972).

Apparently this is by far the most common case in stellar structure, although Paczynski (1972) discusses a few possibilities of real bifurcation points—for instance, when, besides chemical composition, other parameters are involved like the mixing length in convective regions or an equation of state depending on the total mass of the star. Another case could be associated, for instance, with the substitution of convective for radiative equilibrium in some part of the star. However, in this last case, the instability associated with the bifurcation point would be dynamical and could be found only by considering non-radial perturbations.

It seems that most of these "turning" points $D(0) = 0$ correspond to points where s_3 goes through zero by real values and changes sign so that, in general, they separate stable from unstable branches in the linear series. In their investigation of a $10\,M_\odot$ star composed of a hydrogen-rich envelope joined to a helium-rich core by a transition region with a constant gradient of hydrogen with respect to $m(r)$, Kozlowski and Paczynski (1973) find no less than 12 such "turning" points, implying as many as nine different models for the same chemical composition.

However, this approach does not tell us anything about points where the real parts of complex eigenvalues vanish in changing sign unless real and imaginary parts vanish simultaneously. Unfortunately, despite some tentative discussions (Aizenman and Perdang 1971*b*; Gabriel and Noels 1972), we know little on the conditions of occurrence of complex eigenvalues or of the possible physical differences between secular instabilities entering through real or complex eigenvalues. We are also missing significant upper bounds for the real parts of the secular eigenvalues, although it is clear physically that they cannot tend to $+\infty$.

Aizenman and Perdang (1973*a, b*) have also discussed the effects of perturbations of the initial chemical abundances on secular stability and have found instabilities which are typical of the sensitivity of evolutionary tracks to changes in some of the characteristic parameters of the models.

Up to now, we have only considered the secular stability of purely static models in thermal balance, and one may feel that a generalization to actual evolutionary models is indicated. In fact, since the paper of Schwarzschild and Härm (1965), a large part of the numerical results have been obtained by integrating the system of perturbation equations corresponding to this case (cf., for instance, Schwarzschild and Härm 1965, 1967*a;* Rose 1966, 1967, 1970; Hansen *et al.* 1970, 1972; Gabriel 1972). Except for some extra terms in the energy equation, the most important of which is $\delta T(dS/dt)_0$, the differential equations keep the usual form but all the coefficients are slowly variable in time and a separation of the time by a factor e^{st} is no longer possible. Unfortunately, the expected time scale of the motion here is of the order of the actual evolutionary time scale, and asymptotic developments as used by Demaret for the vibrational stability of evolving models are not appropriate.

One may perhaps take the point of view that, if one treats all the coefficients including $(dS/dt)_0$ as frozen, one still gets some indication on the stability for a short time interval. However, if one finds some positive eigenvalues s_3, it is much more difficult to guess how effective the corresponding instability will be, as coupling with other stable modes may soon become dominant. Gabriel (1972), Noels (1972), and Gabriel and Noels (1974) have developed a somewhat different approach in terms of a local linear approximation to the evolution and its acceleration (instability) or deceleration (stability), but it is also subject to strong limitations in time.

As far as the comparison with the general time-dependent Henyey method is concerned, one could still compare the linear iterative process in the latter with the general secular stability problem. The left-hand members of the respective differences equations now become identical if the positive time increment Δt in the Henyey method is set equal to $1/s$. Thus no difficulties will arise in the solution of the set of nonhomogeneous linear equations of the iterative process as long as the s_3 are negative (stability). However if one of the s_3 becomes positive (instability), difficulties will arise if Δt is very close to $1/s_3$. The general recipe is then to shorten Δt; and one knows that thermal pulses connected with secular instabilities have been followed successfully, in this way, by Weigert (1965, 1966), Rose (1966), and others (Giannone and Weigert 1967; Hōshi 1968). That this is possible, even if costly in computer time, is very important since these pulses may allow an interpretation of some puzzling observed phenomena (Christy-Sackmann and Despain 1974) or may give rise at their peaks to new instabilities of the vibrational type (Rose 1967).

Nonradial Perturbations

The General Problem

As a fairly extensive account of the theory of nonradial perturbations has recently appeared (Ledoux 1974), I shall here only underline a few aspects which either have shed new light on the behavior of these modes or which seem susceptible of interesting developments.

The general linear equations for a spherical model in hydrostatic and thermal equilibrium (the only one which we consider here) for which the time may be separated in a factor $e^{i\sigma t}$ can be written

$$\frac{\delta\rho}{\rho} = \frac{\rho'}{\rho} + \frac{\delta r}{\rho}\frac{d\rho}{dr} = -\operatorname{div}\delta r , \tag{22}$$

$$\sigma^2 \delta r - \operatorname{grad}\left(\Phi' + \frac{p'}{\rho}\right) + \frac{r}{r}\,\mathcal{C}\,\frac{\Gamma_1 p}{\rho}\operatorname{div}\delta r$$

$$= \frac{\Gamma_3 - 1}{i\sigma}\frac{1}{\rho}\operatorname{grad}\rho\left(\epsilon - \frac{1}{\rho}\operatorname{div}\boldsymbol{F}\right)' - \frac{i\sigma}{\rho}\operatorname{div}P(\delta r) \tag{23}$$

$$p' + \Gamma_1 p\left(\frac{\rho'}{\rho} + \mathcal{C}\delta r\right) = \frac{\Gamma_3 - 1}{i\sigma}\rho\left(\epsilon - \frac{1}{\rho}\operatorname{div}\boldsymbol{F}\right)' ; \tag{24}$$

or in terms of T and p

$$T' - T\left(\frac{\Gamma_2 - 1}{\Gamma_2}\frac{p'}{p} + s\delta r\right) = \frac{1}{i\sigma C_p}\left(\epsilon - \frac{1}{\rho}\operatorname{div}\boldsymbol{F}\right)' \tag{25}$$

and

$$\nabla^2\Phi' = 4\pi G\rho' ,$$

where a prime denotes an Eulerian perturbation, Φ is the gravitational potential,

$$\epsilon = \epsilon_N - \epsilon_\nu + \frac{1}{\rho}\sum_{ik} P^{ik}\Delta_i v_k ,$$

where the last term represents the heat produced by the viscous stresses. It is convenient to have the factors

$$\mathcal{Q} = \frac{1}{\rho}\frac{d\rho}{dr} - \frac{1}{\Gamma_1 p}\frac{dp}{dr} = \frac{4-3\beta}{\beta}\mathcal{S} + \frac{1}{\bar{\mu}}\frac{d\bar{\mu}}{dr}, \qquad N^2 = -g\mathcal{Q}, \qquad (26)$$

and

$$\mathcal{S} = -\frac{\Gamma_2 - 1}{\Gamma_2}\frac{1}{p}\frac{dp}{dr} - \frac{1}{T}\frac{dT}{dr} \qquad (27)$$

appear explicitly in the equations. $\mathcal{Q}$, which is related to the local Brunt-Väisälä frequency N, reduces essentially to $\mathcal{S}$ for homogeneous chemical composition and is the general argument of the criterion for convection which develops if

$$\mathcal{Q} > 0. \qquad (28)$$

The radial component of the displacement δr, as well as p', ρ', and Φ', can be factorized in the form

$$f'(r, \theta, \phi) = f'(r)P_l{}^m(\cos\theta)e^{im\varphi}, \qquad -l \leq m \leq l.$$

The discussion may be subdivided as in the case of radial perturbations into three parts corresponding to dynamical, vibrational, and secular stability.

Dynamical Stability

This corresponds to the adiabatic approximation obtained in neglecting the nonconservative terms in the right-hand members of equations (23), (24), and (25). If we associate with this simplified problem the four boundary conditions

$$\delta r = 0 \quad \text{in} \quad r = 0$$

$$(\text{in fact,} \quad \delta r \propto r^{l-1} \quad \text{and} \quad p', \rho', \Phi' \propto r^l \quad \text{as} \quad r \to 0),$$

$$\delta p = p' + \delta r \frac{dp}{dr} = 0 \quad \text{in} \quad r = R$$

$$(\Phi'_i)_R = (\Phi'_e)_R \propto \frac{C}{R^{l+1}},$$

$$\left(\frac{d\Phi'_i}{dr}\right)_R + \frac{l(l+1)}{R}(\Phi'_i)_R = -(4\pi G\rho\delta r)_R$$

we are left with a self-adjoint fourth-order problem,

$$\int_0^R \delta r_i \cdot \delta r_k \rho d\mathcal{V} = 0, \qquad i \neq k;$$

and Chandrasekhar (1964a) and Chandrasekhar and Lebovitz (1964) have shown that the eigenvalues obey a variational principle.

The problem is $(2l+1)$ degenerate since m does not appear explicitly in it, and this degeneracy may be lifted totally by rotation and partially by a magnetic field or tidal forces. In fact, $\sigma^2 = 0$ is also a highly degenerate trivial eigenvalue, so that in presence of fields of force devoid of spherical symmetry, the number of distinct eigenvalues may be higher than $(2l + 1)$ for a given l (cf., for instance, Perdang 1968 for a general group theoretical presentation of this question).

It is known (Pekeris 1938; Cowling 1941; Ledoux and Walraven 1958) that for a given l, the eigenvalues fall into three groups, an infinite discrete acoustic spectrum (p modes) with an accumulation point at $+\infty$, an infinite discrete gravity spectrum (g modes) with an accumulation point at zero and an f mode which falls generally in between the two and corresponds to the unique mode of the homogeneous incompressible sphere discussed by Kelvin. This splitting of the spectrum is related to the fact that the parameter σ^2 enters nonlinearly in the problem (cf. Eisenfeld 1968a, b, 1969; Rosencrans 1969).

In all realistic models, the p spectrum has a positive lower bound $(\sigma_p{}^1)^2$, and dynamical instability cannot enter through these p modes.

If $\mathcal{Q}$ is negative everywhere in the star, all the $\sigma_g{}^2$ are also positive; but they become all negative if $\mathcal{Q}$ is everywhere positive. These dynamically unstable (g^-) modes really correspond to the setting in of convection. It can also be shown (Lebovitz 1965, 1966) that $\mathcal{Q} > 0$ in any finite interval, however small, implies negative $\sigma_g{}^2$ and convection in that region.

These properties appear most clearly in the second order system to which the problem reduces when Φ' is neglected and which constitutes a good approximation anyway (cf., for instance, Robe 1968):

$$\frac{dv}{dr} = \left[\frac{l(l+1)}{\sigma^2} - \frac{\rho r^2}{\Gamma_1 p}\right]\frac{p^{2/\Gamma_1}}{\rho}\, w = aw\,, \tag{29}$$

$$\frac{dw}{dr} = [\sigma^2 + \mathcal{Q}g]\frac{\rho}{r^2 p^{2/\Gamma_1}}\, v = bv\,, \tag{30}$$

where

$$v = r^2 \delta r\, p^{1/\Gamma_1}\,, \qquad w = p'/p^{1/\Gamma_1}\,.$$

Discussing the asymptotic equivalent second-order equation in $u = r^2 \delta r$ for g modes (σ^2 small), Ledoux and Smeyers (1966; cf. also Smeyers 1966) showed that, if the star comprises two regions of opposite signs in $\mathcal{Q}$ separated by the turning point $\mathcal{Q} = 0$, the g spectrum itself splits into two parts, one positive and one negative, both with accumulation points in zero. The stable g^+ modes ($\sigma_g{}^2 > 0$) oscillate only in the region $\mathcal{Q} < 0$ and decrease exponentially in the region $\mathcal{Q} > 0$, while the reverse is true of the unstable g^- modes. However, the direct discussion of equations (29) and (30) by Scuflaire (1974a) shows that while the g^- mode may indeed oscillate only in the unstable region ($\mathcal{Q} > 0$), the first g^+ modes may, in some cases, continue to oscillate in an external unstable zone ($\mathcal{Q} > 0$), especially in models of high central condensation.

If $\mathcal{Q}$ changes sign more than once (two or more turning points), Tassoul and Tassoul (1968), on the basis of an asymptotic analysis, suggested that a distinct g spectrum would be associated with each region whose eigensolutions would be negligible in the rest of the star. However, this is a delicate point which deserves a closer study, especially in the physically interesting case where a narrow radiative region is surrounded by two convective regions. For instance, the possible existence in such a case of at least some common unstable g^- modes to the two convection zones could lead to a turnover of the whole region including the radiative zone which would be more efficient than the usual penetrative convection (Saslaw and Schwarzschild 1965; Smeyers 1967; Fowley 1972).

For p modes, the radial component of the displacement tends to be larger than the horizontal component; while the reverse is true of the g modes for which the displacement field in the meridional plane suggests rather a circulation field (cf. Smeyers 1967).

For stellar models of relatively small central condensation (for instance, on the main sequence), the ordering of the modes is simple: the number of nodes of δr, ρ', and p' which is always zero for the f mode becomes one for the first p_1 mode and increases regularly with the order of the mode to which it remains equal. The same is true of the g modes if they are stable (g^+). If they are unstable, the number of nodes in ρ' and p' is still equal to the order of the mode but the number of nodes of δr is lower by one unit.

However, starting with the work of Owen (1957), it became clear that high central condensation of the model could play havoc with these simple rules, extra nodes appearing by pairs first for the f mode and then for higher and higher p and g modes as the central condensation continues to increase (Robe 1968; Dziembowski 1971; Scuflaire 1973). However, for high enough modes the number of nodes becomes regular again.

As shown by Robe (1968), this can be understood in terms of the mobile singularity depending on σ^2 corresponding to the vanishing of the coefficients of the second members of equations (29) and (30). However, a more complete picture based on the theory of the spatial oscillations of solutions of these equations has been recently developed independently by Scuflaire (1974b) and Osaki (1975).

Let us denote these coefficients respectively by a and b. The solutions of equations (29) and (30) may oscillate only in intervals where a and b have opposite signs. If we draw the curves $a = 0$ (for a given l), $b = 0$ in the plane $(x = r/R, \omega^2 = R^3\sigma^2/GM)$, we obtain figures like Figures 2 and 3, respectively, for the polytrope $n = 3$ (various values of l as indicated on the curves $a = 0$) and for the polytrope $n = 4$ ($l = 2$). These curves separate regions of acoustic (A) and gravity (G) modes as indicated.

For the polytrope $n = 3$ ($\rho_c/\bar{\rho} \approx 54$) the values of ω^2 corresponding to the two extrema of the curve $b = 0$ are not very distant, and for lower central condensation the extrema may even vanish. The values of ω^2 for various modes corresponding to $l = 25$ are indicated by horizontal lines on which the positions of the nodes of these same modes are represented by small circles. For $l = 25$, the intersection of the curves $a = 0$, $b = 0$ is higher than the maximum of the curve $b = 0$; and, as expected, all the nodes of the g modes fall in the G region and all the nodes of the p modes in the A region, and their number is equal to the order of the mode. For $l = 2$, the same intersection falls between the extrema but the corresponding ω^2 are essentially outside and the same rules apply (cf. Scuflaire 1974b).

But for the polytrope $n = 4$ ($\rho_c/\bar{\rho} = 3.42 \times 10^3$), the extrema of the curve

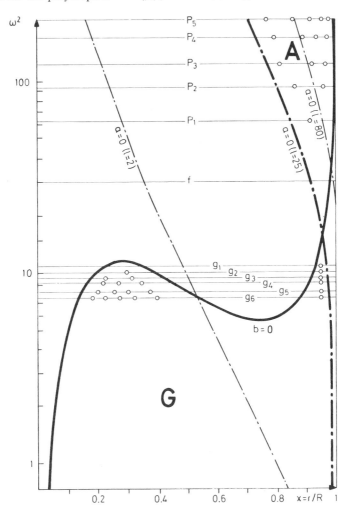

Figure 2. Distribution of the eigenvalues ω^2 and the nodes (*small circles*) for the first few modes of nonradial oscillations of the polytrope $n = 3$, for $l = 25$ in the plane $(x = r/R, \omega^2)$ subdivided in various regions by the curves $a = 0$ (drawn also for $l = 2$ and $l = 80$) and $b = 0$.

$b = 0$ are much sharper and correspond to values of ω^2 separated by quite a large interval and, for $l = 2$, the intersection of $b = 0$ and $a = 0$ falls in between. Many of the actual eigenvalues fall in this interval for $l = 2$, and many of the corresponding eigenfunctions have extra nodes some of which fall in the G region for acoustic modes and in the A region for gravity modes. This is understandable since, for a mode, either p or g, with a ω^2 falling in this interval, a and b, in this case, are of opposite sign in both regions.

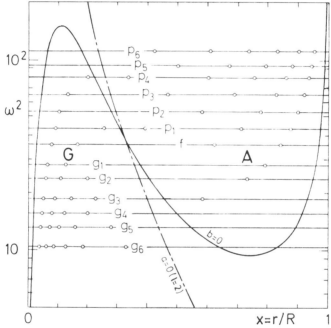

Figure 3. Same as Fig. 2 but for $n = 4$ and $l = 2$.

Of course, for ω^2 outside this interval, the nodes will be confined to their respective regions and will occur again in numbers equal to the order of the mode. But this last rule may now be extended to the first modes with extra nodes: the order of a p mode (g mode) is equal to the number of nodes in the A region (G region) minus the number of nodes in the G region (A region). This also means that, as suggested first by Dziembowski (1971), these first modes have now a mixed character of the gravity type in the central region and of the acoustic type in the external one. All this may also be illustrated on phase diagrams, in principle (v, w) (cf. Scuflaire 1974b), where the representations of p and g modes rotate around the origin in opposite senses while normally the f mode remains in the first quadrant. In the presence of extra nodes the sense of rotation (even the f mode rotates now) changes along the phase path, but the order of a mode p (g) remains equal to the number of intersections with one of the axes, the origin excluded, provided those corresponding to a counter-clockwise (clockwise) rotation are counted positively and those corresponding to a clockwise (counterclockwise) rotation, negatively.

As far as the behavior of the eigenfunctions through the star is concerned, the regular p modes behave very much like radial modes, with amplitudes (say, δr) increasing rapidly toward the surface, the more so the higher the central condensation of the model. However, exceptions with extra nodes in the central region and appreciable amplitudes there could occur when the central condensation is high enough and the p modes acquire a g character in that region. The behavior of the g modes is more difficult to describe in general, but they can reach fairly large amplitudes in the interior regions fairly close to the center (for various examples and references, cf. Ledoux 1974). Of course, for all modes, ρ' and p' (or $\delta\rho$, δp, δT), contrary to the case of radial oscillations, must vanish at the center.

Let us add a few remarks on modes for high degree l of the spherical harmonics. Apart from the work of Smeyers (1970), that of Souffrin *et al.* (1972) concerning the g^- modes associated with a convection zone, and the papers by Wolff (1972*a, b*) treating of a few p modes in the Sun, relatively little is known on their general behavior despite their obvious interest, for instance, for the 5 minute oscillation in the solar atmosphere (cf. Stein and Leibacher 1974).

It is true that p modes for high l values should have appreciable amplitudes only in the very external region, as can be seen on Figure 2 where the band (to the right of $a = 0$) in which they can oscillate regresses toward the surface as l increases. In such cases, the plane approximation nearly always used in this problem is perfectly justified (cf., for instance Michalitsanos 1973).

If this approximation, frequently used also for g^+ modes (MacKenzie, 1971; Thomas *et al.* 1971; Worral 1972), is true for physically trapped modes in a fairly narrow external layer, it may be misleading in the absence of firm grounds for such a space limitation. For instance, Figures 4 and 5 represent the distribution of the amplitudes normalized to 1 at the surface ($x = r/R = 1$) respectively for the p_3 mode ($l = 2, 25, 200$) and a few g modes ($l = 25$) in the polytrope $n = 3$ (Robe and Ledoux 1975). Figure 5 shows clearly that the first few g^+ modes can reach very high amplitudes in the interior fairly close to the center. It seems also that while the frequencies of the p modes increase continually with l, those of the g modes seem to be limited by the hump of $|Ag|$ at $x = 0.28$ in Figure 2 to some value which, for the Sun, would be of the order of 50 minutes.

Vibrational Stability

In the quasi-adiabatic approximation it is easy to generalize the coefficient of vibrational stability (3) to the nonradial case (Simon 1957). For simplicity's

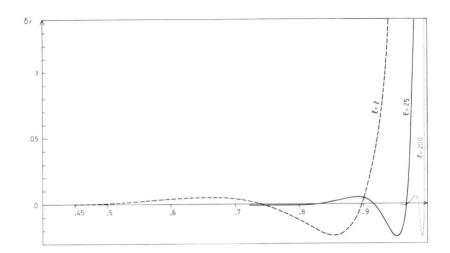

Figure 4. Behavior of δr (p_3 mode) for $l = 2, 25, 200$ in the polytrope $n = 3$.

sake let us drop terms in ϵ_ν and the viscous stresses which may, however, not
be negligible if turbulence is present (Counson *et al.* 1956), and let us assume
that the star is in radiative equilibrium. Regions in convective equilibrium can
be included (Ledoux and Walraven 1958), but their detailed treatment is
always delicate (cf., for instance, Gabriel *et al.* 1974). With the above restrictions, the damping coefficient, for the k mode of the spherical harmonic of
degree l, may be written

$$(\sigma_{q \ a}{}')_l{}^k[2(\sigma_{l,a}{}^k)^2 I] = -\int_0^R \left\{ \left[\left(\frac{\delta T}{T}\right)_{l,a}^k \right]^2 \left(\frac{\epsilon_{N,\rho}}{\Gamma_3 - 1} + \epsilon_{N,T}\right) \right.$$

$$\left. + \left(\frac{\delta T}{T}\right)_{l,a}^k \frac{l(l+1)}{(\sigma_{l,a}{}^k)^2 r^2} \chi_{l,a}{}^k \right\} 4\pi \rho \epsilon r^2 dr$$

$$+ \int_0^R \left(\frac{\delta T}{T}\right)_{l,a}^k \frac{d}{dr} \left\{ L(r) \left[4 \left(\frac{\delta r}{r}\right)_{l,a}^k + (4 - \kappa_T) \left(\frac{\delta T}{T}\right)_{l,a}^k - \kappa_\rho \left(\frac{\delta \rho}{\rho}\right)_{l,a}^k \right. \right.$$

$$\left. \left. + \frac{d/dr(\delta T/T)_{l,a}{}^k}{(1/T)(dT/dr)} - \frac{l(l+1)}{(\sigma_{l,a}{}^k)^2 r^2} \chi_{l,a}{}^k \right] \right\} dr$$

$$+ \int_0^R \left(\frac{\delta T}{T}\right)_{l,a}^k (\delta r)_{l,a}{}^k \frac{l(l+1) 16\pi a c T^4}{3\kappa \rho} S dr$$

$$+ \int_0^R \left(\frac{\delta T}{T}\right)_{l,a}^k \left(\frac{p'}{p}\right)_{l,a}^k \frac{\Gamma_2 - 1}{\Gamma_2} l(l+1) \frac{16\pi a c T^4}{3\kappa \rho} dr , \quad (31)$$

where

$$I = \int_0^R \left[(\delta r_{l,a}{}^k)^2 + \frac{l(l+1)(\chi_{l,a}{}^k)^2}{r^2(\sigma_{l,a}{}^k)^4} \right] 4\pi\rho r^2 dr$$

$$\chi_{l,a}{}^k = \left(\frac{p'}{\rho} + \Phi' \right)^k_{l,a},$$

where Φ' is often negligible and a subscript a denotes solutions of the adiabatic problem.

The essential difference with equation (3) comes from the terms in $l(l+1)$ which correspond to the horizontal gradients created by the nonradial perturbation. In particular, the third term is typical in that respect. It is proportional

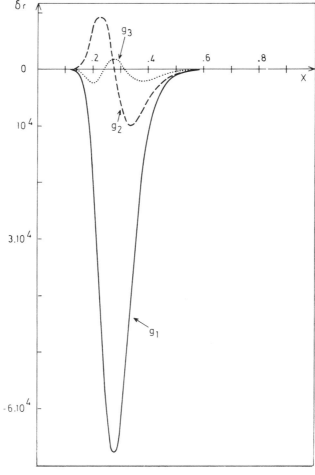

Figure 5. Behavior of δr (g_1, g_2, g_3 modes) for $l = 25$, in the polytrope $n = 3$.

to the argument S of the Schwarzschild's criterion according to which, in case of uniform chemical composition, radiative equilibrium is stable if $S < 0$. In that case, the term considered enhances vibrational stability; but in a super-adiabatic region where $S > 0$, it contributes negatively to σ' and could perhaps, in some cases, lead to vibrational instability as suggested by Spiegel (1964) and Souffrin and Spiegel (1967) for p and g^+ modes, respectively, on the basis of a local discussion.

In the same way, in the presence of a gradient of chemical composition the mean molecular weight $\bar{\mu}$ decreasing toward the exterior, radiative equilibrium remains stable even if $S > 0$ as long as $\alpha < 0$. But, in such case, the third term in (31) again contributes to vibrational instability, as was shown first by Kato (1966) using a local approach. The local treatment may be biased because of the possible stabilizing contribution to σ' of the rest of the star (cf. Gabriel 1969), but a thorough discussion of such cases may be of interest for the physics of semiconvective zones or the total heat transfer across the boundary of a convective region.

The first term in (31) corresponds essentially to the effects of energy generation; and if the latter is highly concentrated around the center, its importance should be smaller than in the radial case since all perturbations here tend to zero at the center. One may then expect that vibrational stability toward nonradial modes will be easily secured by the stabilizing influence of the radiative conductivity (second term in [31]) as confirmed by Wan's discussion (1966). This is particularly true of the p modes considering their behavior described above.

However, as we have also mentioned, g modes at least in some cases are likely to have larger amplitudes in the central region. In particular, g^+ modes may reach large amplitude in a radiative core surrounded by a fairly extensive convective envelope, and this was the reason why vibrational instability toward g^+ modes was first looked for in small-mass stars (Robe *et al.* 1972) and finally found (Noels *et al.* 1974). A similar situation may arise in evolved models with large convective envelopes in the region of the red giants.

On the other hand, Dilke and Gough (1972) suggested that such an instability could also arise in the Sun and could perhaps provide the actual mechanism for the recurrent mixing (Fowler, 1972) which could remove the solar neutrino difficulties. This was again based on a local discussion, and the first attempts at globally testing the idea gave some discordant results (Dziembowski and Sienkievicz 1973; cf. also Ledoux, 1974). However, at present it seems that general agreement (Christensen-Dalsgaard *et al.* 1974; Christensen-Dalsgaard and Gough 1975; Noels *et al.* 1974; Boury *et al.* 1975; Shibahashi *et al.* 1975) has been reached and that early evolutionary solar models (after 2.5×10^8 yr) are vibrationally unstable, at least for the lower g^+ modes corresponding to $l = 1$, and become stable again at an age of the order of 3×10^9 yr. Thus the present conventional model of the Sun is also stable. This can be understood on the basis of the high temperature sensitivity of the ^{3}He-^{3}He link in the p-p chain and the modifications of the eigensolutions for the significant g modes with the Sun's evolution.

However, if mixing results from the first instability, then multiple phases of instability could occur during the Sun's lifetime, and the actual present model could be quite different from the conventional one.

On the other hand, the e-folding time of these vibrational instabilities is still rather long, and one may wonder if they can actually lead to mixing (Ulrich and Rood 1973; Ulrich 1974). In view of the results reported above on g^+ modes corresponding to high l values in the standard model (Fig. 5), one may wonder if the latter could not lead to stronger instabilities. Because of their shorter horizontal wavelengths they may also lead to turbulence more easily and be more efficient for mixing, although this remains a very difficult problem.

These g^+ modes could also play a decisive role in the interpretation of the rapid blue variables (white dwarf variables) both as far as the periods are concerned ($\sigma_g{}^+$ fairly small because α is very close to zero and negative practically through the whole mass) (Osaki and Hansen 1973) and also, if necessary, for the excitation and maintenance of these oscillations, as it seems likely, considering the distribution of the amplitudes of these g modes in such stars, that strong vibrational instability could be associated with some hydrogen burning in the external envelope (Ledoux 1974; Brickhill 1975).

Finally, let us return to the p modes which, as we have seen, are not likely to be energized by nuclear reactions. However, as we have recalled earlier, radial modes can also be excited, for instance in the Cepheids and the RR Lyrae stars, by the Γ and κ mechanisms in the external layers, and one would be tempted to think that this could also be efficient for p modes (Zahn 1968). However, as pointed out by Dziembowski (1971), the central condensation of the appropriate models is already very high and causes the apparition of extra nodes in the central region with an increase of the amplitude (mixed g-behavior) enhancing the conductive dissipation in that region sufficiently to damp the destabilizing influence of the external layers. However, as we have seen, this abnormal behavior of the p modes ceases when $\omega_p{}^2$ is above the critical interval in Figure 2, and this will always occur even for low p modes provided the degree l of the spherical harmonics is high enough. It would be surprising if some of these p modes corresponding to high l values did not become unstable, at least in some stars, perhaps in practically all. In the case of the Sun, the possibility has been considered both for explaining the 5-minute oscillation (Wolff 1972a, b) and for heating the chromosphere and the corona (Ulrich 1970). In other stars they could also play an important role in the structure and the heating of the very external layers and perhaps become agents of mass loss in some cases.

Secular Stability

Little work has been carried on up to now on the subject. To the best of my knowledge, the first investigation is that of Kippenhahn (1967) in which he showed that significant nonradial perturbations compatible with the hydro-static equation exist, for an initially spherical star, only if a gradient of chemical composition is present. He derived a sufficient analytical condition for secular stability of a layer with such a gradient and no energy production. The latter, however, was taken into account in a local discussion (perturbations of small wavelengths) which led to the conclusion that nuclear energy generation has a destabilizing effect on a zone of variable molecular weight. The discussion was somewhat extended by Rosenbluth and Bahcall (1973) with a view especially to solar applications, but up to now it does not seem that one has come across significant secular instabilities toward nonradial perturbations.

The problem may become more important for models in rotation or for models affected by a general magnetic field or tidal effects when new nonradial types of secular instability may result from a combination of mechanical and thermal factors.

Progressive Waves and Associated Transfer Processes

All the studies reported above refer essentially to standing waves and their properties, although nonconservative terms always imply some slight progres-sive character (cf. the variable phase term in equation [15] or the corresponding term in the classical case, Ledoux and Walraven 1958, § 64). In fact, for any stationary stellar pulsation, since the energizing (for instance, by thermo-nuclear reactions in the central region) and the compensating dissipation (for instance, in fairly external layers by radiative conductivity) occur at different places, there must be a net transfer of energy from one place to the other the importance of which, as compared with, for instance, radiative conductivity, has hardly ever been discussed.

At one time (Ledoux 1963a), I thought that, because of the particular dis-tribution of their amplitudes (cf. Fig. 6, appreciable amplitudes essentially at the center and at the surface), high modes of radial pulsations of very short periods might be especially well suited for the purpose. However, it is not yet known whether there are stars in which such modes could be vibrationally un-stable and lead to maintained oscillations.

Of course, the progressive character in the outer layers can be considerably enhanced by the existence of some external medium (chromosphere, corona, interstellar medium) which will allow the filtering out at least of some of the modes (Schatzman 1956; Ledoux and Walraven 1958, § 68β; Schwarzschild and Härm 1959; Simon 1964). If the main pulsation can be maintained by energizing in the deep interior, this provides again an indirect means of transferring energy from the center to the surrounding stellar envelope.

In all these cases of maintained oscillations, the efficiency of the process could probably be roughly estimated simply by taking the positive or the negative part of equation (3) (both of which are supposed equal in absolute values here: stationary pulsation) multiplied by twice the maximum kinetic energy of pulsation for some finite reasonably small limiting amplitude. I am told that Hill (1975), who has discovered oscillations in the Sun with periods of 50 minutes or less (first radial mode or first few p modes of nonradial oscillations for l small), has advocated such extra transfer of energy to decrease the average temperatures in the central region and so relieve the neutrino difficulties—an idea which has been toyed with before, as I discussed in a conference in Montreal in 1972, but about which definite conclusions might be hard to reach.

This is of course very different from the case where a source of excitation

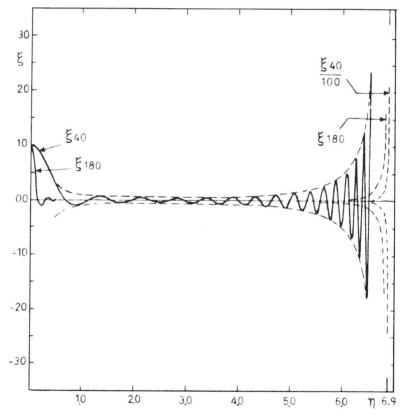

Figure 6. Behavior of the relative amplitude $\xi = \delta r/r$ for high modes (40th and 180th) of radial pulsation in the polytrope $n = 3$.

exists in the star which sends progressive waves outward as, for instance, in the most generally accepted theory of the heating of the solar chromosphere and corona (cf. Kuperus 1969) and in which the energy dissipated at high altitude is really the direct mechanical energy deposited in acoustic waves by the source (i.e., in the Sun, the external convection zone). It may be that similar possibilities in the deep interior, namely, around convective cores, should also receive a little more attention, especially since we know that even slight superadiabaticity will tend to excite and reinforce p and g^+ modes in such a core. One might perhaps also look for more sophisticated sources of perturbations in the central region even if they are very weak, since the excited progressive waves could, in some cases, pick up energy in that region and dissipate it only much farther out in the star. Even if the star is not vibrationally unstable as a whole and even if the excess amount of energy involved is very small, the process could still be efficient because the energy is carried across the star with the velocity of sound, in a very short time compared to the radiative diffusion time. Up to now very few studies have been devoted to the propagation of even linear waves inside a star taking into account the amplification or damping effects of the nonconservative terms.

One can also imagine that special types of nonradial waves could be trapped in regions where vibrational instability could tend to reinforce them continually —high harmonic p modes in the very external layers, for instance, but perhaps also high harmonic g modes at considerable depth in the star in the region of a maximum of $-Ag$ (cf. Figs. 2 and 3). In such cases, one could perhaps build up considerable storage of nonthermal kinetic energy, as strongly advocated by Thomas (1973; also Cannon and Thomas 1977). According to Thomas, it is the leakage of progressive waves from such regions which provides the real basis for the formation of chromospheres, coronae, and stellar winds.

In any case, I believe that these questions will mobilize a good deal of attention in the near future and that perhaps, as suggested in an interesting paper by Souffrin (1975), the time has come for "interior" people and "atmosphere" specialists to come together for a concerted attack on the general problem.

References

Abell, G. O., and Goldreich, P. 1966, *Pub. A.S.P.*, **78**, 232.
Aizenman, M. L., and Cox, J. P. 1974a, *Ap. J.*, **194**, 663.
———. 1974b, *Ap. J.*, **195**, 175.
Aizenman, M. L., and Perdang, J. 1971a, *Astr. and Ap.*, **12**, 232.
———. 1971b, *Astr. and Ap.*, **15**, 200.
———. 1973a, *Astr. and Ap.*, **23**, 209.
———. 1973b, *Astr. and Ap.*, **28**, 327.
Appenzeller, I: 1970a, *Astr. and Ap.*, **5**, 355.
———. 1970b, *Astr. and Ap.*, **9**, 216.

Appenzeller, I., and Fricke, K. 1972a, *Astr. and Ap.*, **18**, 10.

———. 1972b, *Astr. and Ap.*, **21**, 285.

Appenzeller, I., and Kippenhahn, R. 1971, *Astr. and Ap.*, **11**, 70.

Appenzeller, I., and Tscharnuter, W. 1974, *Astr. and Ap.*, **30**, 423.

Arnett, W. D. 1968, *Nature*, **219**, 1344.

———. 1969, *Ap. and Space Sci.*, **5**, 180.

Axel, L., and Perkins, F. 1971, *Ap. J.*, **163**, 29.

Baglin, A. 1967, *Ann. Astrophys.*, **30**, 617.

———. 1968, *Ap. Letters*, **1**, 143.

Baker, N. H. 1968, private communication.

Baker, N. H., and Kippenhahn, R. 1962, *Zs. f. Ap.*, **54**, 114.

———. 1965. *Ap. J.*, **142**, 868.

Baker, N. H., and von Sengbusch, K. 1969, *Mitt. Astr. Gesellschaft*, No. 127, p. 162.

Barkat, Z., Buchler, J. R., and Wheeler, J. C. 1972, *Ap. J.*, **173**, 83.

Biermann, P., and Kippenhahn, R. 1971, *Astr. and Ap.*, **14**, 32.

Boury, A. 1963, *Ann. Astrophys.*, **26**, 354.

———. 1964, *Ap. J.*, **164**, 1322.

Boury, A., Gabriel, M., Noels, A., Scuflaire, R., and Ledoux, P. 1975, *Astr. and Ap.*, **41**, 279.

Boury, A., and Noels, A. 1973, *Astr. and Ap.*, **24**, 255.

Brickhill, A. J. 1975, *M.N.R.A.S.*, **170**, 405.

Buchler, J. R., Wheeler, J. C., and Barkat, Z. 1974, *Ap. and Space Sci.*, **26**, 391.

Cameron, A. G. W. 1962, *Icarus*, **1**, 13.

———. 1970, *Ann. Rev. Astr. and Ap.*, **8**, 179.

Cannon, C. J., and Thomas, R. N. 1977, *Ap. J.*, **211**, 910.

Canuto, V. 1974, *Ann. Rev. Astr. and Ap.*, **12**, 167.

Castor, J. I. 1971, *Ap. J.*, **166**, 109.

Chandrasekhar, S. 1961, *Hydrodynamic and Hydromagnetic Stability* (Oxford: Clarendon Press).

———. 1964a, *Ap. J.*, **139**, 664.

———. 1964b, *Ap. J.*, **140**, 417.

———. 1969, *Ellipsoidal Figures of Equilibrium* (New Haven: Yale University Press).

Chandrasekhar, S., and Lebovitz, N. 1964, *Ap. J.*, **140**, 1517.

Chandrasekhar, S., and Tooper, R. F. 1964, *Ap. J.*, **139**, 1396.

Chiu, H. Y. 1961, *Phys. Rev.*, **123**, 1040.

———. 1964, *Ann. Phys.*, **26**, 364.

Christensen-Dalsgaard, J., Dilke, F. N. W., and Gough, D. O. 1974, *M.N.R.A.S.*, **169**, 429.

Christensen-Dalsgaard, J., and Gough, D. O. 1975, *Proc. 19th Astrophysical Colloquium Liège* (*Mém. Soc. Roy. Sci. Liège*, 6th Ser., **8**, 309).

Christy, R. F. 1962, *Ap. J.*, **136**, 897.

Christy-Sackmann, I.-J., and Despain, K. H. 1974, *Ap. J.*, **189**, 523.

Colgate, S. A., and White, R. H. 1966, *Ap. J.*, **143**, 626.

Conti, P. S., and Burnichon, M.-L. 1975, *Astr. and Ap.*, **38**, 467.

Counson, J., Ledoux, P., and Simon, R. 1956, *Bull. Soc. Roy. Sci. Liège*, **36**, 144.

Cowling, T. G. 1941, *M.N.R.A.S.*, **101**, 367.

Cox, J. P. 1963, *Ap. J.*, **138**, 487.

———. 1974*a*, *Ap. J. (Letters)*, **192**, L85.

———. 1974*b*, *Rept. Progr. Phys.*, **37**, 563–598.

Cox, J. P., Davey, W. R., and Aizenman, M. L. 1974, *Ap. J.*, **191**, 439.

Cox, J. P., Hansen, C. J., and Davey, W. R. 1973, *Ap. J.*, **182**, 885.

Dallaporta, N. 1971, in *Supergiant Stars, Proc. 3d Colloquium on Astrophysics*, ed. M. Hack (Trieste: Osservatorio Astronomico).

Davey, W. R., and Cox, J. P. 1974, *Ap. J.*, **189**, 113.

Demaret, J. 1974*a*, *Bull. Acad. Roy. de Belgique*, Cl. Sci., **60**, 183.

———. 1974*b*, *Ap. and Space Sci.*, **31**, 305.

———. 1975*a*, *Ap. and Space Sci.*, **33**, 189.

———. 1975*b*, *Proc. 19th Astrophysical Colloquium Liège* (*Mém. Soc. Roy. Sci. Liège*, 6th Ser., **8**, 161).

———. 1975*c*, private communication.

Demaret, J., and Ledoux, P. 1973, *Astr. and Ap.*, **23**, 111.

Dilke, F. W. W., and Gough, D. O. 1972, *Nature*, **240**, 262.

Dziembowski, W. 1971, *Acta Astr.*, **21**, 289.

Dziembowski, W., and Sienkievicz, R. 1973, *Acta Astr.*, **23**, 273.

Eisenfeld, J. 1968*a*, *J. Math. Anal. Appl.*, **23**, 58.

———. 1968*b*, *J. Math. Anal. Appl.*, **26**, 357.

———. 1969, *J. Math. Mech.*, **18**, 991.

Ferrini, F. 1975, *Ap. and Space Sci.*, **32**, 231.

Feshchenko, S. F., Shkil', N. I., and Kikolenko, L. D. 1967, *Asymptotic Methods in the Theory of Linear Differential Equations* (New York: American Elsevier).

Fowler, W. A. 1964, "The General Instability of Massive Stars," Joint Discussion on Radio Galaxies, 12th IAU Meeting, Hamburg.

———. 1972, *Nature*, **238**, 24.

Fowley, W. M. 1972, *Ap. J.*, **180**, 483.

Fricke, K. J., and Strittmatter, P. A. 1972, *M.N.R.A.S.*, **156**, 129.

Gabriel, M. 1964, *Ann. Astrophys.*, **27**, 141.

———. 1967, *Ann. Astrophys.*, **30**, 745.

———. 1969, *Astr. and Ap.*, **1**, 321.

———. 1972, *Astr. and Ap.*, **18**, 242.

Gabriel, M., and Ledoux, P. 1967, *Ann. Astrophys.*, **30**, 975.

Gabriel, M., and Noels-Grötsch, A. 1968, *Ann. Astrophys.*, **31**, 167.

Gabriel, M., and Noels, A. 1972, *Astr. and Ap.*, **20**, 455.

———. 1974, *Astr. and Ap.*, **30**, 339.

Gabriel, M., Scuflaire, R., Noels, A., and Boury, A. 1974, *Bull. Acad. Roy. Belgique*, Cl. Sci., 5th Ser., **60**, 866.

Giannone, P., and Weigert, A. 1967, *Zs. f. Ap.*, **67**, 41.

Glansdorff, P., and Prigogine, I. 1971, *Structure, stabilité et fluctuations* (Paris: Masson).

Hansen, C. J. 1972, *Astr. and Ap.*, **19**, 71.

——. 1974, ed., *Physics of Dense Matter, IAU Symposium No. 53* (Dordrecht: Reidel).

Hansen, C. J., Cox, J. P., and Herz, M. A. 1970, *Bull. AAS*, **2**, 319.

——. 1972, *Astr. and Ap.*, **19**, 144.

Hansen, C. J., and Spangenberg, W. H. 1971a, *Ap. J.*, **163**, 653.

——. 1971b, *Ap. J.*, **168**, 71.

Härm, R., and Schwarzschild, M. 1972, *Ap. J.*, **172**, 403.

Hayashi, C. 1966, *Ann. Rev. Astr. and Ap.*, **4**, 171.

Heintzmann, H., and Hillebrandt, W. 1975, *Astr. and Ap.*, **38**, 51.

Henyey, L., and L'Ecuyer, J. 1969, *Ap. J.*, **156**, 549.

Henyey, L., and Ulrich, R. K. 1972, *Ap. J.*, **173**, 109.

Hill, H. A. 1975, Communication to the Working Group on Convection, Cambridge, England, 1975 June.

Hōshi, R. 1968, *Progr. Theor. Phys.*, **39**, 957.

Howard, W. M., Arnett, W. D., and Stanford, E. 1972, *Ap. J.*, **175**, 201.

Hoyle, F. 1946, *M.N.R.A.S.*, **106**, 343.

Iben, I. 1971, *Ap. J.*, **166**, 131.

——. 1972, *Ap. J.*, **178**, 433.

Imshennik, V. S., and Nadyazhin, D. K. 1974, *Astr. Sci. Inform.*, **29**, 27.

Itoh, N. 1969, *Progr. Theoret. Phys.*, **41**, 1211.

Jeans, J. H. 1928, *Astronomy and Cosmogony* (Cambridge: Cambridge University Press), sections 104–111.

Kähler, H. 1972, *Astr. and Ap.*, **20**, 105.

Kähler, H., and Weigert, A. 1974, *Astr. and Ap.*, **30**, 431.

Kantorovich, L. V., and Krylov, V. I. 1958, *Approximate Methods of Higher Analysis*, transl. C. D. Benster (Netherlands: P. Noordhoff).

Kato, S. 1966, *Pub. Astr. Soc. Japan*, **18**, 374.

Kato, S., and Unno, W. 1967, *Pub. Astr. Soc. Japan*, **19**, 1.

Kippenhahn, R. 1967, *Zs. f. Ap.*, **67**, 271.

——. 1970, *Astr. and Ap.*, **8**, 52.

Kippenhahn, R., Thomas, H.-C., and Weigert, A. 1966, *Zs. f. Ap.*, **64**, 373.

Kozlowski, M. 1971, *Ap. Letters*, **9**, 65.

Kozlowski, M., and Paczynski, B. 1973, *Acta Astr.*, **23**, 65.

Kuperus, M. 1969, *Space Sci. Rev.*, **9**, 713.

Larson, R. B. 1973, *Fund. Cosmic Phys.*, **1**, 1.

Larson, R. B., and Starrfield, S. 1971, *Astr. and Ap.*, **13**, 190.

Lauterborn, D. 1972, *Astr. and Ap.*, **19**, 473.

——. 1973, *Astr. and Ap.*, **24**, 421.

Lauterborn, D., Refsdal, S., and Roth, M. L. 1971a, *Astr. and Ap.*, **13**, 119.

Lauterborn, D., Refsdal, S., and Stabell, R. 1972, *Astr. and Ap.*, **17**, 113.

Lauterborn, D. Refsdal, S., and Weigert, A. 1971b, *Astr. and Ap.*, **10**, 97.

Lebovitz, N. R. 1965, *Ap. J.*, **142**, 229.

——. 1966, *Ap. J.*, **146**, 946.

Ledoux, P. 1941, *Ap. J.*, **94**, 537.

――――. 1945, *Ap. J.*, **102**, 143.

――――. 1958, in *Handbuch der Physik*, Vol. **51**, p. 605.

――――. 1960, *Bull. Acad. Roy. Belgique*, Cl. Sci., 5th Ser., **46**, 429.

――――. 1963*a*, *Bull. Acad. Roy. Belgique*, Cl. Sci., 5th Ser., **49**, 286.

――――. 1963*b*, *Stellar Stability and Stellar Evolution*, *Rend. d. Scuola Int. d. Fisica "E. Fermi,"* XXVIII Corso 1962, in *Star Evolution*, ed. L. Gratton (New York: Academic Press).

――――. 1965, in *Stars and Stellar Systems*, Vol. **8**, ed. L. Aller and D. McLaughlin (Chicago: University of Chicago Press), p. 499.

――――. 1969, in *La Structure interne des étoiles*, XIe Cours de Perfectionnement de l'Association Vaudoise des Chercheurs en Physique, Saas-Fee.

――――. 1974, in *Stellar Instability and Evolution*, *IAU Symposium No. 59*, ed. P. Ledoux, A. Noels, and A. W. Rogers (Dordrecht: Reidel).

Ledoux, P., and Pekeris, C. L. 1941, *Ap. J.*, **94**, 245.

Ledoux, P., and Sauvenier-Goffin, E. 1950, *Ap. J.*, **111**, 611.

Ledoux, P., and Smeyers, P. 1966, *C. R. Acad. Sci. Paris*, **262**, 841.

Ledoux, P., and Walraven, Th. 1958, in *Handbuch der Physik*, Vol. **51**, p. 353.

Lee, T. D. 1950, *Ap. J.*, **111**, 625.

Liepholz, H. 1970, *Stability Theory* (New York and London: Academic Press).

Mackenzie, J. F. 1971, *Astr. and Ap.*, **15**, 450.

Mariska, J. T., and Hansen, C. J. 1972, *Ap. J.*, **171**, 317.

Marshall, M. P., and Van Horn, H. M. 1973, *Ap. J.*, **182**, 517.

Mestel, L. 1952, *M.N.R.A.S.*, **112**, 583.

Meurice, P. 1964, Etude de la stabilité vibrationelle d'étoiles en contraction gravifique, Th. de Licence, Université de Liège.

Michalitsanos, A. G. 1973, *Earth and Extraterr. Sci.*, **2**, 125.

Murai, T. 1974, *Pub. Astr. Soc. Japan*, **26**, 323.

Murphy, J. O. 1967, private communication.

Narita, S., Nakano, T., and Hayashi, C. 1970, *Progr. Theoret. Phys.*, **43**, 942.

Noels-Grötsch, A. 1967, *Ann. Astrophys.*, **30**, 349.

Noels, A. 1972, *Astr. and Ap.*, **18**, 350.

Noels, A., Boury, A., Scuflaire, R., and Gabriel, M. 1974, *Astr. and Ap.*, **31**, 185.

Noels, A., and Gabriel, M. 1973, *Astr. and Ap.*, **24**, 201.

Noels, A., Gabriel, M., Boury, A., Scuflaire, R., and Ledoux, P. 1975, *Proc. 19th Astrophysical Colloquium Liège* (*Mém. Soc. Roy. Sci. Liège*, 6th Ser., **8**, 317).

Noels-Grötsch, A., Boury, A., and Gabriel, M. 1967, *Ann. Astrophys.*, **30**, 13.

Okamoto, I. 1967, *Pub. Astr. Soc. Japan*, **19**, 384.

Opoien, J. W., and Grossman, A. S. 1974, *Astr. and Ap.*, **37**, 335.

Osaki, Y. 1966, *Pub. Astr. Soc. Japan*, **18**, 384.

――――. 1975, *Pub. Astr. Soc. Japan*, **27**, 237.

Osaki, Y., and Hansen, C. J. 1973, *Ap. J.*, **185**, 277.

Owen, J. W., 1957, *M.N.R.A.S.*, **117**, 384.

Paczynski, B. 1972, *Acta Astr.*, **22**, 164.

Paczynski, B., and Ziołkowski, J. 1968, *Acta. Astr.*, **18**, 255.

Papaloizou, J. C. 1973*a*, *M.N.R.A.S.*, **162**, 143.

Papaloizou, J. C. 1973*b*, *M.N.R.A.S.*, **162**, 169.
Pekeris, C. L. 1938, *Ap. J.*, **88**, 189.
Perdang, J. 1968, *Ap. and Space Sci.*, **1**, 355.
———. 1975*a*, *Adv. Chem. Phys.*, **32**, 207.
———. 1975*b*, *Ap. and Space Sci.*, **36**, 111.
Prigogine, I., and Glansdorff, P. 1965, *Physica*, **31**, 1242.
Rakavy, G., and Shaviv, G. 1967, *Ap. J.*, **148**, 803.
———. 1968, *Ap. and Space Sci.*, **1**, 347.
Refsdal, S., and Weigert, A. 1970, *Astr. and Ap.*, **6**, 426.
Robe, H. 1968, *Ann. Astrophys.*, **31**, 475.
Robe, H., and Ledoux, P. 1975, *Bull. Acad. Roy. Sci. Belgique*, Cl. Sci., **61**, 198.
Robe, H., Ledoux, P., and Noels, A. 1972, *Astr. and Ap.*, **18**, 424.
Rose, W. K. 1966, *Ap. J.*, **146**, 838.
———. 1967, *Ap. J.*, **150**, 193.
———. 1970, *Ap. J.*, **159**, 903.
Rosenbluth, M. N., and Bahcall, J. N. 1973, *Ap. J.*, **184**, 9.
Rosencrans, S. 1969, *J. Math. Anal. Appl.*, **25**, 616.
Roxburgh, I. W. 1967, *Nature*, **215**, 838.
Russell, H. N. 1925, *Nature*, **116**, 209.
Sandage, A. R., and Schwarzschild, M. 1952, *Ap. J.*, **116**, 463.
Saslaw, W. S., and Schwarzschild, M. 1965, *Ap. J.*, **142**, 1468.
Sastri, V. K., and Simon, N. R. 1973, *Ap. J.*, **186**, 997.
Sastri, V. K., and Stothers, R. 1974, *Ap. J.*, **193**, 677.
Schatzman, E. 1956, *Ann. Astrophys.*, **19**, 45.
———. 1958, *White Dwarfs* (Amsterdam: North-Holland).
Schechter, R. S. 1967, *The Variational Method in Engineering* (New York: McGraw-Hill).
Schönberg, M., and Chandrasekhar, S. 1942, *Ap. J.*, **96**, 161.
Schwarzschild, M., and Härm, R. 1959, *Ap. J.*, **129**, 637.
———. 1965, *Ap. J.*, **142**, 855.
———. 1967*a*, *Ap. J.*, **150**, 961.
———. 1967*b*, Communication at 12th IAU General Assembly, Prague.
Scuflaire, R. 1973, private communication.
———. 1974*a*, *Astr. and Ap.*, **34**, 449.
———. 1974*b*, *Astr. and Ap.*, **36**, 107.
———. 1975, Ph.D. thesis, Université de Liège.
Shibahashi, H., Osaki, Y., and Unno, W. 1975, *Pub. Astr. Soc. Japan*, **27**, 401.
Simon, N. R. 1970, *Ap. J.*, **159**, 859.
———. 1971, *Ap. J.*, **164**, 331.
Simon, N. R., and Sastri, V. K. 1972, *Astr. and Ap.*, **27**, 39.
Simon, N. R., and Stothers, R. 1970, *Astr. and Ap.*, **6**, 183.
Simon, R. 1957, *Bull. Acad. Roy. Belgique*, Classe Sci., **43**, 610.
———. 1964, *Astrophys. Norv.*, **9**, 113.
Smeyers, P. 1966, *Ann. Astrophys.*, **29**, 539.
———. 1967, *Bull. Soc. Roy. Sci. Liège*, **36**, 357.

————. 1970, *Astr. and Ap.*, **7**, 204.

Smith, R. L., and Rose, W. K. 1972, *Ap. J.*, **176**, 395.

Souffrin, P. 1975, Wave propagation in solar atmosphere (preprint).

Souffrin, P., Grisvard, P., and Zerner, M. 1972, *Astr. and Ap.*, **17**, 309.

Souffrin, P., and Spiegel, E. 1967, *Ann. Astrophys.*, **30**, 985.

Spiegel, E. 1964, *Ap. J.*, **139**, 959.

Stein, R. F., and Leibacher, J. 1974, *Ann. Rev. Astr. and Ap.*, **12**, 407.

Stellingwerf, R. F. 1974, *Ap. J.*, **192**, 139.

Stellingwerf, R. F., and Cox, J. P. 1972, *Astr. and Ap.*, **19**, 8.

Stothers, R. 1976, *Ap. J.*, **204**, 853.

Stothers, R., and Simon, N. R. 1970, *Ap. J.*, **160**, 1019.

Talbot, R. J. 1971*a*, *Ap. J.*, **163**, 17.

————. 1971*b*, *Ap. J.*, **165**, 12.

Tassoul, M., and Tassoul, J.-L. 1968, *Ann. Astrophys.*, **31**, 251.

Thom, R. 1972, *Stabilité structurelle et Morphogenèse* (New York: Benjamin).

————. 1974, *Structural Stability and Morphogenesis*, transl. D. H. Fowler (Reading, Mass.: Addison-Wesley).

Thomas, J. H., Clark, P. A., and Clark, A. 1971, *Solar Phys.*, **16**, 51.

Thomas, L. H. 1931, *M.N.R.A.S.*, **91**, 122, 619.

Thomas, R. N. 1973, *Astr. and Ap.*, **29**, 297.

Thompson, J. M. T. 1969, *Z. Angew. Math. and Phys.*, **20**, 797.

————. 1975*a*, *Phys. Letters*, **51A**, 201.

————. 1975*b*, *Nature*, **254**, 392.

Thompson, J. M. T., and Hunt, G. W. 1973, *A General Theory of Elastic Stability* (New York: Wiley).

Tolman, R. C. 1939, *Ap. J.*, **90**, 541, 568.

Toma, E. 1972, *Astr. and Ap.*, **19**, 76.

Ulrich, R. K. 1970, *Ap. J.*, **162**, 993.

————. 1974, *Ap. J.*, **188**, 369.

Ulrich, R. K., and Rood, R. T. 1973, *Nature*, **241**, 111.

Unno, W. 1965, *Pub. Astr. Soc. Japan*, **17**, 205.

————. 1968, *Pub. Astr. Soc. Japan*, **20**, 356.

Van der Borght, R. 1968, *Bull. Acad. Roy Belgique*, Cl. Sci., 5th Ser., **54**, 1159.

————. 1969, *Australian J. Phys.*, **22**, 497.

Van Horn, H. M., Richardson, M. B., and Hansen, C. J. 1972, *Ap. J.*, **172**, 181.

von Sengbusch, K. 1975, *Proc. 19th Astrophysical Colloquium Liège (Mém. Soc. Roy. Sci. Liège*, 6th Ser., **8**, 189).

Vauclair, G. 1971, *Ap. Letters*, **9**, 161.

Vila, S. C. 1971, *Ap. J.*, **163**, 543.

Wan, F. S. 1966, Ph.D. thesis, Australian National University.

Weigert, A. 1965, *Mitt. Astr. Gesellschaft*, **25**, 61.

————. 1966, *Zs. f. Ap.*, **64**, 395.

Weinberger, H. F. 1968, *J. Math. Anal. Appl.*, **21**, 506.

Wheeler, J. A. 1966, *Ann. Rev. Astr. and Ap.*, **4**, 393.

Wheeler, J. C., Hansen, S. T., and Cox, J. P. 1968, *Ap. Letters*, **2**, 253.

Whitney, C. 1956, *Ann. Astrophys.*, **19**, 34.

Wolff, C. L. 1972*a*, *Ap. J.*, **176**, 883.

————. 1972*b*, *Ap. J. (Letters)*, **177**, L87.

Wood, P. R. 1974, *Ap. J.*, **190**, 609.

Worral, G. 1972, *Ap. J.*, **172**, 749.

Zahn, J. P. 1968, *Ap. Letters*, **1**, 209.

Zel'dovich, Ya. B., and Novikov, I. D. 1971, *Relativistic Astrophysics*, ed. K. S. Thorne and W. D. Arnett (Chicago: University of Chicago Press).

Zhevakin, S. A. 1953, *Astr. Zh.*, **30**, 161.

————. 1963, *Ann. Rev. Astr. and Ap.*, **1**, 367.

Ziebarth, K. 1970, *Ap. J.*, **162**, 947.

3

The Influence of Rotation on Stars and Stellar Systems

Jeremiah P. Ostriker

Introduction

The general problem of determining the equilibrium and stability of rotating self-gravitating objects is ancient. It has attracted mathematicians from Maclaurin in the eighteenth century to Chandrasekhar in the twentieth; along the way Poincaré, Riemann, Jeans, Darwin, and many others have left their names attached to significant solutions and theorems. However, despite this history, Lord Kelvin was able to write (in Thomson and Tait 1912)

> No one seems yet to have attempted to solve the general problem of finding all the forms of equilibrium which a mass of homogeneous incompressible fluid rotating with uniform angular velocity may assume. Unless the velocity be so small that the figure differs but little from a sphere, the problem presents difficulties of an exceedingly formidable nature.

Thus, if even the simplest ($\rho = $ const., $\Omega = $ const.) problem is difficult, one may imagine the complexities to be encountered in studying "real" rotating stars or stellar systems. The problem defined by Kelvin and further restricted to the case in which the zero-pressure surface is described by an ellipsoid has been studied in exhaustive detail. Chandrasekhar and co-workers, particularly N. Lebovitz, have made substantial contributions to this problem, and the best review is contained in the former's book *Ellipsoidal Figures of Equilibrium* (1969).

In the last decade the development of electronic computers has permitted the study of compressible, differentially rotating, nonspherical equilibrium figures by approximate numerical techniques. Fortunately, the solutions found for these more realistically prescribed models have striking family resemblances to the classical exactly solved problem, so it is extremely useful to first review the problem of Maclaurin.

Polytropic Sequences

Equilibrium

Consider a fluid self-gravitating object with given total mass M and angular momentum J. Construct a first model within which the density, ρ, is very small

Jeremiah P. Ostriker is at the Princeton University Observatory, Princeton, New Jersey.
This work was supported in part by National Science Foundation grant MPS74-18970.

and constant and the rotation is described by a uniform angular velocity Ω. In the limit $\rho \to 0$ it follows that $\Omega \to 0$ and a spherical shape is obtained. Define this to be the first member of a sequence. We may consider this object a polytropic sphere of index $n = 0$ if we follow the convention of relating the pressure to the density by $P = \text{const. } \rho^{(n+1)/n}$. Then we can construct successive members of the $n = 0$ sequence in the following way. We imagine the density of each fluid element in the original configuration to increase by a given factor, the angular momentum of each element remaining constant (no viscosity). Then it can be shown that the new configuration will also be a uniform, uniformly rotating, spheroid. Successive members will rotate more and more rapidly; the eccentricity of the surface also increases monotonically along such a sequence, finally approaching unity as the density approaches infinity. It is convenient to label each member of the sequence by a dimensionless parameter measuring, in an integrated fashion, the importance of rotation. We define a quantity equal to the ratio of kinetic energy of rotation to gravitational self energy

$$t \equiv |T_{\text{rot}}/W_{\text{grav}}| = \left[\frac{(3 - 2e^2)\sin^{-1} e - 3e(1 - e^2)^{1/2}}{2e^2 \sin^{-1} e}\right], \tag{1}$$

where the relation between t and eccentricity e is derived from the known properties of the $(n = 0)$ Maclaurin sequence. If it is noted that the internal energy U_{int} is always positive, the requirement that an object satisfy the equilibrium virial theorem $2T_{\text{rot}} - W_{\text{grav}} = -2U_{\text{int}} < 0$ implies that $0 < t < \frac{1}{2}$. Figures 1a, 1b, and 1c show the dependence of rotation velocity, axis ratio, and central density on the parameter t. One seemingly trivial point is important: *there exists an equilibrium configuration for every allowed value of t $(0 < t < \frac{1}{2})$.* That is, there is no limiting value of $t < \frac{1}{2}$ where objects "rotationally shed matter" or "break up due to rotational forces." This is obviously not a result of the way we have defined the sequence. Had we imagined the sequence constructed of objects having fixed density and mass but increasing angular momentum, they would of course map identically on to the $(n = 0)$ curves of Figure 1 since the plotted quantities are dimensionless. A maximum is reached in the curve for $\langle \Omega \rangle / (4\pi G \rho_c)^{1/2}$ at $t = 0.2379$, but this does not limit the objects in any way; it simply reflects the fact that, for rapidly rotating stars $(t \gtrsim \frac{1}{4})$, the moment of inertia about the rotation axis increases with J faster than J itself does, resulting in a decrease of $J/I \equiv \langle \Omega \rangle$.

Now we are ready to consider generalized polytropic sequences defined by two numbers (n, n') representing the distributions of density and angular momentum. The first member of a given sequence is (in the limit $\rho \to 0$) a nearly spherical polytrope of index n (cf. Chandrasekhar 1939). It is assumed to be slowly rotating with rotation constant on cylindrical surfaces. The angular momentum distribution [number of mass elements with specific angular momentum in the interval $j(m) \to j(m) + dj$] is assumed to be the same as in a uniformly rotating polytrope of index n'. Thus the Maclaurin sequence is the $(0, 0)$ generalized polytropic sequence. Subsequent members of a given sequence are constructed by imagining an increase in the central density, the density increasing for all other elements in such a manner that it remains in polytropic equilibrium. Each element conserves its angular momentum, and rotation is constrained to be constant on cylindrical surfaces. Naturally, the equipotential surfaces are no longer similar spheroids (cf. Fig. 2). Generalized zero-viscosity sequences for astrophysically interesting values of n (e.g., 3/2, 3) have been computed by Bodenheimer and Ostriker (1973) using the self-consistent-field method developed by Ostriker and Mark (1968). Some of the results, plotted in Figures 1 and 2, show that the generalized sequences appear to be in every way analogous to the classical $(0, 0)$ sequence. In particular, no "rotational mass ejection" was ever found, a fact that can be understood on the basis of simple physical arguments. In the limit of rapid rotation $(t \rightarrow \frac{1}{2})$, centrally concentrated configurations must and do approach a Keplerian angular velocity distribution. The angular velocity distribution is of course a calculated quantity

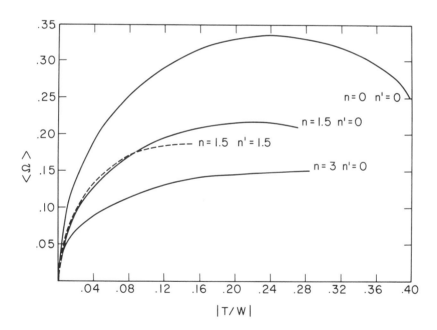

Figure 1a. The average angular velocity $\langle \Omega \rangle = J/I$ for the generalized polytropic sequences, plotted as a function of $|T/W|$. I is the moment of inertia about the rotation axis. The unit of $\langle \Omega \rangle$ is $(4\pi G \rho_c)^{1/2}$.

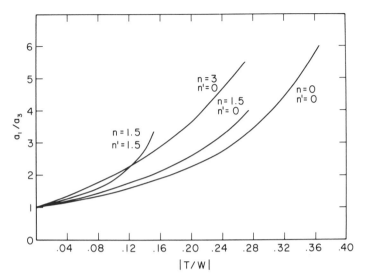

Figure 1b. The ratio of equatorial and polar radii for four polytropic sequences plotted as a function of $|T/W|$.

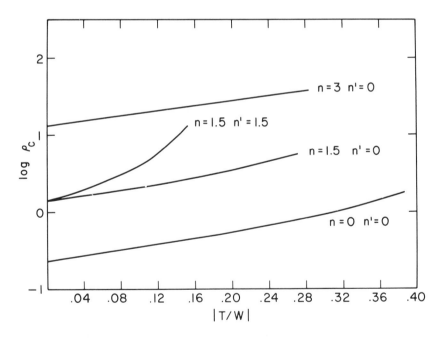

Figure 1c. The logarithm of the central density, in the unit M/a_1^3, for the generalized polytropic sequences plotted as a function of $|T/W|$.

which cannot be prescribed *ab initio*, just as in the solar system one may always order the spacing of the planets to satisfy any proposed $j(m)$ relation but $\Omega(r)$ is fixed at the Keplerian relation.

Stability of the Kelvin Modes

Although satisfying the force balance equations does not appear to in any way limit the total angular momentum of self-gravitating objects having pre-scribed distributions of specific angular momentum, it is entirely possible that the more rapidly rotating members of the sequences are unstable to small perturbations, and so would be expected to have only a transitory existence. Fluid motions associated with several different modes tend to be stabilized by rotation. Convective modes are the classic example of this, but the situation is somewhat complicated (cf. Lebovitz 1967). Rotation has a destabilizing effect,

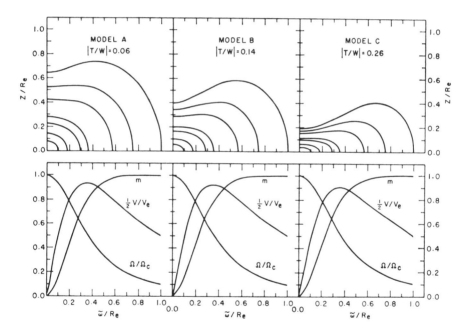

Figure 2. Detailed structure of three models along the $(n, n') = (3, 0)$ sequence, il-lustrating the effects of an increase in t. The radius in the equatorial plane is indicated by $\tilde{\omega}$; that along the axis of rotation, by z. R_e = total equatorial radius. Upper por-tions give equidensity contours of densities 0.8, 0.5, 0.2, 0.1, 0.01, 10^{-3}, and 0 times the central value. Lower portions give the ratio of angular velocity Ω to the central value Ω_c, the fraction m of the total mass interior to the corresponding cylindrical surface about the rotation axis, and the ratio of the circular velocity V to the surface value V_e (taken from Bodenheimer and Ostriker 1970).

however, on certain of the lowest nonaxisymmetric Kelvin modes. All of the $l = 2$, $m = (2, 1, 0, -1, -2)$, $n = 0$ modes have frequency $\sigma_K{}^2 = \frac{4}{5}(W/I)$ in the degenerate case of no rotation. Ledoux (1945) has shown that the above expression for $\sigma_K{}^2$ is an excellent approximation for inhomogeneous stars ($n \neq 0$) and exact for the homogeneous ($n = 0$) case. Rotation splits these modes and, at $t = 0.1376$, one of the $m = \pm 2$ modes reaches zero frequency (cf. Fig. 3a, which mode depends on whether one works in the laboratory or rotating frame). The point of neutral stability is a manifestation of the fact that, at this point along the sequence, an infinitesimal perturbation can be found which will transform the spheroid into a lower energy triaxial Jacobi or Dedekind ellipsoid having the same density, mass, and angular momentum. For rotation with $0.1376 < t < 0.2738$, Maclaurin spheroids are unstable in the presence of viscosity (Roberts and Stewartson 1963) or gravitational radiation (Chandrasekhar 1970) on the dissipative time scale. Above $t = 0.2738$, dynamical instability sets in via the same modes (cf. Lebovitz 1961) and the further existence of the equilibrium sequence is of no real interest. At present it is not at all clear

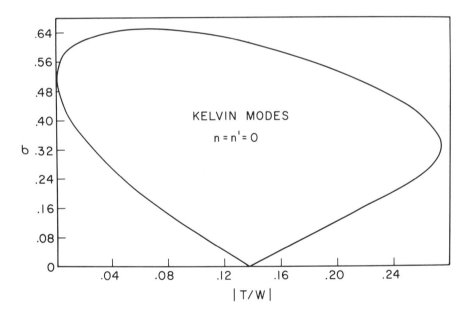

Figure 3a. Frequencies of the Kelvin modes belonging to $l = 2$, $m = \pm 2$ as a function of rotation parameter t for the sequence of Maclaurin spheroids. In all graphical and tabular results the frequencies are measured in the laboratory frame and are given in the unit $(4\pi G\rho_c)^{1/2}$.

what the consequences of dynamical instability would be. In the time-dependent calculations of Rossner (1967) and Fujimoto (1968) very elongated objects are generated which, in turn, are probably unstable to higher harmonics leading to fission.

The virial tensor techniques for studying stability, which are exact for homogeneous ellipsoids, were generalized by Tassoul and Ostriker (1968) and were found to be an efficient, approximate method for studying the stability to the Kelvin modes of rapidly rotating inhomogeneous stars. Some of the results (from Ostriker and Bodenheimer 1973) are shown in Figure 3b. A point of bifurcation was found along all sequences, and along all sequences a point of dynamical instability was found which could be studied to large enough values of the rotation parameter t. The results show that the points of bifurcation t_b (Table 1) are remarkably independent of (n, n'), always occurring quite close to 0.138; the points of dynamical instability t_d are similarly close to 0.27. Various tests applied in the above quoted papers indicated that the approximate methods used were probably accurate to within several percent. Subsequently, however, Friedman and Shutz (1975) and Hunter (1975) have shown by exact analytical techniques that the appearance of a point of bifurcation

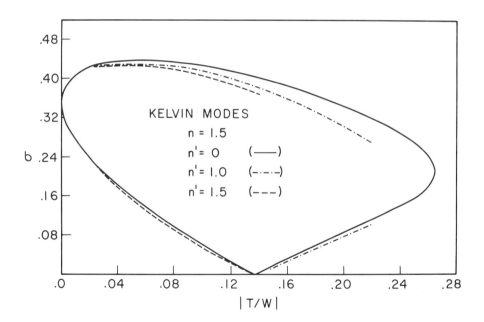

Figure 3b. Same as Fig. 3a, for polytropic sequences of $n = 3/2$ with three different distributions of angular momentum.

along a sequence as determined by the virial tensor method is a sufficient (but not necessary) condition for instability. That is, the true points of bifurcation occur for smaller values of t than those quoted in Table 1.

Stellar Systems

Equilibrium

In a steady axisymmetric stellar system, the distribution of velocities and positions must be only a function of the energy, the angular momentum about the symmetry axis, and a third integral if such exists (cf. Contopoulos in this volume). This prescription allows so much freedom that there is very little one can say qualitatively about the nature of rotating stellar systems. For example, the fact that the pressure tensor is not generally isotropic implies that there is

Table 1 Properties of Models at Point of Bifurcation

n	0	3/2	3/2	3/2	3	3
n'	0	0	1	3/2	0	3/2
$t_b(n, n') \equiv \|T/W\|_b$	0.1376	0.1377	0.1375	0.1375	0.1381	0.1348
Virial test	0	0.00002	0.00001	0.0003	0.0023	0.048
$t_b(n, n') - t_b(0, 0)$.	0	+0.0001	−0.0001	−0.0001	+0.0005	−0.0028

no necessary relation between rotation and flattening; the classical illustration of this point is the spherical system that is made to rotate rapidly (while remaining spherical) by simply reversing the velocity vectors of all stars with one sign of angular momentum about a chosen axis. Since two-body relaxation is a small effect in most galactic scale systems, initial conditions and violent relaxation must determine the structure. Recent work and references can be found in Gott (1973), where it was shown that large ellipticities are difficult to produce by the above mentioned processes alone.

Stability to Nonaxisymmetric Modes

We will restrict our discussion of stability to a consideration of the same types of modes which were found to destabilize rapidly rotating fluid systems. Numerical experiments by Prendergast and co-workers (cf. Miller *et al.* 1970) and Hohl (1971) had shown that flat cold systems tended to rapidly develop random motions at the expense of ordered rotation. Systems with stellar velocity distributions like that found in the solar neighborhood where $(v_{random}/v_{rotation})^2$ $\sim 10^{-2}$ did not persist in this state. A recent investigation by Miller (1976) shows that these results were not due to the numerical imperfections of n-body experiments. To test specifically for the existence of the bar instability, Ostriker and Peebles (1973) examined three-dimensional (but cold) simulated stellar systems. They found that cold disks ($t > 0.14$) were invariably unstable, with barlike modes developing on the rapid dynamical time scale. In models which contained no stabilizing halo component, rough stability was reached in less than one rotation period when t was decreased to values $t \gtrsim 0.14$. Models with added (hot) spherical components were more stable; and when the spherical mass was comparable to the disk mass and the overall value of t reduced to less than the critical value $= 0.14$, no instabilities were observed. A summary of the earlier n-body experiments is found in Table 2. Finally, analytic studies of

Table 2 Values of $t \equiv T_{\mathrm{mean}}/|W|$

Study	t_{crit}
Maclaurin spheroid (fluid, exact)	$= 0.1376$
Generalized polytropes (fluid, approximate)	$\approx 0.137 \pm 0.002$
n-body, flat (Hohl 1971) ($n = 10^5$, approximate)	≈ 0.141
n-body, flat (Miller 1971) ($n = 1.25 \times 10^5$, approximate)	$\approx 0.130\text{--}0.135$
n-body, 3-D (Ostriker and Peebles 1973)* ($n = 150\text{--}500$, approximate)	$\approx 0.14 \pm 0.02$

* Average and standard deviation of t at $\tau = 1$ for the 12 models discussed in this paper.

cold flat fluid systems by Bardeen (1975) and of stellar systems having harmonic potentials by Kalnajs (1972) have shown similar results. A hot disk or halo component is necessary to stabilize a cold disk; otherwise the latter tends to be unstable to barlike modes.

Conclusion

It appears that equilibrium stellar and fluid systems may exist with arbitrarily large amounts of angular momentum (given fixed mass and density), so that all values of the rotation parameter $0 < t < \frac{1}{2}$ are allowed. However, in all systems so far studied, more or less violent barlike instabilities set in for $t > 0.14$ reflecting the fact that, when angular momentum is large, it becomes possible for a system to reduce its total energy by assuming a triaxial shape, increasing its moment of inertia and thereby reducing its kinetic energy of rotation.

References

Bardeen, J. 1975, *IAU Symposium No. 69*, Besançon, France.
Bodenheimer, P., and Ostriker, J. P. 1970, *Ap. J.*, **161**, 1101.

Bodenheimer, P., and Ostriker, J. P. 1973, *Ap. J.*, **180**, 159.

Chandrasekhar, S. 1939, *Introduction to the Study of Stellar Structure* (Chicago: University of Chicago Press).

———. 1969, *Ellipsoidal Figures of Equilibrium* (New Haven: Yale University Press).

———. 1970, *Ap. J.*, **161**, 561.

Friedman, J. L., and Shutz, B. F. 1975, *Ap. J.*, **199**, 157.

Fujimoto, M. 1968, *Ap. J.*, **152**, 523.

Gott, J. R. 1973, *Ap. J.*, **186**, 481.

Hohl, F. 1971, *Ap. J.*, **168**, 343.

Hunter, C. 1975, preprint.

Kalnajs, A. 1972, *Ap. J.*, **175**, 63.

Lebovitz, N. 1961, *Ap. J.*, **134**, 500.

———. 1967, *Ap. J.*, **150**, 203.

Ledoux, P. 1945, *Ap. J.*, **102**, 145.

Miller, R. H. 1971, *Ap. and Space Sci.*, **14**, 73.

———. 1976, preprint.

Miller, R. H., Prendergast, K. H., and Quirk, W. J. 1970, *Ap. J.*, **161**, 903.

Ostriker, J. P., and Bodenheimer, P. 1973, *Ap. J.*, **180**, 171.

Ostriker, J. P., and Mark, J. W.-K. 1968, *Ap. J.*, **151**, 1975.

Ostriker, J. P., and Peebles, P. J. E. 1973, *Ap. J.*, **186**, 467.

Roberts, P. H., and Stewartson, K. 1963, *Ap. J.*, **137**, 777.

Rossner, L. F. 1967, *Ap. J.*, **149**, 145.

Tassoul, J. L., and Ostriker, J. P. 1968, *Ap. J.*, **154**, 613.

Thomson, W., and Tait, P. G. 1912, *Treatise on Natural Philosophy* (Cambridge: Cambridge University Press).

4 The Influence of Magnetic Fields on Stars and Stellar Systems

Leon Mestel

Magnetic fields are observable directly from the Zeeman effect in some dense galactic gas clouds, in a restricted group of early-type stars, and in the Sun. We have indirect evidence for a general galactic field from the polarization of starlight, from the Faraday rotation of radio waves, and from the background synchrotron radiation. Analysis of radio sources, of X-ray sources, and of pulsars is inconceivable without magnetism. I shall attempt to summarize in depth just some of the problems of stellar magnetism, of magnetic gas clouds and star formation, and of pulsar electrodynamics.

Stellar Magnetism—Dynamo or Fossil?

From the start we emphasize that we do not restrict the discussion to stars with observable magnetic fields. A surface field that is too weak to yield a measurable integrated Zeeman broadening may yet be a manifestation of a large-scale field deep down which can have important effects over the leisurely time scale of normal stellar evolution, especially with regard to stellar rotation. In fact, a vitally important problem is precisely the relation between the total magnetic flux F_t and that fraction F_s which manages to emerge from a surface hemisphere.

Let me begin by referring to two fundamental papers, both by Cowling. In the first (1934), there appeared the original and most celebrated antidynamo theorem: the impossibility of offsetting the ohmic decay of fields topologically similar to axisymmetric fields from the kinetic energy of bulk motion of the fluid. In the second paper (1945), Cowling pointed out that although fields of simple structure must decay, the e-folding time of a large-scale stellar field is longer than the stellar lifetime, partly because of the fairly high conductivity, but primarily because of the smallness of the currents required to maintain fields with cosmical length scales. Thus a large-scale stellar field could be a "fossil"—a slowly decaying relic of the field present when the star formed.

The fear that Cowling's theorem might be a precursor of a universal antidynamo theorem was dispelled by the publication of rigorous existence theorems, and indeed we now have a plethora of dynamo models. A pioneering

Leon Mestel is at the Astronomy Centre, University of Sussex, England.

paper by Parker (1955) contained the seminal idea for much subsequent work (Krause *et al*. 1971). Nonuniform rotation—as observed in the Sun—is pictured as generating a toroidal field component B_t from an initial poloidal field $(B_p)_0$. To complete the dynamo cycle, we require that there be other motions—essentially nonaxisymmetric—which twist the field B_t in such a way as to yield a new poloidal component B_p. Parker noted how Coriolis forces acting on convective motions would give the field such a twist. The discovery that the polar field of the Sun apparently reverses during the solar cycle is strong, though perhaps not overwhelming, evidence (cf. Piddington 1972) that at least part of the solar field is periodically built up, destroyed, and reversed by mass motions in the subphotospheric convective zone. Indirect support comes from observations of late-type stars in clusters of different ages. There is a systematic decrease of angular velocity with increasing age (Kraft 1967), correlated with a steady decline in calcium activity (Wilson and Woolley 1970). In some cases there is tentative observational evidence for periodic calcium activity, suggestive of a solar cycle (Wilson 1968–1973). This is satisfactory: one would expect both expanding coronae and solar dynamo action in stars of similar type, with consequent removal of angular momentum by a magnetically controlled stellar wind. Moreover, the strength of the field—as indicated by the calcium activity—should decline as the star's rotation is reduced.

The recent advances in dynamo theory lead us naturally to ask whether the fields of the strongly magnetic stars are also due to dynamo action. These stars are all of early type, without strong outer convective zones, so we would not expect to see something similar to the solar cycle. But they do have convective cores, so it is not unreasonable to postulate dynamo action deep down, though without knowing the details of the rotation law or of any laminar circulation within the core we cannot say whether the dynamo is steady or oscillatory. It is also possible to build dynamos by using just laminar circulation and differential rotation in the radiative envelope, including models of oblique rotators (Krause 1971). However, one should note that most of the dynamo work so far is kinematic, using *prescribed* velocity fields such as nonuniform rotation. Comparatively little has been done on how the field's growth is limited, presumably by the nonlinear reaction of the magnetic forces on some vital part of the motion. We want to be able to predict both the nonuniform rotation field and the relation $F_t(\Omega)$ between the total magnetic flux F_t and the angular velocity. It is not even clear yet that the solar equatorial acceleration has a purely hydrodynamical explanation, or whether the magnetic torques play an essential rote. Further, the slowness of the circulation normally to be expected in a stellar radiative zone (even in the outer layers) would suggest that any field built by dynamo action dependent on circulation in radiative zones would be rather weak (cf. Mestel 1961). Extrapolation of our limited current knowledge on dynamos to the strongly magnetic stars seems highly risky.

In fact, I wish in any case to argue that it would be premature to abandon the idea of fossil stellar magnetism, especially for the strongly magnetic group among the Ap stars. I feel that Cowling's proposition should be restated: one should rather ask why the fields of the observably magnetic stars are not much stronger than they are, and simultaneously why these stars are restricted to a subclass of stars in one part of the Hertzsprung-Russell diagram. Indeed, a superficial study suggests that the observed galactic magnetic field would be an impediment to the condensation of protostars; and when condensations do form, one might guess that the total flux trapped within is far above the upper limit suggested by plausible inward extrapolation of observed fields. We return to this problem later; for the moment we are concerned with the possibility that some (or even all) stars may possess fossil fields. Any primeval field possessed by the Sun would probably be expelled from the convective envelope and largely concentrated in the radiative core. It would tend to couple the rotation of the core with that of the envelope, and it may affect solar evolution in other ways, but it is difficult to see how it would be unambiguously detectable at the surface. However, early-type stars have at most very weak subphotospheric convective zones in essentially radiative envelopes, so that a primeval field expelled from the core should—other things being equal—be observable. So it is prima facie not surprising that there exists a class with observably strong fields—up to 3.5×10^4 gauss.

Further, there are stringent tests, theoretical and observational, which an adequate theory of stellar magnetism must pass. To carry conviction, a dynamo explanation of stellar magnetism must be able to account for the most striking features of the data. It is now well attested that the observably magnetic stars —and indeed the whole class of Ap stars, and the Am stars—are almost all slower rotators than normal A stars; and within the class of magnetic stars there is now evidence (Landstreet 1975; Wolff 1975) of a gross correlation between stronger effective field and *lower* rotation. It is this *anticorrelation* between surface magnetism and stellar rotation which is so surprising: intuitively one expects that the more rapid the rotation, the greater the corresponding differential rotation and circulation speeds, and so the more efficient the dynamo. And in addition to this statistical anticorrelation, there are other observations that seem difficult to fit into a dynamo picture. Why is there a class of slowly rotating stars near type A0 which show no magnetic fields and spectral peculiarities (Deutsch 1967); and why does one find, for example, two stars with very similar Zeeman curves and spectra, and yet differing in rotation by a factor 10 or more?

None of these arguments is conclusive. As we shall see below, the interpretation of the observations is intimately linked with the question of how much of the total internal flux F_t, whether fossil or dynamo-maintained, emerges from the surface at a given epoch in the star's life. To give a definitive answer we shall also need to know much more about the dynamics of dynamo action, in particular the criteria which fix the dependence $F_t(\Omega)$ for a star of given structure. But it is at least a plausible hypothesis that the flux F_t possessed by a star may be an extra parameter, determined by conditions at the birth of the star and its subsequent history, but not physically linked with the star's rotation.

Stability: External and Internal Fields

The fossil theory is based on Cowling's estimate of the ohmic decay time in a medium *at rest*. If a domain is superadiabatic and therefore spontaneously unstable, then all but the very strongest fields will be tangled up by the convective motions and probably expelled. But in a subadiabatic (radiative) zone the density-temperature field is stable for motions with a component in the direction of gravity, so it has been tacitly assumed that a magnetic field that is not too strong will persist (apart from slow ohmic decay). However, it can be shown (Wright 1973; Markey and Tayler 1973) that the simplest magnetic structures are themselves dynamically unstable: for example, a purely poloidal field—such as Cowling's slowest decaying dipolar mode, with just one O-type neutral point—is subject to an analog of the "kink" instability familiar to workers in laboratory plasma physics. The unstable modes have displacements ξ nearly perpendicular to the local gravitational field g, so that the strong adiabatic stabilizing term $\propto (\xi \cdot g) \times [\xi \cdot (\nabla \rho / \rho)]$ is negligible. The stability theory is worked out only to the first order in ξ, but there is no obvious qualitative reason why the motions should be halted in the nonlinear domain. A purely toroidal field is likewise unstable, but one can construct fields of mixed topology, with toroidal flux linking poloidal loops, which are stable against the obvious disturbances. According to Wright a toroidal flux about one-quarter as large as the poloidal will stabilize the poloidal field against kink modes. Dynamo-built fields automatically have this type of structure, and it is at least arguable that the instabilities will so distort a purely poloidal primeval field that ohmic diffusion allows the field to adopt a stable topology.

It is always much more difficult to prove stability against any possible set of displacements than to demonstrate instability by a judicious choice of one set, and we cannot be sure that fields stable against kinking are not subject to more subtle modes. We shall provisionally assume that a nontrivial class of dynamically stable fields exists; but if this were not the case, so that within a short time scale *all* fossil fields spontaneously convert their energy into kinetic energy (ultimately to be dissipated), then we would be forced back to dynamo regeneration for all stellar fields.

Most of this recent stability work has implicitly assumed the field to be "weak," in the sense that the magnetic energy is much less than the star's gravitational energy. The star is therefore only mildly distorted from the spherical, and in fact the instabilities are essentially topological. It would be of interest to know if dynamical stability considerations prevent the existence of any bodies—stars or gas clouds—in which magnetic forces balance more than a small fraction of the gravitational forces.

Once a dynamically stable field has been constructed, one needs to worry about nonadiabatic, dissipation-dependent instabilities. An example of this is "magnetic buoyancy." Consider a thin toroidal loop in pressure equilibrium with its surroundings, so that

$$p_i + \frac{B^2}{8\pi} = p_e \tag{1}$$

where p_i, p_e are respectively the thermal pressures within the loop and in the surrounding gas. To be in dynamical equilibrium, the gas within the tube must be at about the same density as the ambient gas, so that the pressure differences are maintained by a lower temperature within the loop. The consequent heat inflow must cause the loop to rise, but in a thermal time scale. In a radiative zone, magnetic buoyancy is essentially an "Eddington-Sweet" type of flow: the time scale is at least of the order of the Kelvin-Helmholtz time, and may very well be longer. In general, the forces exerted by any large-scale field cause a disturbance to the pressure-temperature field, with a consequent imbalance in heat flow: in standard notation for a radiative envelope

$$\nabla \cdot F \equiv -\nabla \cdot \left(\frac{4}{3} \frac{acT^3}{\kappa\rho} \nabla T \right) \neq 0 . \tag{2}$$

The consequent buoyancy forces lead to a circulation, which in turn tends to drag and so distort the field. It is possible to choose special magnetic field structures which in fact do satisfy $\nabla \cdot F = 0$ (circulation-free systems), but it is not clear that they would themselves be "stable" against spontaneous transformation into fields that do yield thermally driven circulation. It could be that even if a field is dynamically stable, if it is too strong over the bulk of the star it will generate a circulation which is able (well within the star's lifetime) to drag the flux into regions of lower conductivity, there to be destroyed. Perhaps this is how a newly formed star loses during the pre–main-sequence phase any excess flux it has inherited.

In fact, any discussion of thermally driven circulation cannot ignore the effect of centrifugal forces, which prima facie are likely to dominate over the magnetic forces over the bulk of the star. The circulation due to uniform rotation (Sweet 1950) has a large-scale quadrupolar form; and if it flows inexorably, it will tend to concentrate flux deep down into the regions of high density. This is seen most simply if the field is also basically quadrupolar, and a steady state has been reached with $\rho v/B$ constant (Chandrasekhar 1956; Mestel 1961);

as the circulation speed for uniform rotation does not vary much over the bulk of the star (Sweet 1950), B/ρ is approximately constant. This suggests that if the circulation persists, and if the total flux F_t is more or less constant, then very little flux will emerge from the surface: a star may be in reality quite strongly magnetic, yet may appear normal. Conversely, if a star is to appear magnetic, we require that any circulation be either suppressed or very slow, at least in the surface regions if not over the bulk of the star. And it should be noted that even in slow rotators the circulation can be quite important in the low-density surface regions, especially if there is a modest degree of nonuniform rotation (Baker and Kippenhahn 1959).

With these considerations in mind, one is led to look for conditions which allow a star to be visibly magnetic. In the first studies Davies (1968) and Wright (1969) postulated that the magnetic forces are strong enough to ensure that the thermally driven circulation is killed: the radial structure of a dipolar field is chosen so that

$$(\nabla \cdot F)_\Omega + (\nabla \cdot F)_B = 0 \qquad (3)$$

where the suffixes refer to the disturbances to the thermal field due respectively to centrifugal and magnetic forces. The results are formulated as an answer to the question: given the total flux F_t, then as Ω varies, how much flux F_s does condition (3) allow to emerge from a surface hemisphere? It turns out that if Ω is supposed to increase, then F_s steadily declines, until at a finite value of Ω, F_s drops to zero. With Ω still higher, no circulation-free solutions can be found, but there do exist self-consistent solutions with circulation (Mestel and Moss 1977). Deep down in the star, the flow is essentially the Eddington-Sweet circulation, with the magnetic field merely acting to preserve the uniform rotation against the advection of angular momentum by the circulation. In the outer layers the density is low enough for the magnetic forces to make a significant contribution to $\nabla \cdot F$. The surface flow is nearly horizontal. Because of finite resistivity—which plays an essential role—some flux does emerge, but it is far less than that predicted, for example, by the principal eigenmode of Cowling's decay equation in a stationary medium. The work has been extended to more realistic fields with the mixed poloidal-toroidal structure essential for dynamical stability (Moss 1977a).

These results are prima facie favorable to the fossil theory of stellar magnetism. They do show why we should expect rapid rotators to be without observable fields, and why there should be an anticorrelation between rotation and surface field, as recently reported by Landstreet (1975) and Wolff (1975), while still leaving the total flux F_t as a free parameter not tied to the star's rotation. However, one should note that the same tendency to concentrate the field into dense regions will apply even if the field is dynamo-built: the difference is that a dynamo-built field has a flux F_t that is itself a function of Ω. We can parametrize this by $F_t \propto \Omega^n$, with n presumably positive, so that for larger Ω the magnetic as well as the centrifugal forces increase. It appears (Mestel and Moss 1977) that if the index $n < 1$, then the increase of the magnetic forces with Ω is insufficient to offset the stronger Eddington-Sweet circulation, and the anticorrelation between F_s and Ω would again be qualitatively explicable. However, if $n \geq 1$, the more rapidly rotating stars should show systematically stronger surface flux, contrary to observation, and the case for the fossil theory would be strengthened. Note that one should then conclude not that dynamo action is not occurring deep down in A stars (and indeed in others with convective cores), but rather that the flux F_s we see at the surface is not part of this dynamo-maintained field, but is the emerging part of a fossil flux F_t that is unrelated physically to the star's rotation.

Clearly, we need a better understanding of both the kinematics and dynamics of dynamo action, especially the back-reaction of a growing magnetic field on the rotation field. It should be noted that a "fossil" field need not be a relic of the galactic field in the gas from which the star formed; it could be a relic of a field built up by dynamo action in an earlier phase of the star's life. The question at issue is whether the fields of the strongly magnetic stars are being maintained by *contemporary* dynamo action. Theory must explain also why, of the slowly rotating A stars, only some of the Ap stars and apparently none of the Am stars show strong surface fields. It is just this bewildering complexity of the observational data which leads us to favor a fossil theory, in which the total magnetic flux possessed by a star (at least those of early spectral type) is an extra parameter, not related physically to the rotation Ω. However, unambiguous interpretation of the observations is likely to depend on the time lag between dynamo generation of a field deep down and its manifestation at the surface, and on whether a dynamo-maintained flux F_t is a single-valued function of Ω.

Magnetic Braking

We have emphasized the effect of the slow but inexorable rotation-dependent circulation in trapping magnetic flux within an early-type star. However, it is also very plausible that a stellar magnetic field coupling the star with its surroundings will transfer angular momentum, and so slow up the rotation: we envisage a two-way interaction between rotation and magnetism. The braking process that has been studied in most detail involves a stellar wind. In the absence of any magnetic control, outflowing gas would remember the angular momentum it had when it left the star, so that it would acquire a strongly nonuniform rotation field. The consequent twisting of magnetic field lines generates torques which try and restore uniform rotation. Intuitively, one expects approximate corotation with the star to be maintained as long as the gas velocity v is well below the Alfvén speed v_A; and indeed detailed analysis (Mestel 1966a, 1968a; Weber and Davis 1967) shows that the transport of angular momentum by both outflowing gas and magnetic stresses is equivalent to assuming strict corotation out to the Alfvénic surface S_A. The exact shape of S_A depends on the structure of the field, and this in turn is not known except as part of the solution of the whole problem; however, one can write

$$ -\frac{dh}{dt} \equiv -k^2 R^2 M \dot{\Omega} = -\frac{dM}{dt} \Omega R_A{}^2 \tag{4} $$

where R_A is a mean Alfvénic radius, and h is the stellar angular momentum written in terms of a radius of gyration kR. For the Sun and other late-type stars the dynamo-built magnetic field is presumably a function of Ω, so that $R_A = R_A(\Omega)$, and equation (4) predicts an algebraic law of variation of Ω with time (Spiegel 1968; Skumanich 1972). Extrapolation back via late-type stars in the young Hyades and Pleiades clusters yields a zero-age main-sequence rotation some 10–20 times the present solar rotation (Ostriker 1972). This is still much below the limit at which centrifugal force and gravity balance at the surface: it appears that the Sun and other late-type stars began their main-sequence lives as "slow" rotators. Now studies on the breakup of diffuse gas clouds (cf. below) suggest that protostars enter the opaque, pre–main-sequence phase rotating rapidly, with centrifugal force comparable with gravity. If so, then the Sun must have lost most of its angular momentum during the Hayashi phase. Perhaps the violent mass-loss inferred from observations of T Tauri stars implies also a greatly increased angular momentum loss, presumably again via a dynamo-built magnetic field. Alternatively, one can revive the idea (Alfvén 1954; Hoyle 1960) that the missing solar angular momentum is largely stored in the orbital motion of the planets (especially Jupiter) around the Sun, with magnetic stresses again being the means of transfer. More detailed study on this problem is called for, both on the stability of the magnetic field structures postulated, and indeed on the plasma models used, which seem sometimes to have ad hoc properties (Alfvén and Arrhenius 1973).

It is tempting to try to use a similar braking process to explain the systematically slower rotations of the magnetic early-type stars. If the field is a fossil, and also if any dragging of the field lines beneath the surface is too slow to be significant, then the external flux will be essentially independent of Ω. If also the mass loss is primarily due to thermal pressure and only marginally to "centrifugal wind" effects (Mestel 1968a), then equation (4) predicts an exponential law of braking. I am uncomfortable with such a law: the fear is that unless the time constant is just right, either the star will lose virtually no angular momentum, or it will end up with a rotation period orders of magnitude longer than even the several years found by Preston (1970) for a few magnetic stars. I much prefer a process which has a built-in self-adjustment. In fact, one can reasonably doubt whether a main-sequence early-type star acquires a corona hot enough to drive a thermal wind; and centrifugal winds are efficient only for rapid rotators (Mestel 1968a). During the comparatively short pre-main-sequence phase the star's outer convection zone would presumably compress a primeval field into the central regions, so one would not expect more braking than for other stars of the same mass.

A more promising model appeals not to a stellar wind but to its opposite—gravitational accretion. Consider a sequence of nonmagnetic stars, with the temperature $\bar{T}$ of the corona as a parameter. When $\bar{T} \approx 10^6$ K, then we know that the pressure at "infinity" predicted by the condition of hydrostatic support much exceeds the pressure of the interstellar medium; in the absence of a "lid," the corona expands, forming a thermally driven stellar wind (Parker 1963). If $\bar{T} \approx 10^5$ K, the two pressures are roughly equal, and a static corona is possible; if $\bar{T} \approx 10^4$ K, the scale-height is so short that the pressure exponentiates to almost zero very near the star, and the interstellar gas will tend to flow into the star's potential hole. Now suppose the star has a dipolar magnetic field. The gas now tries to establish hydrostatic equilibrium along the field lines, while the magnetic field resists distortion from a nearly curl-free or force-free structure. If $\bar{T} \approx 10^6$ K, then gas still flows out along field lines emanating from near the poles; but if $\bar{T} \approx 10^4$ K, then the density again exponentiates to near zero a short distance from the star. The vital difference is that now the magnetic field interferes with accretion: prima facie, the inflow is stopped near a new Alfvénic radius r_A, defined by

$$\frac{B^2}{8\pi} \approx \frac{\rho v^2}{2} \approx \frac{\rho G M}{r_A} . \tag{5}$$

However, the resulting system, with interstellar gas hanging on the boundary of a virtually empty magnetosphere, is likely to be Rayleigh-Taylor unstable: gas will slide into troughs in between magnetic planes. As a consequence the gas will pick up angular momentum from the rotating star via magnetic pressure gradients. The subsequent flow of the gas depends on the ratio $\eta \equiv \Omega^2 r_A^3 / GM$. If $\eta > 1$, then at least some of the gas will be shot out by the centrifugal force, and the star will lose angular momentum until $\eta \approx 1$, after which the gas can flow in, and braking ceases. The efficiency of the process and the final rotation period depend only weakly on the accretion rate and so on the interstellar density. Rough estimates (Mestel 1975) yield an e-folding time of $\sim 10^7$ years and a typical asymptotic rotation period of a few days. Thus we have a process which has a natural cutoff, given by $\eta \approx 1$, and with promising numbers; it merits more careful study, especially on the details of the instability (cf. Arons and Lea 1976).

The same magnetic tensions which yield a net torque about the star's rotation axis will in general geometries exert smaller but nonzero torques about the two perpendicular axes: only in highly symmetrical systems—for example, those with axial or equatorial symmetry—will these components vanish (Mestel 1968*b*). Their effect is to cause the instantaneous axis of rotation to precess through the star, and simultaneously for the magnetic axis to rotate in space, while the angular momentum vector stays invariant in direction. For the one case studied in detail—with the rather implausible assumptions of braking by a thermal wind, and a field which is a small departure from a split monopole—it was found that the axis of rotation looks for the region of the star where the field is strongest (Mestel and Selley 1970). One can find qualitative arguments why this may be a more general result. An adequate theory of the magnetic stars must explain why interpretation of the observations using the very plausible oblique rotator model (Deutsch 1958, 1970) seems to require a nonrandom distribution of angles of obliquity χ between the magnetic and rotation axes. According to Preston (1971) there is a marked preference for χ small or large—near alignment or near perpendicularity. (Doubts have been expressed by Borra [1974] on the reality of the small χ cases, but Landstreet [1975] has confirmed at least one such case.) It may be that an explanation can be found by combining this precessional effect with the distribution of surface flux, itself presumably determined by the internal hydrodynamics of the star.

Magnetic Fields and Stellar Structure

Probably the most powerful influence of an internal magnetic field—dynamo-maintained or fossil—on stellar structure and evolution is again via its effect on the rotation field. During the leisurely time scale of all but the late phases of stellar evolution, even a weak field will maintain something near uniform rotation, at least within nonturbulent regions. In particular, during the contraction of a burned-out core we expect the magnetic field lines to transfer most of the core angular momentum to the expanding envelope, so that both white dwarfs and pulsars should be slow rotators, in the sense that $\Omega^2 R^3/GM \ll 1$—even the Crab pulsar with its present-day 1/30 second period. Estimates for the braking of the Crab indicate that it was a slow rotator even at birth—a point particularly emphasized by Arnett (1975). I am sorry that this is the case: like Kip Thorne, I would like pulsars to rotate rapidly, so that effects characteristic of general relativity—gravitational radiation, dragging of inertial frames—would appear much more strongly; but regret is untinged by surprise. Likewise, it is difficult to see how a strong differential rotation between the radiative core of the Sun and the base of the convective envelope could survive the presence of even a weak field trapped within the core.

Direct effects of the field on stellar structure are difficult to show unambiguously. The bulk effects are probably masked by those of the similar but normally stronger centrifugal forces. In low-density surface regions one might expect magnetic forces to dominate when the magnetic energy density exceeds the centrifugal. However, what tends to happen is for the field to adjust itself to a nearly curl-free or force-free structure: the energy densities are not a good measure of the local forces (e.g., Wright 1969). Magnetic interference with convection looks more promising. The turbulent velocities v_t predicted by standard mixing-length theory in the absence of a magnetic field are normally highly subsonic. Thus the turbulent energy density is much less than the thermal, so that a "weak" magnetic field would seem able, prima facie, to interfere with the heat transport. However, it is certainly wrong to claim that such a field will suppress the turbulence and convert the zone into a radiative zone. Reduction of the efficiency of heat transport will lead to a corresponding increase in the superadiabatic temperature gradient and so to stronger velocities. To suppress the convection would require a field of energy $B^2/8\pi \approx \rho a^2$, where a is the sound speed (Gough and Tayler 1966; Moss and Tayler 1969; Mestel 1970) and there is certainly no evidence for a field of anything like this strength. However, a weaker field may still be able to interfere with the convection. One could picture, for example, field lines leaking from the radiative core into the solar convection zone, and forcing up the turbulent energy until it is capable of expelling the field once more, after which the turbulence would revert to its normal energy density. It is not clear what effect such quasi-periodic behavior would have on evolution.

If the turbulence builds up its own field by dynamo action, then one is again faced with the problem of the asymptotic field strength. It is possible that in addition to the large-scale fields built up by the periodic solar dynamo, there are smaller scale fields with stresses comparable with the Reynolds stresses of the turbulence. If so, then by affecting the extent of zones of instability the field could again influence stellar evolution.

A primeval field with its axis inclined to the rotation axis at a general angle χ can be shown to cause quasi-periodic internal motions. The oblique rotator is a body with three unequal axes of inertia. If it were a rigid body, its motion could be described as a combination of the Eulerian nutation about the magnetic axis, and a rocking motion of the obliquity angle χ, both with the frequency $\omega \approx \Omega(\mathfrak{M}/|\mathfrak{B}|)$, where $\mathfrak{M}$ and $\mathfrak{B}$ are respectively the magnetic and gravitational energies (Mestel and Takhar 1972). However, because of the requirement that the star remain in equilibrium without exerting nonhydrostatic stresses, there is in addition a field of (nearly divergence-free) oscillatory "ξ-motions," again with frequency ω. In a rapidly rotating star their amplitudes are large enough to cause significant mixing of material between convective and radiative zones; and even in a slow rotator, there may be sufficient departure from strict periodicity for some mixing to occur via a random walk.

Such effects depend on persistence of the field of ξ-motions. Any dissipation of energy—by radiative conduction, turbulent viscosity, ohmic resistance—

will cause a secular change in the star. A star of given angular momentum h has kinetic energy of rotation $h^2/2I$, where I is the moment of inertia about the instantaneous axis of rotation: as noted already by Spitzer (1958), dissipation of the energy of dynamically driven internal motions must cause the star to try to rotate about the axis of maximum moment of inertia. If the star is dynamically oblate about the magnetic axis, then dissipation will cause $\chi \to 0$; if dynamically prolate, then $\chi \to \pi/2$. In general, a poloidal field tends to make the star oblate, while a toroidal field tends to cause prolateness. We noted earlier that a mixed poloidal-toroidal structure is a necessary condition for dynamical stability. Putting together the different strands in the argument, we arrive at a possible reason for the preference for small or large χ (Preston 1971); the decisive criterion is the relative strength of the poloidal and toroidal fluxes, which decides whether the star is dynamically oblate or prolate.

However, the argument depends on the dissipation time scale being sufficiently short. Rough arguments suggest that ohmic dissipation may be efficient enough, provided the field lines extend to the low-conductivity surface regions. Thus the possibility exists that in a late-type star containing a trapped magnetic field in the radiative core, an initial obliquity and the consequent ξ-motions may persist through the star's lifetime, whereas in an early-type star with magnetic field lines penetrating to the surface, χ may reach one or other asymptotic value in a comparatively short time. However, it is too early to say whether or not this dissipative process is more significant for the obliquity problem than the magnetic torque process outlined in the last section. And one should note that a purely kinematic process, in which the Eddington-Sweet circulation distorts a magnetic field of given obliquity, may be capable of accounting for Preston's observations (Mestel and Moss 1977; Moss 1977b).

These studies are somewhat frustrating, as they are difficult to relate unambiguously to observation. However, the observations of pulsating stars offer a much more sensitive test of theory. At one time it was thought that RR Lyrae had an observable magnetic field; and although this now seems doubtful, there is indirect evidence from the cyclical variation with a period of several weeks in the shape of the light curves (Christy 1974). The obvious interpretation is that the period is one of rotation, so that the star presents a changing aspect to us because of a magnetic structure that lacks symmetry about the rotation axis. In the low-density surface regions the spherically symmetric gravitationally modified sound waves become mixed sound-Alfvén waves with properties dependent on the local inclination of the magnetic field to the radius vector. The analysis of this problem is likely to be formidable but rewarding.

Magnetic Gas Clouds and Star Formation

Opinion seems to have converged toward an estimate of 3–5×10^{-6} gauss for the large-scale galactic magnetic field. Time does not permit discussion of the role of the field in the gross dynamics of the galactic gas, such as the Rayleigh-Taylor-Parker instability (Parker 1966) and its nonlinear development (Mouschovias 1974), or the restriction on cosmic-ray streaming to the Alfvén speed (Kulsrud and Pearce 1969), with its important consequences for our ideas on cosmic-ray generation. We pass straight to the problems of self-gravitating clouds, forming, for example, by thermal instability in gas permeated by the galactic field. With flux-freezing a good approximation, condensation with a component of motion across the field will distort the field and increase its strength, so generating magnetic forces that tend to oppose further contraction of the cloud. The condition for indefinite contraction to occur nevertheless is most easily found from the Chandrasekhar-Fermi virial theorem (1953), which yields the *mass-flux relation*: the cloud must have a mass greater than M_c, which is related to the total flux F within the cloud by

$$F^2 = \frac{9\pi^2}{5} kGM_c^2 \tag{6}$$

where k is a numerical factor of order unity, dependent on geometrical details. If M is less than but comparable with M_c, then the cloud is not massive enough to collapse indefinitely; instead, it achieves a state of magneto-thermal-gravitational equilibrium, with magnetic and gravitational forces approximately balancing in two dimensions, and pressure balancing gravity along the field lines (Mestel 1965; Strittmatter 1966; Parker 1973, 1974; Mouschovias 1976a, b). These models may very well be relevant to observations (Verschuur 1969a, b; Mestel 1969) of Zeeman splitting of the 21 cm line in moderately dense gas clouds, implying a *local* field strength of 10^{-5} or 2×10^{-5} gauss; for in at least some cases the estimated masses of the clouds are indeed comparable with M_c.

If M exceeds M_c, then the magnetic field is unable to prevent indefinite collapse. We are then led to ask: How does the field affect the *fragmentation* problem—the breakup of the cloud into masses of stellar order? A superficial study suggests that the field is just a nuisance. If the cloud is supposed to contract isotropically, then at no stage will subcondensations be able to form: a local density fluctuation trying to amplify under its self-gravitation would generate magnetic forces which would halt and reverse the condensation. For this reason, early studies (Mestel and Spitzer 1956) looked for conditions under which the flux-freezing constraint would be broken: we wanted flux-loss to occur "rapidly"—i.e., in a time short compared with the basic dynamical time scale, determined by gravitation—and at comparatively low densities. Our estimates were in fact too optimistic: the frictional coupling between the neutral gas on the one hand, and the ionized component (inductively coupled to the magnetic field) on the other, is larger than was thought, both because the collision cross section turns out to be higher (Osterbrock 1961), and because cosmic rays keep the ionization level comparatively high. It is likely, however, that in some or even most cases there is a phase in which flux loss does become rapid. At moderately high densities the cloud is still diffuse enough for the temperature to be far too low for thermal ionization; ionizing photons and cosmic rays can be absorbed in the outer regions of the cloud, and the ionized component can then decline by rapid attachment to dust grains. But at lower densities, flux loss is slow, and theories of star formation must encompass this. I shall therefore discuss a scenario in which the magnetic field remains dynamically significant.

The first point to emphasize is that the presence of a large-scale magnetic field is not in itself a bar to breakup of the cloud. The assumption of *isotropic* collapse—which leads to a "magnetic $\gamma = 4/3$"—is highly implausible; it is far more likely that there will be preferential flow of gas down the field lines, and it can then be shown (Mestel 1965) that fragmentation can occur. For example, if a cloud of mass M and flux F flattens parallel to the magnetic axis into an oblate spheroid, then the condition that a subcondensation can separate out gravitationally is simply reducible to the condition that the cloud has a mass greater than the critical mass $M_c(F)$, given by equation (6). Thus provided the cloud is massive enough to be able to contract against magnetic opposition, then by the same criterion it can fragment into subcondensations, which, however, will also be "strongly magnetic," in the sense that the magnetic and gravitational forces will be comparable. This is in striking contrast to even the strongest observably magnetic stars, for which the magnetic forces are inferred to be only a weak perturbation over the bulk. A clear distinction must be kept between the conditions for subcondensations to be able to form, and the additional requirement that by the time stars have reached the main sequence they have all become "weakly magnetic."

We can also make a virtue of necessity, and again use the magnetic field as a means of redistributing angular momentum. A simple calculation shows that if a gas cloud has the angular momentum to be expected because of the galactic rotation alone, and if subsequent contraction occurs with each element of gas conserving its angular momentum, then it is very difficult to see how a fragment can reach anything like stellar densities: the centrifugal forces of spin become much too large (Hoyle 1945; Mestel 1965). Conditions are even worse if one includes the angular momentum associated with the galactic turbulence. These difficulties may disappear if a coupling process transfers angular momentum from a condensation to the surrounding gas. And if a fragment is magnetically "strong," then the characteristic time for magnetic braking, which is just the time of travel of an Alfvén wave through the fragment, is comparable with the gravitational free-fall time, so that braking is clearly dynamically significant. However, it seems unlikely that the braking will be so efficient as to reduce the centrifugal force to a small fraction of gravity (Gillis et al. 1974, 1977). It is usually more reasonable to picture centrifugal forces—in both the cloud as a whole and in the fragments—as remaining close to gravity, so that contraction is determined not by gravitational forces alone, but by the rate at which angular momentum is removed: as noted earlier, we expect protostars to begin their pre–main-sequence lives *rotating rapidly*.

If the angular momentum vector of the gas cloud is more or less parallel to the magnetic field, then the centrifugal forces will not interfere with the flow down the field lines, and our picture of fragmentation in spite of flux-freezing will survive. However, if the two vectors are inclined at a large angle—as prima facie one would expect if the angular momentum of the cloud is essentially due to the galactic rotation—then it appears that fragmentation may sometimes be prevented by the centrifugal forces, which will inhibit flow down the field. As long as the magnetic field lines of the cloud remain a locally distorted part of the galactic field, then transport of angular momentum from the cloud will lead to isotropic contraction but not to breakup. Thus in this geometry a modest degree of flux loss would help. At this point we focus attention on the problem of the detailed structure of the field. As the cloud loses angular momentum and contracts, the magnetic field becomes steadily more and more distorted. Detailed study (Mestel 1966*b*) shows that as a consequence the field exerts strong local pinching forces, which build up a high-density zone. Sooner or later the constraint of strict flux-freezing will give way in this zone, and most of the cloud field lines will detach from the background field (Mestel and Strittmatter 1967). The cloud field will then look like the field of a local dipole, with a ring of O-type neutral points. So far, the cloud has still not lost any flux, so the dilemma on fragmentation remains. However, we have already noted that any purely poloidal field with O-type neutral points is dynamically unstable, even within subadiabatic stellar zones: one simply had to consider modes of motion perpendicular to the gravitational field, so that the adiabatic stabilization is killed. The same instability is found within a diffuse gas cloud, with the difference that one now has no need to restrict the motions to be perpendicular to g. By definition, for a diffuse cloud the time of approach to thermal balance is short compared with the dynamical time, so that there are many more modes of instability than for a stellar field. Making large extrapolations into the nonlinear domain, one can argue that once the field has begun to develop a structure with O-type neutral points, it will spontaneously acquire such small length scales that the field-freezing constraint will again be relaxed. Energy is fed from the largest scale elements into smaller and smaller scales until dissipative processes lead to flux destruction, but at a rate determined by the dynamical instabilities of the large-scale field. As the cloud steadily loses flux, it can flatten along the rotation axis, and (strongly magnetic) subcondensations can now form; they in turn will contract isotropically as they lose angular momentum to the rest of the cloud, unless and until the same process of field-detachment followed by instabilities starts up again. The significant feature is that flux loss is not "rapid": it is not flux loss that determines the evolution of the cloud, but the gross dynamical properties of the field.

We thus argue that the galactic magnetic field does not prevent contraction and breakup of a rotating cloud, and in fact plays a vital role by redistributing angular momentum. However, the field is certain to affect the details of the fragmentation process, especially when the magnetic and rotation axes are inclined at a large angle. In the simplest fragmentation models (Hoyle 1953; Mestel and Spitzer 1956) the systematic breakup of a cloud is halted when the bulk of the gravitational energy released during the contraction of a blob is no longer retained as the random kinetic energy of the fragments, but is all dissipated and trapped as the internal heat of the final protostars. This occurs when the gas becomes so opaque that the heat of compression generated cannot be radiated away quickly enough. The opacity of cold interstellar matter is so low that this argument predicts a typical stellar mass of well below the solar mass. This suggests that while opacity may be the dominant physical property that fixes the average stellar mass, something else intervenes during the formation of the minority of more massive stars (including perhaps the Sun); and it may be that the magnetic field is the extra factor. Certainly, in our picture of the evolution of a cold rotating magnetic cloud, with B and Ω inclined at a large angle, the field interferes strongly with breakup of the cloud. As long as flux-freezing holds, then loss of angular momentum leads to steady contraction, with the gravitational energy released supplying the increasing rotational kinetic energy and magnetic energy, but not to breakup; it is only when the magnetic field structure changes—leading to instabilities and to "slow" flux loss—that we envisage some fragmentation occurring. It remains as a challenge to theorists to predict a mass function for stars forming according to this picture.

Hydromagnetic instabilities occur because flux-freezing imposes such a severe constraint on the field that it prefers to relax toward a lower energy state, converting its excess energy into other forms. When flares occur on the Sun, we observe the conversion of a small fraction of the energy released into high-speed particles. One is led to wonder whether something similar occurs in a gas cloud (a suggestion hinted at in the early papers of Woltjer on cosmical hydromagnetics). A lot of magnetic energy is being destroyed; if only a small fraction is converted into *locally produced* cosmic rays, it could still make a significant contribution to the ionization balance of the cloud, especially at the higher densities when penetration of the cloud by galactic cosmic rays is limited. Is it possible that flux loss is kept *slow* during star formation because the degree of ionization is maintained at a moderately high level through·this generation of a local cosmic ray flux, at the cost of a modest rate of flux destruction? If so, it will be a supreme example of eating one's cake and having it: enough flux is lost to enable some fragmentation to occur, but the field remains strong enough to deal with the angular momentum problem.

Once a fragment has become an opaque protostar, its rate of contraction is no longer determined by dynamical considerations, but is determined by the rate of heat loss. If our scheme with strongly magnetic bodies forming at each stage does extend all the way to the opaque regime, then there has to be a dramatic reduction of flux well before the star reaches the main sequence. And in fact we may again appeal to hydromagnetic instabilities, which will now no longer lead, for example, to flattening and breakup within a dynamical time scale; instead, the protostar will be transformed from a strongly to a weakly magnetic body. Possibly when the total magnetic energy has become small, then the instabilities will transform the field into a dynamically stable structure, as discussed earlier.

Pulsar Electrodynamics

For the last topic, we turn to the pulsar problem. We accept the canonical model: an obliquely rotating magnetized neutron star, with an angular velocity Ωk that defines the frequency of the pulsing. The rate of loss of rotational energy from the Crab pulsar, as inferred from the observed rate of period increase, agrees well with the known energy emitted per second by the Crab Nebula in the form of synchrotron radiation, but is several orders of magnitude larger than the energy emitted in the pulses. We may in general anticipate that whereas the pulses are the diagnostic by which the pulsar phenomenon is discovered, from the point of view of energetics the zero-order problem is to understand how the magnetic field converts the macroscopic kinetic energy of rotation into relativistic particle energy. Conceivably, a solution of this "pulsar magnetosphere" problem could yield as a bonus a hint as to the location of the pulse-emitting particles.

The simplest model (Ostriker and Gunn 1969) assumes the surroundings to be a strict electrodynamic vacuum, with no charges and hence no sources of the electromagnetic field. With appropriate boundary conditions, both at infinity (no incoming wave) and at the pulsar surface (continuity of the normal component of B and the tangential component of E), the external electromagnetic field is that of a classical Maxwell-Hertz wave, carrying away energy and angular momentum to infinity. If the magnetic field is assumed mainly dipolar, and if the angle of obliquity is not too small, the classical formula immediately yields the oft quoted value of 10^{12} gauss for the field at the pulsar surface. The wave is able to accelerate individual charges to highly relativistic energies, but it is assumed implicitly that the density of such charges is too low for the particle current to be comparable with the displacement current. The light cylinder, of radius $\bar{\omega}_c \equiv c/\Omega$, separates the near zone, where quasi-static, induction fields dominate, from the wind zone, where the radiation fields dominate. The factor $(1 - \Omega^2\bar{\omega}^2/c^2)$ appears in the differential equations, but in an innocuous place: the light cylinder is *not* a singularity of the vacuum equations.

The assumptions implicit in the vacuum model were challenged by Goldreich and Julian (1969). They studied the steady, axisymmetric, aligned rotator, but their arguments are to some extent extendable (Mestel 1971; Cohen and Toton 1971) to the oblique rotator. The near-zone vacuum solutions in general have a strong electric field component $E_{\parallel}$ along B. Such a field will not persist if there are available sufficient free charges: a plasma tends rapidly to become polarized so as to reduce $E_{\parallel}$ to a much smaller value—an excellent example of Le Chatelier's principle. Within the highly conducting pulsar surface $E_{\parallel}$ is certainly almost zero; thus if the vacuum conditions hold outside, there will be a discontinuity in the normal component of E, and so also a surface charge density, which, however, is subject to large unbalanced electrical forces. Goldreich and Julian argued that these charges would therefore be pulled out of the pulsar, converting the surroundings from a vacuum into a perfectly conducting domain containing a *charge-separated* "plasma." In zones where the field lines emanating from the pulsar close well within the light cylinder, the plasma can corotate with the star; but along field lines which cross the light cylinder, steady-state conditions require that corotation be supplemented by motion along the field lines, in the right sense to ensure that the total velocity in the inertial frame is subluminous. The picture that emerged was of an electrically driven wind, with distinct streams of electrons and ions emanating from different regions of the pulsar surface.

For the oblique rotator the appropriate generalization of a steady state is the "quasi-steady" state, in which changes in time as seen in the inertial frame are a consequence merely of the rotation with angular velocity Ω of the non-axisymmetric structure (Mestel 1971; Endean 1972). In either case, the Goldreich-Julian assumptions imply an electric field

$$E = -\frac{(\Omega k \times r) \times B}{c} \tag{7}$$

maintained by a charge density

$$\rho_e = \frac{\nabla \cdot E}{4\pi} = -\frac{\Omega k}{2\pi c} \cdot [B - \tfrac{1}{2} r \times (\nabla \times B)] . \tag{8}$$

In a normal, nonrelativistic plasma ρ_e is the difference between the electron and ion densities; the Coulomb forces are so strong that they ensure that this net charge density is "small" in the sense that convection current is negligible compared with conduction current, and electrical force density negligible compared with magnetic. But if the plasma is strictly charge-separated, then if $\rho_e > 0$, it must be equal to the ion density, and if $\rho_e < 0$, to the electron density. In either case one finds (Mestel 1971, 1973)

$$\frac{B^2}{8\pi\rho c^2} \approx \frac{\omega_g}{\Omega} \gg 1 , \tag{9}$$

where ρ is the local mass density, and ω_g a typical (nonrelativistic) Larmor frequency. The point is that the amount of charge needed by Coulomb's law to transform the vacuum electric field into the hydromagnetic field (7) is so small that its Einstein rest-energy density is much less than the magnetic energy density. It is this which suggests that one should make a first attempt to describe the pulsar magnetosphere as an "infinitely conducting vacuum." The charge density (8) implies a current density of order $\rho_e \Omega \bar{\omega}$, which near the light cylinder is certainly important as a source of the field; and in fact for the oblique case the material current and the displacement current are comparable. Thus the medium can certainly not be treated as an electrodynamic vacuum; but *provided the relativistic γ factors do not become too large*, the dynamical effects of the plasma may still be negligible. If so, the condition (7)—which itself depends strictly on zero rest-mass—may be supplemented by the *relativistic force-free* condition

$$\rho_e E + \frac{j \times B}{c} = 0 \,, \tag{10}$$

where the current density j includes a contribution due to motion of charges along the field lines. In a strictly charge-separated plasma, the conditions for neglect of inertia in equations (7) and (10) are identical. In a mixed plasma, condition (10) will break down before (7); however, the inequality (9) is so strong that one would need to add a very large density in the form of neutral plasma before $B^2/8\pi\rho c^2$ would approach unity.

Equations (7) and (10) jointly yield

$$\left(j - \frac{\rho_e(\Omega k \times r) \times B}{c} \right) = \frac{c}{4\pi} \psi B \,, \tag{11}$$

where ψ is a scalar. Substitution from Maxwell's equations transforms (11) (for both steady and quasi-steady cases) into

$$\nabla \times \tilde{B} = \psi B \,, \tag{12}$$

where

$$\tilde{B} = \left[B_{\bar{\omega}}\left(1 - \frac{\Omega^2 \bar{\omega}^2}{c^2}\right), B_\phi, B_z\left(1 - \frac{\Omega^2 \bar{\omega}^2}{c^2}\right) \right] \tag{13}$$

in cylindrical polar components (Mestel 1973; Endean 1974). In the steady, axisymmetric case, equations (12) and (13) reduce to the form derived by Michel (1973), Scharlemann and Wagoner (1973), and Julian (1973).

There are in fact several reasons for doubting whether the physical assumptions that lead to equations (12) and (13) are adequate to describe the whole magnetosphere. The assumption of strict charge separation leads to paradoxes (already apparent in the Goldreich-Julian paper) which have been elucidated by Okamoto (1974). These difficulties stem from trying to posit *a priori* the directions in which charges flow. The conditions determining the charge density and current density are different, and the restriction to a pure *convection* current

$j = \rho_e v$—implicit in a charge-separated plasma—is too severe a constraint. The assumption that charges flow away from the pulsar but never in the reverse direction is not obvious (quite apart from the difficulties concerning ionic emission from the pulsar surface emphasized by Ruderman [1971]). But it is difficult to see how a two-stream system beyond the light cylinder can be made consistent with both the plasma condition (7) and the velocity-of-light limitation: one or other species will need to break away from the field lines at each locality.

The familiar pulsar magnetospheric picture with open field lines is inspired by the analogy with the solar wind, which has field lines which far from the Sun are pulled out by the frozen-in gas so as to be nearly radial. But this depends explicitly on the energy in the wind being ultimately dominant over the energy in the field: and in fact it is at the Alfvénic surface that the magnetic and kinetic energies do become comparable. Further, radial field lines exert strong pinching forces at the magnetic equator, where the magnetic field changes sign, and so require implicitly a plasma pressure in the zero-field region to hold them apart. In the analogous axisymmetric pulsar problem, one must take account of the electrical as well as the magnetic forces. For a solution of equations (12) and (13) to describe a genuinely force-free field and yet also to have an open structure, the electromagnetic stress vector at the equator must be not just continuous but actually zero; for since the field is changing sign, there must again be a region of zero field, so that any nonvanishing stress would require a relativistically hot plasma to balance it. None of the published models appear to satisfy this condition. Claims that the domain in which nonelectromagnetic forces are significant can be banished arbitrarily far beyond the light cylinder seem therefore unfounded: the light cylinder is the analog of the Alfvénic surface in the nonrelativistic problem.

In fact, the picture with open field lines is forced on us only because of the kinematic constraint of field-freezing, together with the requirement that particle speeds are subluminous, so that pure corotation beyond the light cylinder is forbidden. But if, e.g., inertia terms become large enough, so that field-freezing can be relaxed, then there is no objection to field lines that close beyond the light cylinder: the field structure can have a linear velocity greater than c, as long as the particles lag behind. Axisymmetric computations by Kuo-Petravic et al. (1975), which admittedly involve artificially large inertia terms and a nonphysical diffusion introduced for numerical reasons, do indeed yield a more satisfactory picture, with field lines closing behind the light cylinder.

It is perhaps not surprising that in the axisymmetric case it appears impossible to study the relativistic analog of hydromagnetic braking without explicit introduction of finite inertia: after all, it is the gas which must ultimately carry the energy and angular momentum given up by the pulsar. However, in the oblique case we know that Maxwell's displacement current ensures that energy and angular momentum are radiated away through a surrounding vacuum: as the wave propagates, more and more of the vacuum acquires the angular momentum density $r \times [(c/4\pi)E \times B]$. We now consider the effect on the wave of the plasma with the density (8), that maintains the zero rest-mass electric field (7). Does the Goldreich-Julian charge-density combined with the force-free conditions kill the Gunn-Ostriker wave? Or does the wave propagate more or less as if the plasma were not there? Or do the boundary conditions again force one to admit that the zero rest-mass assumption is an untenable zero-order approximation?

For an answer, we need to solve equations (12) under the most general conditions. So far, only very special cases have been studied (Mestel 1973; Mestel *et al.* 1976). The crucial difference from the analogous vacuum problem is that the light cylinder $\bar{\omega}_c$ appears now as a *singularity* of the differential equations, so that it is impossible to construct an outgoing wave, analogous to the Hankel functions $H^{(1)}$. In some cases one can find standing wave solutions, implying zero total flow of energy across the light cylinder; but such solutions clearly require a reflector at infinity. With a realistic boundary condition at infinity (no incoming wave) the result is a *reductio ad absurdum* of the assumptions leading to equations (12), in particular the constraint $E_\parallel = 0$. If we were able to solve the initial value problem, we would again find that the system does not reach a quasi-steady state until the electric fields have accelerated particles to energies high enough for nonelectromagnetic forces (including inertial forces) to be significant, at least locally. Low-frequency waves propagate away from the pulsar, but in so doing they necessarily accelerate charges to high energy. But equally the particle current is as important as the displacement current, and the purely dipolar wave is likely to be an oversimplification. Particles will no longer be tied to field lines, so it is no longer obvious that any counterstreaming particles required for a steady state must have come from the nebula (cf. Jackson 1975); and again there is no objection to field loops closing beyond the light cylinder.

The superficially most simple generalization of the zero rest-mass equations (12) includes just the relativistic inertia terms. However, it should be noted that γ may be forced up to values for which radiation damping is important. Further, the component $E_\parallel$ will accelerate particles of opposite sign in opposite directions; hence in any region of mixed species, a sizable $E_\parallel$ will set up a relativistic relative drift, almost certainly producing a two-stream instability (Coppi and Trèves 1972). A more realistic model would probably have both ions and electrons with nearly the same γ, and with a strong anomalous resistivity coupling the two streams together, and allowing diffusion across the field lines. It will be of interest to know how much of the low-frequency wave would survive these essentially dissipative processes.

Conclusion

Astrophysics is applied physics par excellence. It is difficult to think of any branch of physics which does not have relevance to some astrophysical problem. However, one is only too conscious of how difficult it is to be sure that one's work is really a description of a phenomenon up there, rather than a (hopefully) accurate piece of physical or mathematical analysis. Given axioms A, we can endeavor to show that conclusions B follow (or are at least plausible). But how often can we be sure that A describe accurately or fully the object—star, gas cloud, pulsar, galaxy—under study, even if B are apparently in agreement with observation? Perhaps we should always mentally preface each paper by the phrase: "A study toward the understanding of . . . ." However, we should not begrudge later generations some part in the writing of the definitive treatise on the structure and history of the universe. And let us recall the advice of the second century Rabbi Tarphon, who wrote, admittedly à propos of his own very different world-picture: "It is not thy duty to complete the labor, but neither art thou free to desist therefrom"—a sentiment surely appropriate to a meeting in honor of Chandra, who certainly has never felt he had the right to desist from pushing back the frontiers of astronomy.

References

Alfvén, H. 1954, *On the Origin of the Solar System* (Oxford: Oxford University Press).

Alfvén, H., and Arrhenius, G. 1973, *Ap. and Space Sci.*, **21**, 117.

Arnett, D. W. 1975, private communication.

Arons, J., and Lea, S. M. 1976, *Ap. J.*, **207**, 914.

Baker, N., and Kippenhahn, R. 1959, *Zs. f. Ap.*, **48**, 140.

Borra, E. F. 1974, *Ap. J.*, **187**, 271.

Chandrasekhar, S. 1956, *Ap. J.*, **124**, 232.

Chandrasekhar, S., and Fermi, E. 1953, *Ap. J.*, **118**, 116.

Christy, R. F. 1974, private communication.

Cohen, J. M., and Toton, E. T. 1971, *Ap. Letters*, **7**, 21.

Coppi, B., and Trèves, A. 1972, *Cosmic Plasma Physics*, ed. K. Schindler (New York: Plenum Press).

Cowling, T. G. 1934, *M.N.R.A.S.*, **94**, 39.

———. 1945, *M.N.R.A.S.*, **105**, 166.

Davies, G. F. 1968, *Australian J. Phys.*, **21**, 294.

Deutsch, A. J. 1958, in *Electromagnetic Phenomena in Cosmic Physics*, ed. B. Lehnert (Cambridge: Cambridge University Press), p. 209.

———. 1967, in *Magnetic and Related Stars*, ed. R. C. Cameron (Baltimore: Mono Book Corp.), p. 181.

———. 1970, *Ap. J.*, **159**, 985.

Endean, V. G. 1972, *Nature Phys. Sci.*, **237**, 72.
———. 1974, *Ap. J.*, **187**, 359.
Gillis, J., Mestel, L., and Paris, R. B. 1974, *Ap. and Space Sci.*, **27**, 167.
———. 1977, in preparation.
Goldreich, P., and Julian, W. H. 1969, *Ap. J.*, **157**, 869.
Gough, D. O., and Tayler, R. J. 1966, *M.N.R.A.S.*, **133**, 85.
Hoyle, F. 1945, *M.N.R.A.S.*, **105**, 302.
———. 1953, *Ap. J.*, **118**, 513.
———. 1960, *Quart. J. R. Astr. Soc.*, **1**, 28.
Jackson, E. A. 1975, preprint.
Julian, W. H. 1973, *Ap. J.*, **183**, 967.
Kraft, R. P. 1967, *Ap. J.*, **150**, 551.
Krause, F. 1971, *Astr. Nach.*, **293**, 187.
Krause, F., Rädler, K.-H., and Steenbeck, M. 1971, *The Turbulent Dynamo* (transl. P. H. Roberts and M. Stix) (Boulder: NCAR).
Kulsrud, R. M., and Pearce, W. P. 1969, *Ap. J.*, **156**, 445.
Kuo-Petravic, L. G., Petravic, M., and Roberts K. V. 1975, *Ap. J.*, **202**, 762.
Landstreet, J. D., 1975, Report to IAU-Colloquium **32**, *Physics of Ap Stars*, Vienna.
Markey, P. A., and Tayler, R. J. 1973, *M.N.R.A.S.*, **163**, 77.
Mestel, L. 1953, *M.N.R.A.S.*, **113**, 716.
———. 1961, *M.N.R.A.S.*, **122**, 473.
———. 1965, *Quart. J. R. Astr. Soc.*, **6**, 161 and 265.
———. 1966a, Liège Colloquium on *Gravitational Instability*.
———. 1966b, *M.N.R.A.S.*, **133**, 265.
———. 1968a, *M.N.R.A.S.*, **138**, 359.
———. 1968b, *M.N.R.A.S.*, **140**, 177.
———. 1969, Report to 1968 Asilomar Conference, *Plasma Instabilities in Astrophysics*, ed. D. A. Tidman and D. G. Wentzel (New York: Gordon and Breach).
———. 1970, *Mém. Soc. R. Sci. Liège*, 5th Ser., **19**, 167.
———. 1971, *Nature Phys. Sci.*, **233**, 149.
———. 1973, *Ap. and Space Sci.*, **24**, 289.
———. 1975, *Mém. Soc. R. Sci. Liège*, 6th Ser., **8**, 79.
Mestel, L., and Moss, D. L. 1977, *M.N.R.A.S.*, **178**, 27.
Mestel, L., and Selley, C. S. 1970, *M.N.R.A.S.*, **149**, 197.
Mestel, L., and Spitzer, L., Jr. 1956, *M.N.R.A.S.*, **116**, 583.
Mestel, L., and Strittmatter, P. A. 1967, *M.N.R.A.S.*, **137**, 95.
Mestel, L., and Takhar, H. S. 1972, *M.N.R.A.S.*, **156**, 419.
Mestel, L., Westfold, K. C. and Wright, G. A. E. 1976, *M.N.R.A.S.*, **175**, 257.
Mestel, L., and Wright, G. A. E. 1977, in preparation.
Michel, F. C. 1973, *Ap. J.*, **180**, 207.
Moss, D. L. 1977a, *M.N.R.A.S.*, **178**, 51.
———. 1977b, *M.N.R.A.S.*, **178**, 61.
Moss, D. L., and Tayler, R. J. 1969, *M.N.R.A.S.*, **145**, 217.

Mouschovias, T. Ch. 1974, *Ap. J.*, **192**, 37.
———. 1976*a*, *Ap. J.*, **206**, 753.
———. 1976*b*, *Ap. J.*, **207**, 141.
Okamoto, I. 1974, *M.N.R.A.S.*, **167**, 457.
Osterbrock, D. E. 1961, *Ap. J.*, **134**, 270.
Ostriker, J. P. 1972, in *On the Origin of the Solar System*, ed. H. Reeves (Paris: C.N.R.S.), p. 154.
Ostriker, J. P., and Gunn, J. E. 1969, *Ap. J.*, **157**, 1395.
Parker, D. A. 1973, *M.N.R.A.S.*, **163**, 41.
———. 1974, *M.N.R.A.S.*, **168**, 331.
Parker, E. N. 1955, *Ap. J.*, **122**, 293.
———. 1963, *Interplanetary Dynamical Processes* (New York: Interscience).
———. 1966, *Ap. J.*, **145**, 811.
Piddington, J. H. 1972, *Solar Phys.*, **22**, 3.
Preston, G. W. 1970, in *IAU Symposium on Stellar Rotation*, ed. A. Slettebak (Dordrecht: Reidel).
———. 1971, *Pub. Astr. Soc. Pac.*, **83**, 571.
Ruderman, M. A. 1971, *Phys. Rev. Letters*, **27**, 1306.
Scharlemann, E. T., and Wagoner, R. V. 1973, *Ap. J.*, **182**, 951.
Skumanich, A. 1972, *Ap. J.*, **171**, 565.
Spiegel, E. A. 1968, in *Highlights of Astronomy*, ed. L. Perek (Dordrecht: Reidel).
Spitzer, L., Jr. 1958, in *Electromagnetic Processes in Cosmical Physics*, ed. B. Lehnert (Cambridge: Cambridge University Press), p. 169.
Strittmatter, P. A. 1966, *M.N.R.A.S.*, **132**, 359.
Sweet, P. A. 1950, *M.N.R.A.S.*, **110**, 548.
Verschuur, G. L. 1969*a*, *Ap. J. (Letters)*, **155**, L155.
———. 1969*b*, *Ap. J.*, **156**, 861.
Weber, E. J., and Davis, L., Jr. 1967, *Ap. J.*, **148**, 271.
Wilson, O. C. 1968–1973, Hale Observatories Annual Reports.
Wilson, O. C., and Woolley, R. 1970, *M.N.R.A.S.*, **148**, 463.
Wolff, S. C. 1975, private communication.
Wright, G. A. E. 1969, *M.N.R.A.S.*, **146**, 197.
———. 1973, *M.N.R.A.S.*, **162**, 339.

5   Stellar Dynamics   George Contopoulos

Introduction

Stellar dynamics was developed as a by-product of stellar kinematics, in order to explain the observed patterns in the distribution of stellar velocities. The main subjects of the early stellar dynamics were the theory of star streaming, including Schwarzschild's ellipsoidal distribution, and the differential rotation of our Galaxy, following the work of Oort and Lindblad.

The first to develop stellar dynamics as a discipline in itself, by bringing forth and trying to solve its own theoretical problems, was Chandrasekhar. In his *Principles of Stellar Dynamics* Chandrasekhar (1942) isolates the two outstanding problems of stellar dynamics, which "appear to have an interest for general dynamical theory even apart from the practical context in which they arise". These problems refer on one hand to the solutions of the collisionless Boltzmann equation (called also Liouville's equation) and on the other hand to the time of relaxation and the evolution of star clusters. These are typical and basic problems of what we call today collisionless and collisional stellar dynamics.

I became interested in stellar dynamics myself in the early 1950s, and, as I was quite isolated and inexperienced, I tried to generalize Chandrasekhar's treatment of Liouville's equation. I started full of optimism—I even published a short paper on the subject—but then I came to the real difficulties. I reached a system of 20 partial differential equations of second order with respect to three independent variables, and finally I had to give up. All I learned was the following principle: "Don't try to go beyond Chandrasekhar by following the same road. Your only hope is to find a different road." As a consequence I tried a different approach. I was impressed by the fact that some problems of stellar dynamics are closely related to celestial mechanics. The fact that I was isolated turned out to be in my favor, because I did not realize how out of fashion this approach was at that time. Thus, I started a study of galactic orbits, and I was fortunate enough to encounter the "third" integral.

The new interest in stellar dynamics started around 1960 in three main directions. On one hand the theory of orbits gave a large spectrum of problems connected with the behavior of the integrals of motion, the general topological properties of orbits, etc. The interest in these problems extended well beyond

George Contopoulos is in the Astronomy Department at the University of Athens.

astronomy, to several other branches of physics, like dynamics in general, plasma physics, the physics of molecules and of solids, statistical mechanics, etc.

The second kind of problem is collisionless dynamics, namely, the study of the collisionless Boltzmann equation together with Poisson's equation. I should mention here the problems connected with the equilibrium solutions of spherical or axisymmetric stellar systems, the study of modes in galaxies—e.g., bar modes or spiral modes—and the collisionless evolution of stellar systems, especially in cases of violent relaxation, for which a new statistics like that of Fermi and Dirac has been proposed by Lynden-Bell. Of special interest in this area is the theory of spiral density waves of Lin, Shu, Kalnajs, and their associates, and its recent nonlinear developments.

The third area refers to the collisional evolution of stellar systems. In this area belong phenomena taking place in time scales of the order of the time of relaxation, or larger. Such is the escape of stars from a cluster, the formation of binaries, and the evolution of the nucleus.

In recent years much impetus was given to stellar dynamics by the use of large electronic computers. Computers have been used extensively in calculations of stellar orbits and integrals of motion, in simulations of galaxies by Prendergast, Miller, Hohl, etc., as well as in calculations of the evolution of star clusters by Hénon, Aarseth, Spitzer, and others.

Thus, at the present moment there is a lot of activity going on in the area of stellar dynamics. It is not possible to review this whole area here. Therefore, I will report only on some particular developments that have wider implications for physics in general. These are connected with the existence of integrals of motion, besides the classical integrals of energy, angular momentum, etc.

Integrable versus Ergodic Systems

We know that in a rather flat galaxy we have not two but three integrals of motion: the energy, the angular momentum, and the "third" integral (which in the lowest approximation is the energy in the z-direction). More generally in many systems of N degrees of freedom we can find under certain conditions N integrals of motion. As we will see below, this has far reaching consequences not only for dynamics, but also for statistical mechanics and maybe also for quantum mechanics.

Let us consider a very simple example, namely, the density of the gas inside this room. Let us idealize the walls as perfect reflectors and disregard gravitational fields and any other "external" interaction. We know that the density of the gas in this room is constant everywhere. Why is this so? Because the probability of having any particle inside either of two equal volumes is the same. This, in turn, is due to the fact that the orbit of every particle goes everywhere and the particle stays equal times in equal volumes. In other words, the motion of every particle is ergodic.[1]

The ergodic assumption for noninteracting particles moving with constant velocity within a room with parallel perfect reflecting walls can be proved very easily. In general, an orbit in such a room fills the whole space; i.e. it is everywhere dense.

The only exceptions are the periodic orbits. For example, we can imagine all the particles of this room moving up and down with constant velocity in such a way that they are always on a (moving) plane. Such a system is not ergodic at all, and the density is zero except on the plane of the particles. However, one can prove that such configurations have a probability zero. In fact, the whole set of periodic orbits has a measure zero.

A much more general theorem was proved in 1963 by the Russian mathematician Sinai. Sinai proved the ergodic hypothesis for the hard-sphere gas. The particles of such a gas are assumed to move along straight lines until a collision occurs, either between two particles, or between a particle and the wall. The collisions are elastic, and they produce the ergodicity of the system. Thus Sinai's theorem justifies the application of the usual methods of statistical mechanics to a hard-sphere gas.

The question is now: What happens in other, more general cases? The hard-sphere gas is peculiar in the sense that the particles interact only during the collisions, and the forces are then infinitely repulsive. What happens if we have attractive forces, or both attractive forces (for large distances) and repulsive forces (for short distances)?

Consider, e.g., a large number of planets of infinitesimal mass moving in the same plane around the Sun. It is obvious that, unless we have some very unlikely resonant configuration, the differences in the periods of the planets will bring about a random distribution of their position vectors (Fig. 1). However, the distribution of the major axes and eccentricities will not change in time, because each planet is not influenced by the others. The motion is not ergodic on the energy surface. We may only say that it is ergodic on the subspace of the angle variables.

These examples lead us to the basic problem. What is the most representative case in dynamics—the hard-sphere gas, or the noninteracting planetary system? And how general is this latter case? In particular, what will happen if the masses of the planets are different from zero? In this way we come to the famous problem of the stability of the solar system. Is it possible that the mutual perturba-

---

[1] "Ergodic" is a combination of two Greek words *ergon* (energy) and *hodos* (path). It means that the path of a particle goes through the (arbitrary) neighborhood of every point of the energy surface. (With this definition "ergodic" is equivalent to what used to be called "quasi-ergodic" in the past.)

tions of the planets will force the planets to escape, or to collide with each other or with the Sun?

The answer to this problem was given a few years ago by Arnold (1963). If the eccentricities and the masses of the planets are sufficiently small, then for most initial conditions the solar system is stable; i.e., the changes in the major axes and eccentricities are small.

A peculiar characteristic of this theorem is that it is proved for a set of initial conditions that is nowhere dense. That is, the theorem is not valid whenever the mean motions of the planets are in resonance or near resonance. In the case of two planets with mean motions ω_1 and ω_2 $(0 < \omega_1/\omega_2 < 1)$ the above condition requires that ω_1/ω_2 be sufficiently far from all rationals. This means that from the set of numbers between 0 and 1 we have to exclude all rational num-

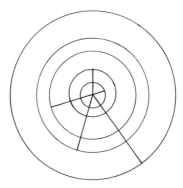

Figure 1. The position vectors of a set of infinitesimal planets distributed at random.

bers, *together with their neighborhoods*. One may wonder whether anything is left in this way. However, if one chooses the neighborhoods of the rationals sufficiently small, the total set of the excluded regions has also a small measure.

In fact let the neighborhood of the rational n/m $(n < m)$ be ϵ/m^3. There are $m - 1$ rationals with denominator m; therefore the sum of their neighborhoods is less than ϵ/m^2. The sum S of all neighborhoods, for all m, is smaller than $\epsilon\Sigma m^{-2} = \frac{1}{6}\pi^2\epsilon \approx 1.6\epsilon$. Thus, the excluded set is small, and we can be sure that in the majority of cases the theorem of Arnold is applicable and the system is nonergodic.

Another example of a nonergodic system is the model of a nonlinear string considered by Fermi, Pasta, and Ulam (1955). The model consists of a number of particles on a straight line interacting not only with elastic (linear) forces, but also with a small nonlinear force. This problem can be reduced to that of N *coupled* harmonic oscillators (modes).

If the coupling constant is zero, we have N uncoupled oscillators, and there is no energy transfer from one mode to the other. However, if the coupling is different from zero, then Fermi, Pasta, and Ulam expected that there would be a mixing of the modes; i.e., the energy would be shared by all the modes in an ergodic way. Thus, they performed a numerical experiment with a computer (one of the first computer experiments ever made) to verify this hypothesis. Instead of that, the numerical experiments have shown that there is no ergodicity at all. If the energy is initially put in one mode, only a few modes are excited, and the whole situation is quasi-periodic. This result was quite unexpected and has led to a large amount of numerical and theoretical work exploring this field. As an example I may mention the discovery of the solitons and their applications in solid-state physics.

It is of interest to remark here that if Fermi, Pasta, and Ulam had used a larger initial energy, they would have found the ergodic behavior they expected and the interesting nonergodic results would have been unnoticed.

Such a case occurred recently with a Lennard-Jones potential. This potential is attractive for large distances but repulsive for very short distances:

$$V = 4\epsilon\left[\left(\frac{1}{r}\right)^{12} - \left(\frac{1}{r}\right)^{6}\right]. \tag{1}$$

As this potential seems to represent fairly well the interaction between the particles of a gas, or a solid, it has been applied to many model problems in recent years.

The first numerical experiments (Stoddard and Ford 1973) indicated that a system of particles interacting with a Lennard-Jones potential is always ergodic, while Galgani and Scotti (1972) have shown that no ergodicity appears for energies smaller than a certain threshold.

A similar situation seems to occur in most cases of actual interest. There is a more or less abrupt threshold, such that for energies smaller than this threshold the motions are quasi-periodic, while for larger energies the motions are ergodic.

It is of interest to explain this transition from quasi-periodic to ergodic motion in some detail. The situation that has been studied most extensively is that of two coupled oscillators. In such a case the Hamiltonian can be written in action-angle variables in the form

$$H \equiv \omega_1 I_1 + \omega_2 I_2 + \epsilon(\text{coupling terms}) = E, \tag{2}$$

where E is the total energy, while the coupling terms are time independent and trigonometric in the angle variables.

If the coupling constant ϵ is zero, the energies of both oscillators $E_1 = \omega_1 I_1$ and $E_2 = \omega_2 I_2$ are separately conserved. If the coupling term is small, there is some energy exchange between the two oscillators, but this is small and changes periodically in time. One can find a generalization of the energy of one oscillator,

$$\Phi_1 = E_1 + \text{higher order terms} \tag{3}$$

which is a formal second integral of motion, besides the total energy. This is usually called an "adelphic"[2] integral (Whittaker 1937), or a "third" integral because it was found also in the case of galactic orbits of stars as an integral beyond the energy and angular momentum (Contopoulos 1960).

The higher-order terms of the formula (3) can be given by a computer program (Contopoulos 1966; Gustavson 1966) for any given Hamiltonian of the form (2). Thus, we can calculate Φ_1 up to the terms of any given degree.

If the coupling constant is small, we find that the calculated quantity Φ_1 is better and better conserved as we add higher and higher order terms. Thus, Φ_1 behaves like an analytic integral of motion.

However, we know that Φ_1 is not, in general, an analytic integral. In fact, if the energy of the system becomes large (or if the coupling constant becomes large for constant energy), then the divergence of the series Φ_1 is quite obvious. Any truncated value of Φ_1 oscillates, and the inclusion of further higher-order terms does not improve the situation. In such a case we can check that the orbits behave much as in an ergodic system.

Why do we have this transition from ordered motion (where the orbits are, in fact, quasi-periodic) to a stochastic, or ergodic, motion?

The reason is well understood today. It is what we call an "interaction of resonances" (Chirikov 1967; Contopoulos 1967; Rosenbluth, Sagdeev, and Taylor 1966; Walker and Ford 1969).

Interaction of Resonances

We know that the initial formula (3) of the "third" integral is not valid near resonances. The extent of each resonant region is given by the theory (Contopoulos 1965, 1967). That is, if the ratio of the frequencies ω_1/ω_2 is near the rational number n/m, the area of the resonant region is proportional to $E^{(n+m-4)/4}$ for fixed ϵ.[3] Thus, if n and m are large, the resonant regions are extremely small. On the other hand, if n and m are small, the resonance phenomena become quite prominent.

If we add all resonant regions together, we find a sum of the order of

$$E^{(1+\omega_1/\omega_2)/4}$$

(for $\omega_1 < \omega_2$).[4]

---

[2] "Adelphic" comes from the Greek word *adelphos* (brother). This integral was named so by Whittaker because it "stands in a much closer relation to the integral of energy than do the other integrals." However, in some resonant cases the new integral is quite different from the energy. Such is the case of two coupled oscillators with ratio of frequencies ω_1/ω_2 equal to 1 or 2. In these cases the energies of the two oscillators are not even approximately conserved.

[3] It is easy to show that a change of ϵ is equivalent to a change of E together with a change of scale.

[4] We know (Contopoulos 1965) that for a given ratio of frequencies $\omega_1/\omega_2 < 1$ the range of values of n/m, near ω_1/ω_2, that produce resonant phenomena is of $O(E)$. Thus, we may

If the energy E is small, this quantity is small and the various resonant regions are well isolated. However, if E increases, the resonant regions increase and they interact with each other.

The interaction of resonances starts near the unstable periodic orbits. It is of interest to see how the unstable orbits produce ergodicity. Let us consider the Hamiltonian

$$H \equiv \omega_1 I_1 + \omega_2 I_2 + f_0(I_1, I_2) + f_{12}(I_1, I_2) \cos(\theta_1 - 2\theta_2) = E, \qquad (4)$$

where $(I_1, I_2, \theta_1, \theta_2)$ are conjugate action-angle variables. Such is the problem near the inner Lindblad resonance of a galaxy, where the epicyclic frequency ω_1 is the double of the rotational frequency ω_2. This Hamiltonian contains only

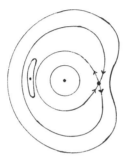

Figure 2. Invariant curves in a 2/1 resonance case (e.g., the inner Lindblad resonance of a galaxy).

one combination of the angle variables, namely, $(\theta_1 - 2\theta_2)$. Because of that it can be checked that the combination

$$J_2 = I_2 + 2I_1 \qquad (5)$$

is an exact integral of motion. Thus, this problem is an integrable one.

Let us consider now the intersections of an orbit by a fixed line through the origin (e.g., $y = 0$). At every point of intersection we find the values of x and $\dot{x}$, and plot in a diagram the successive points $(x, \dot{x})$. Such points were called consequents by Poincaré (1889). All the consequents that correspond to the same orbit lie on a curve, which is called an invariant curve. If ω_1/ω_2 is near 2/1, the total set of invariant curves is given in Figure 2. We notice that there are three invariant points, two stable and one unstable. The invariant points represent periodic orbits. In Figure 3 we give a few examples of periodic orbits in a

---

state that we have $\Delta n = QEm$ resonant regions of multiplicity m. We have roughly $n \approx m\omega_1/\omega_2$; thus the total area of the resonant regions is of order

$$\Sigma Q m E^{m(1+\omega_1/\omega_2)/4} \approx QE^{(1+\omega_1/\omega_2)/4}/[1 - E^{(1+\omega_1/\omega_2)/4}]^2,$$

i.e., roughly of $O[E^{(1+\omega_1/\omega_2)/4}]$.

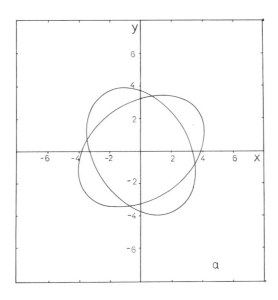

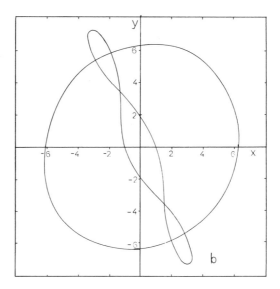

Figure 3. Periodic orbits in a model of our Galaxy composed of a Barbanis-Woltjer (1967) spiral field superimposed on a Contopoulos-Strömgren (1965) axisymmetric model. The spiral force is about 5% of the axisymmetric force. The orbits are given for two values of the Hamiltonian corresponding to the following values of the radius of the circular orbit of the axisymmetric field: (a) r_c = 3.8 kpc, (b) r_c = 6.4 kpc.

realistic model of our Galaxy. Some resonant orbits extend quite far from the inner Lindblad resonance, which occurs at 3.7 kpc.

Orbits near a stable periodic orbit form invariant curves surrounding the corresponding invariant point. In general there is only one invariant curve through each point of the plane, except for the unstable invariant points. The two invariant curves emanating from an unstable point are called asymptotic curves, because if we start an orbit at a point of such a curve the successive consequents approach asymptotically the unstable point as $t \rightarrow \infty$, or as $t \rightarrow -\infty$.

We notice, in Figure 2, that the asymptotic curves separate the invariant curves that surround the first and the second stable periodic point. For this reason the asymptotic curves are called also separatrices. Outside the outer

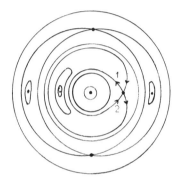

Figure 4. Approximate form of the invariant curves when we have two resonant terms in the Hamiltonian (corresponding to the resonances 2/1 and 3/2). The invariant points that correspond to other (secondary) resonances are not shown in this figure.

branch of the asymptotic curve there are nonresonant invariant curves surrounding all three invariant points.

If we add now one further trigonometric term in the Hamiltonian (4), whose argument is not a multiple of $(\theta_1 - 2\theta_2)$, say

$$f_{23}(I_1, I_2) \cos (2\theta_1 - 3\theta_2), \qquad (6)$$

we find a number of new phenomena. The most conspicuous is the appearance of two further periodic orbits (if ω_1/ω_2 is between 3/2 and 2, and the energy E is large enough), each represented by two invariant points, corresponding to the resonance 3/2. The new stable invariant points are surrounded by closed invariant curves, called islands (Fig. 4). However, the most important phenomenon is that now the asymptotic curves emanating from the unstable invariant points are no longer smooth curves but intersect each other in a most intricate way. For example, the asymptotic curves 1 and 2 from the original unstable invariant point form not one curve, as they seem to do in Figure 4, but two different curves; each curve forms loops of larger and larger amplitude, intersecting the other curve an infinite number of times, but never intersecting itself (Fig. 5). If we take the successive consequents of an orbit starting near an unstable periodic orbit, they look scattered in a practically random way in a certain region (Fig. 6). This is what we call a "dissolution" of the invariant curves. This "dissolution" is the main manifestation of nonintegrability. We

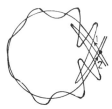

Figure 5. The actual form of the asymptotic curves emanating from an unstable invariant point. For small E the oscillations of the asymptotic curves are so small that they are inconspicuous.

Figure 6. The distribution of the consequents corresponding to an orbit starting near an unstable invariant point.

may say that the oscillations of the asymptotic curves and the corresponding dissolution is due to the interaction of two different resonances in the Hamiltonian (in the present case of the resonances 2/1 and 3/2).

The regions of randomness are called also "zones of instability" (Birkhoff 1927). In these regions we have an infinite number of new periodic orbits, most or all of them unstable. Every unstable periodic orbit of this type has its own asymptotic curves intersecting those of the original orbit in an even more complicated way. Thus, we have an interaction of several resonances.

If the energy E is small, the region where we have dissolution of the invariant curves is very small. For small enough E we cannot distinguish the oscillations of the asymptotic curves, which look smooth as in Figure 4.

As E increases, we see some dissolution near the unstable invariant points, but the most important resonances (2/1 and 3/2) are well isolated by closed invariant curves. Thus, the zones of instability are well separated from each other.

If, however, E increases still further, the resonant regions increase and finally the asymptotic curves of the two main resonances intersect each other in a complex way, producing an important and quite conspicuous interaction of resonances. In particular the points of intersection of the asymptotic curves of different resonances (heteroclinic points) are starting points of orbits that ap-

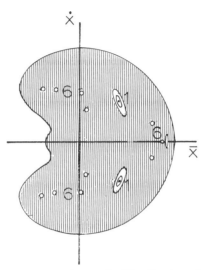

Figure 7. An "ergodic sea", corresponding to the Hamiltonian of Fig. 8, for $\epsilon = 4.5$. The consequents of all orbits starting in the shaded region are scattered practically at random in it. However, there are some "islands of stability" containing good (closed) invariant curves, surrounding the invariant points of stable periodic orbits intersecting the x-axis once (two orbits) or 6 times (two orbits).

proach two quite dissimilar unstable periodic orbits corresponding to the resonances 2/1 and 3/2.

In such a case the points representing an orbit are said sometimes to fill an "ergodic sea" (Fig. 7). Of course, we cannot prove that the distribution of the points in this "sea" is really random—in fact, we often find there small islands of stability (Fig. 7)—but for many practical purposes we can apply there stochastic arguments, especially if the ergodic sea covers most of the available space.

Another way of looking at the interaction of resonances is the following. If we consider the invariant curves around a stable invariant point, we can define on each of them a rotation number, i.e., the average angle between successive position vectors from the stable invariant point, joining successive consequents. In Figure 8 we give, in a particular case, the rotation number, r, versus the distance of each invariant curve, $\bar{x} = \omega_1 x$, from the origin (along the axis $\dot{x} = 0$) for various values of ϵ, and constant energy E. It is seen that these "rotation curves" have plateaus in the neighborhood of the stable invariant points (extending as far as the corresponding islands), while they have apparent discontinuities near the unstable invariant points. The discontinuity near the unstable point $r = \frac{2}{3}$ is inconspicuous for $\epsilon = 2$ (it would be zero in an inte-

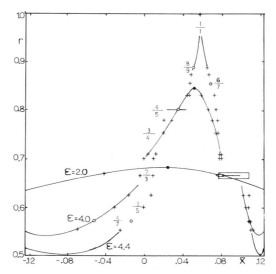

Figure 8. The rotation number r of invariant curves in the Hamiltonian $H \equiv \frac{1}{2}(\dot{x}^2 + \dot{y}^2 + \omega_1^2 x^2 + \omega_2^2 y^2) - \epsilon x y^2 = E$, for $\omega_1^2 = 1.6$, $\omega_2^2 = 0.9$, $E = 0.00765$, and various values of ϵ, vs. $\bar{x} = \omega_1 x$ (for $\dot{x} = 0$). Periodic orbits are shown by circles (stable) or crosses (unstable). Large resonant regions appear at $r = 2/3$ and $r = 4/5$ (the latter for $\epsilon = 4.4$ only).

grable case) but it increases abruptly if ϵ becomes about 4 or larger. For $\epsilon = 4.4$ most of the space is covered by the ergodic sea. For most orbits no rotation number can be defined, and only a skeleton of periodic orbits remains, to remind one of the rotation curve that appeared for small ϵ.

Systems of Many Degrees of Freedom

Up to now we have treated only problems of two dimensions. However, the problems become more difficult if we have more than two degrees of freedom.

A dynamical system of N degrees of freedom which is near an integrable case has, in general, N formal integrals of motion including the energy. We can construct these integrals in the form of series. As in the case of two degrees of freedom, the series are not convergent, in general; and they are not valid near resonances. We have again a set of integral surfaces, corresponding to frequencies "far from resonances." However, there is an important difference between this problem and the problem of two degrees of freedom. The integral surfaces are N-dimensional in a space of $2N - 1$ dimensions (the surface of constant energy). Therefore, if $N > 2$, these surfaces do not separate the $(2N - 1)$-dimensional space. For example, if $N = 3$, the energy surface is five-dimensional but the integral surfaces are three-dimensional. Therefore, they do not impose topological restrictions to the motions of particles that do not belong to the integral surfaces.

The situation is similar to that of a room filled with perpendicular strings going through every point with coordinates x and y irrational, and sufficiently far from all rationals (thus, we do not have such strings whenever x, or y, is rational, or near a rational). These strings cover most of the volume of this room. However, if one is slender enough, he can move freely along lines where one coordinate is rational, and thus reach the neighborhood of every point in this room.

A similar phenomenon is believed to occur in most dynamical systems (Arnold and Avez 1967), including the N-body problem.

Thus, we come to a very curious situation. If a particle is on an integral surface, its motion is nonergodic, in the $(2N - 1)$-dimensional space. This is the case for the majority of phase space (if the energy E is small). However, if we allow for some irregular forces, most systems become ergodic; or, if the energy surface extends to infinity, most particles eventually escape. In the case of the solar system Arnold concludes that the system is, in fact, unstable, although for the great majority of the initial conditions it is stable. The instability is produced by irregular forces (like the resistance of the interplanetary gas).

A number of people tried to find this Arnold diffusion by numerical experiments. However, the results up to now have been negative, or doubtful. In particular, the work of Froeschlé (1970) on the three-dimensional restricted three-body problem and that of Hadjidemetriou (1975) on the general three-body problem give evidence that this diffusion does not exist, or is not operative in quite important cases. Hadjidemetriou reduced the planar three-body problem to a system of three degrees of freedom, and found several families of periodic orbits for various masses, and in particular for the case of three equal masses. Then he studied the behavior of nonperiodic orbits by taking the initial conditions at various distances from a periodic orbit. In many cases such orbits define good integral surfaces, indicating the existence of three integrals of motion. No Arnold diffusion appears, even for very long times.

An example of such a surface is given in Figure 9. We have placed there the successive points of intersection of an orbit in the space $(x_1, \dot{x}_1, x_3, \dot{x}_3, y_3, \dot{y}_3)$ by the plane $y_3 = 0$.[5] If there are three integrals of motion, all these points lie on a two-dimensional invariant surface that corresponds to the one-dimensional invariant curve of a problem of two degrees of freedom. In Figure 9 the successive points of intersection are near each other, and we can join them by a continuous line. The time between two crossings of the plane $y_3 = 0$ defines an average period P_1. There are about 50 points on a loop [marked (1) in Fig. 9], defining a second period $P_2 \approx 50 P_1$. We see that the loop does not close but has a slow precession on the invariant surface, which is like a distorted cylinder,

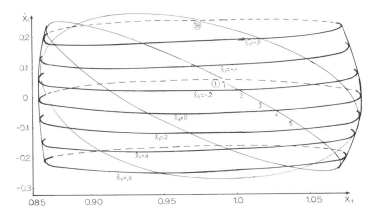

Figure 9. An invariant surface corresponding to a particular orbit in the planar general three-body problem. It contains the successive intersections of the orbit by the plane $y_3 = 0$ (consequents). The various level curves $\dot{x}_3 = 0$, $\dot{x}_3 = 0.2$, etc., are projected on the plane $(x_1, \dot{x}_1)$. The successive consequents like 1, 2, 3, 4, 5, are joined by a continuous line; this line forms loops like those labeled (1) and (38).

covering this cylinder completely after about 250 loops. Thus, we have a third period P_3 equal to about $250 P_2$. The existence of these three well separated periods is another indication that we have three integrals of motion in this problem, which act as adiabatic invariants.

If we go further away from the periodic orbit, we find that the particles escape, sometimes after many revolutions. One may argue that this escape is due to Arnold diffusion and that there is always a similar diffusion, i.e., the particles always escape, but the time scale is extremely large. This may be so, but it seems rather improbable. In fact, according to Arnold the time scale of the diffusion is of $O(e^{1/\sqrt{\epsilon}})$. (This expression, incidentally, shows why this phenomenon cannot be predicted by a perturbation technique.) Here ϵ is a measure of the departure of the dynamical system from a separable system, where no Arnold diffusion exists. Hadjidemetriou has studied systems composed of a

[5] The masses m_1 and m_2 are always on the x-axis, and the origin is their center of mass. Thus, there are only three independent coordinates and the corresponding velocities.

binary and a third body at distances from 2 to 10 times larger than the dimensions of the binary. This case can be considered as a perturbation of an integrable system, namely of a two-body problem (assuming the binary as one point). The relative perturbations in the potential are rather large, of order $\frac{1}{3}$ to $1/10$. One would expect that in such cases the Arnold diffusion would be prominent, but it is not so.

One can make the argument more quantitative in the following way. In one case it was found that, if an orbit deviates by $\epsilon = 0.2$ from a stable periodic orbit (of size ~ 1), it produces a good invariant surface, while for a deviation $\epsilon = 0.3$ the particle escapes. Now $e^{1/\sqrt{0.2}} \approx 9.4$, while $e^{1/\sqrt{0.3}} \approx 6.2$. Thus, if the particle escapes (i.e., it goes beyond a certain limiting distance with large enough velocity) in the last case in time T, the same phenomenon should occur in the first case in a time scale $T' \approx 1.5\ T$. However, the numerical experiments indicate that no escape occurs for times much larger than T'.

In order to explore the existence or nonexistence of Arnold diffusion in various cases, we have started a study of the regions near unstable invariant points in more than two dimensions. Arnold expects that the asymptotic surfaces from the unstable points extend very far, intersecting the asymptotic surfaces of practically all other unstable points. However, this is doubtful. The equations of motion may provide further restrictions, which are not yet well understood. Thus, the study of the asymptotic surfaces near unstable invariant points would be quite useful in understanding the dynamics of systems of many degrees of freedom.

The general three-body problem is a typical problem of more than two degrees of freedom, and it has attracted recently much interest. The discovery of many families of periodic orbits (Hénon 1974; Hadjidemetriou 1975) has been the starting point of a systematic effort to understand the totality of orbits. It is expected that this work will allow a separation of the phase space in regions of stability, ergodicity, and escape, as was done a few years ago by Hénon in the case of the restricted three-body problem. Similar phenomena appear in the case of the N-body problem also; e.g., Hadjidemetriou demonstrated the existence of families of periodic orbits in the general N-body problem.

It is well known that the equations of motion of the N-body problem can be reduced to $6N - 10$ first-order differential equations, with the help of the classical integrals of motion. According to a theorem of Moser (1967) the stable periodic orbits are followed, in general, by nearby quasi-periodic orbits with $3N - 5$ frequencies, which lie on $(3N - 5)$-dimensional surfaces (tori). These surfaces are represented asymptotically by the new integrals, which are formal series of the same form as the "third" integral described above. However, the forms of the new integrals are different near every stable periodic orbit. Thus, it remains to be studied how large the stable regions are that surround the various periodic orbits where these integrals are applicable. This study would also give the ergodic, or stochastic regions in the N-body problem, where one can apply statistical arguments. No such care has been used in the statistical mechanical schemes proposed so far for N-body systems.

I believe that some unexplained phenomena and difficulties encountered both in collisionless and in collisional stellar dynamics are due to the appearance of such integrals of motion. I would like to mention two examples.

First, as an example of a collisionless problem let us consider the applicability of Lynden-Bell's (1967) statistics. Lynden-Bell found that a stellar system in violent relaxation obeys a statistics that is similar to that of Fermi and Dirac. Several numerical experiments have been made to verify the predictions of the new statistics. In many cases the agreement between theory and numerical calculations is good up to a certain energy, but bad beyond this limit. It has been argued that this is due to the fact that the initial mixing was not violent enough. However, Lecar and his associates found (Lecar and Cohen 1972, and references there) that there is no clear correlation between violence of relaxation and degree of applicability of Lynden-Bell statistics. In some cases the most violently relaxing systems have the greatest deviations from the predictions of Lynden-Bell statistics.

It is known that the analytic integrals can be incorporated in Lynden-Bell's statistics (Lynden-Bell 1967). However, the new integrals behave differently in different regions of the phase space. Thus, the orbits may be ergodic in some regions, while in other regions they may belong to a practically dense set of integral surfaces. Thus, in order to use the new integrals as arguments in the distribution function, we must specify the regions of their validity.

Second, in the case of collisional N-body systems, there are several indications (see, e.g., Miller 1973; Ipser 1974) that in many cases the "thermodynamic" arguments may not be applicable to such systems. For example, the "entropy" of an N-body system may increase or decrease, and no maximum entropy states seem to appear.

These problems are closely connected with the question of what part of the phase space is available for the motions in every case. Therefore, an understanding of the restrictions imposed by the nonclassical integrals of motion is of great importance in this area.

The Case $N \rightarrow \infty$

Several people have investigated whether various many-body systems behave as nearly integrable or nearly ergodic. For this purpose one usually studies the increase of the distance of two points in phase space that start very near each other. If the increase is linear in time, the system is practically integrable; if the increase is exponential, the system is practically ergodic.

The main problem is what happens when $N \rightarrow \infty$. It has been assumed (see, e.g., Ford 1975) that the transition energy would go to zero for large N. This means that as $N \to \infty$ all systems would be ergodic. The argument is that the interactions between resonances are much more numerous for large N than for small N. However, the extensive work of Galgani and Scotti (1972, 1975) in Milan has shown that this is not the case.

For small enough energies the system of N particles is quasi-periodic. Galgani and Scotti (1975) applied the computer program of Contopoulos (1966) in constructing formal integrals of motion for systems of N particles interacting through a Lennard-Jones potential. For small energies the truncated integrals are conserved very well. If one takes two nearby systems and calculates their deviation in phase space, this is linear in time. On the other hand, the devia-

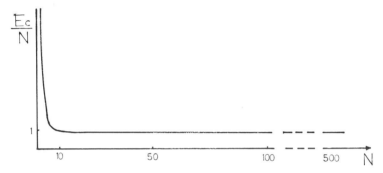

Figure 10. The transition energy per particle, E_c/N, as a function of N (number of particles, or degrees of freedom).

tions become exponential in time if the energy is above a certain transition energy, E_c, and the systems become stochastic. The transition energy per particle, E_c/N follows the curve of Figure 10. For small N the quantity E_c/N decreases with N, but for $N \geq 8$ it remains stable at about 1. This kinetic energy is only about 4 percent of the energy necessary for the particles to escape from the potential well, but it is clearly finite and does not tend to zero.

Experiments were made with N up to 500 and the results give a fairly constant transition energy per particle $E_c/N \approx 1$. I would claim that $N = 500$ is a good representation of $N = \infty$.

Of course, this is not a proof, and there will always be doubters, but it is remarkable that while some time ago practically no one dared to doubt seriously stochasticity for $N \to \infty$, now the doubts are on the other side. At any rate, there is already enough evidence for our case to justify the exploration of the consequences of nonstochasticity for $N \to \infty$.

These consequences are really remarkable. One can derive by classical methods several results usually considered as basically quantum-mechanical (Galgani 1975). In particular one can derive by classical methods an analog of Planck's law. In order to do that, one has to interpret the threshold energy E_c as the zero-point energy of thermodynamics.

The idea of using the zero-point energy to derive Planck's law of radiation by classical methods was introduced already in 1913 by Einstein (Einstein and Stern 1913). Einstein had tried previously (Einstein and Hopf 1910) to derive a radiation law, in a cavity full of resonators interacting with each other, by assuming tacitly that the zero-point energy is zero. The law derived in this way was the Rayleigh-Jeans law. With the introduction of the zero-point energy the result of the calculation was Planck's law. This result was rederived independently by Boyer (1969).

It is obvious that research in this field opens many exciting possibilities. This is why we have started a collaboration with the group in Milan in order to explore some basic problems in this area—e.g., the role of the various integrals of motion and the regions of their applicability, the importance of the resonances in producing stochasticity, the reason of the lack of Arnold diffusion, etc. It is only after we have a good understanding of these questions that we can attack the fundamental problems of the foundations of statistical mechanics and quantum mechanics.

The examples that I gave above show the progress made in recent years in stellar dynamics and in areas related to it.

Chandrasekhar's role in this field has been that of a pioneer. For several years after the publication of his book the subject looked too difficult and perhaps too esoteric to attract much interest. However, the situation gradually changed. As time passed, more and more research workers entered this field. Today the areas related to stellar dynamics are full of interesting problems, full of new results, and full of young people working in them.

References

Arnold, V. I. 1963, *Uspekhi Math. Nauk*, **18**, 91.

Arnold, V. I., and Avez, A. 1967, *Problèmes Ergodiques de la Mécanique Classique* (Paris: Gauthier-Villars).

Barbanis, B., and Woltjer, L. 1967, *Ap. J.*, **150**, 461.

Birkhoff, G. D. 1927, *Dynamical Systems* (Providence: American Mathematical Society).

Boyer, T. H. 1969, *Phys. Rev.*, **182**, 1374.

Chandrasekhar, S. 1942, *Principles of Stellar Dynamics* (Chicago: University of Chicago Press).

Chirikov, B. V. 1971, *Nucl. Phys. Inst. Siberian Section USSR Acad. Sci.*, Rept. No. 262 (CERN Transl. 71–40).

Contopoulos, G. 1960, *Zs. f. Ap.*, **49**, 273.

———. 1965, *A.J.*, **70**, 526.

———. 1966, *Ap. J. Suppl.*, **13**, 503.

————. 1967, *Bull. Astr.*, Ser. 3, **2**, 223.

Contopoulos, G., and Strömgren, B. 1965, *Tables of Plane Galactic Orbits* (New York: NASA Institute for Space Studies).

Einstein, A., and Hopf, L. 1910, *Ann. Phys.*, **33**, 1105.

Einstein, A., and Stern, O. 1913, *Ann. Phys.*, **40**, 551.

Fermi, E., Pasta, J., and Ulam, S. 1955, Los Alamos Sci. Lab., Rept. LA-1940.

Ford, J. 1975, in *Fundamental Problems in Statistical Mechanics*, Vol. **3**, ed. E. D. Cohen (Amsterdam: North-Holland).

Froeschlé, C. 1970, *Astr. and Ap.*, **4**, 115.

Galgani, L. 1975, *Scientia*, **110**, 483.

Galgani, L., and Scotti, A. 1972, *Rev. Nuovo Cimento*, **2**, 189.

————. 1975, private communication.

Gustavson, F. G. 1966, *A.J.*, **71**, 670.

Hadjidemetriou, J. 1975, *Celest. Mech.*, **12**, 155.

Hénon, M. 1974, *Celest. Mech.*, **10**, 375.

Hénon, M., and Heiles, C. 1964, *A.J.*, **69**, 73.

Ipser, J. R. 1974, *Ap. J.*, **193**, 463.

Lecar, M., and Cohen, L. 1972, in *Gravitational N-Body Problem*, ed. M. Lecar (Dordrecht: Reidel), p. 262.

Lynden-Bell, D. 1967, *M.N.R.A.S.*, **136**, 101.

Miller, R. H. 1973, *Ap. J.*, **180**, 759.

Moser, J. 1967, *Math. Ann.*, **169**, 135.

Poincaré, H. 1899, *Les Méthodes Nouvelles de la Mécanique Celeste*, Vol. **3** (Paris: Gauthier-Villars).

Rosenbluth, M. N., Sagdeev, R. Z., and Taylor, J. B. 1966, *Nucl. Fusion*, **6**, 297.

Sinai, Ya. 1963, *Soviet Math. Dokl.*, **4**, 1818.

Stoddard, S., and Ford, J. 1973, *Phys. Rev. A*, **8**, 1504.

Thirring, W. 1970, *Zs. f. Phys.*, **235**, 339.

Walker, G. H., and Ford, J. 1969, *Phys. Rev.*, **188**, 416.

Whittaker, E. T. 1937, *Analytical Dynamics of Particles and Rigid Bodies* (4th ed.; Cambridge: Cambridge University Press).

6 Mathematics of Radiative Transfer — T. W. Mullikin

Introduction

The mathematical equations of radiative transfer have been known for many years. The complexity (reality) of models studied has increased with computation power of modern computers (Hunt 1971; Lenoble 1974). Another strong influence on models of planetary atmospheres is the observational capabilities of artificial satellites (Kurijan 1973). The role of mathematical analysis has shifted somewhat from the analytical work of Hopf in the 1930s on the Milne equation to questions of obtaining asymptotic formulae and of improving efficiency and accuracy of numerical algorithms or Monte Carlo statistics.

In this contribution in honor of Professor Chandrasekhar we return to the problem of anisotropic scattering in a gray, homogeneous, plane-parallel atmosphere. Our reasons for studying this idealized problem are twofold. First is our desire to relate part of Chandrasekhar's analytical work (1950) in radiative transfer to the classical work of Hopf (1934) and to some recent mathematical work on operator factorizations developed extensively by Krein (1958) and Gohberg and Feldman (1974).[1] Our second reason is that, even with modern computers, numerical computations for thick atmospheres and strongly peaked phase functions are difficult. We formulate some equations which we hope may be useful for such problems.

Hopf (1934) presented several mathematical results for radiative transfer equations including existence and uniqueness results as well as estimates of solutions. For the Milne problem for isotropic scattering, he made use of the Fourier transform, the Wiener-Hopf factorization, and analytic function theory to give explicit solutions.

During the period 1940 to 1950 different methods were developed by Ambartsumian (1952) and Chandrasekhar (1950) for the study of equations of radiative transfer, including Chandrasekhar's extension to account for polarization of light by use of Stokes vectors. By means of invariance principles, equations were obtained for reflection and transmission functions S and T. These nonlinear equations replace the linear transfer equation and give informa-

---

T. W. Mullikin is in the Division of Mathematical Science, Purdue University, West Lafayette, Indiana.

[1] References are made to books rather than to original papers whenever possible.

113

tion about reflection and transmission by an atmosphere without requiring a determination of the radiation field within the atmosphere.

For homogeneous, plane-parallel atmospheres of optical thickness τ_1, invariance principles lead to initial-value problems for nonlinear integrodifferential equations for S and T as functions of τ_1 and angular variables (Chandrasekhar 1950, chap. 8). These equations have been used for numerical computations (Bellman *et al.* 1963). For the scalar transfer equation with a phase function having a finite spherical harmonic expansion and for the vector transfer equation for Rayleigh scattering, the equations for S and T were reduced by Chandrasekhar to systems of nonlinear integral equations for functions ψ_l^m and ϕ_l^m of one variable in which τ_1 enters as a parameter. Before the advent of modern computers this reduction made possible the solution of many interesting problems.

Typical of the systems of nonlinear integral equations obtained by such methods is the one for Chandrasekhar's X and Y functions,

$$X(\mu) = 1 + \frac{\mu \varpi_0}{2} \int_0^1 \frac{X(\mu)X(\nu) - Y(\mu)Y(\nu)}{\mu + \nu} \, d\nu \,,$$

$$Y(\mu) = \exp\left(-\tau_1/\mu\right) + \frac{\mu \varpi_0}{2} \int_0^1 \frac{Y(\mu)X(\nu) - Y(\mu)Y(\nu)}{\mu - \nu} \, d\nu \,.$$

For $\tau_1 = \infty$ these reduce to the single equation

$$H(\mu) = 1 + \frac{\mu \varpi_0}{2} \int_0^1 \frac{H(\mu)H(\nu)}{\mu + \nu} \, d\nu \,.$$

Questions of existence and uniqueness of solutions to these equations were studied by Busbridge (1960) and Mullikin (1962). It had been known that these equations fail to have a unique solution for conservative scattering ($\varpi_0 = 1$). We showed that this is true for $0 < \varpi_0 \leq 1$. The mathematical technique essentially amounted to a return to the approach of Hopf together with the use of singular integral equations whose theory has been used in radiative transfer by Sobolev (1956). Additional equations were obtained in the form of linear constraints sufficient to specify the desired solution.

In this work we return to the scalar transfer equation for a general continuous, positive, phase function. We use some results for the inhomogeneous Milne problem studied by Maslennikov (1968) for even more general phase functions. We show that if we replace Chandrasekhar's H, X, and Y functions by operator-valued functions, reflection and transmission operators can be expressed in terms of these basis operators. By Fourier transform methods we obtain the operator version of the Wiener-Hopf factorization and nonlinear equations for H-operators. Uniqueness questions are also studied and linear constraints obtained. These are complex equations, but they summarize in a compact form results that are difficult to arrive at when spherical harmonic expansions are used to obtain Chandrasekhar's $\psi_l{}^m$ and $\phi_l{}^m$ equations.

The same approach could be developed for the study of the vector transfer equation accounting for polarization. Basic to any of these problems seems to be the operator factorization in terms of H-operators. These also play a fundamental role in applications of the Case method of singular eigenfunction expansions (Case and Zweifel 1967; Siewert 1972).

An Equation of Radiative Transfer

We seek to determine a scalar field I defined, for a specific frequency, on the product of a subset of R^3 and the unit sphere, S^2. The number $I(r, \Omega)$ is the specific intensity which determines the steady-state flow of radiant energy through a medium in a region in three space in direction Ω (Chandrasekhar 1950).

We restrict to a plane-parallel scattering atmosphere in a steady state. Depth into this medium is expressed in terms of optical-depth variable τ measured normal to one face, and we shall treat both finite $(0 \leq \tau \leq \tau_1 < \infty)$ and half-space $(0 \leq \tau < \infty)$ atmospheres. The material density and mass absorption coefficient are incorporated in the definition of τ. We consider the face of the atmosphere at $\tau = 0$ to be exposed to a uniform radiation field constant in time and from a specific direction relative to the τ-axis, and the face at $\tau = \tau_1$ (or ∞) to be free of inflowing radiation.

We investigate the diffuse radiation field in which absorbed radiation which is not subsequently scattered into other directions is ignored. The specific intensity for the diffuse field depends on the single optical-depth variable τ and direction vector Ω. It satisfies the linear Boltzmann equation

$$\mu \frac{\partial I}{\partial \tau} + I = J \ . \tag{2.1}$$

The source function J is taken to be

$$J(\tau, \Omega, \Omega_0) = p(\Omega, \Omega_0)\mu_0{}^{-1} \exp\left[-\tau\mu_0{}^{-1}\right]$$

$$+ \int p(\Omega \cdot \Omega')I(\tau, \Omega', \Omega_0)d\omega' \ , \tag{2.2}$$

$d\omega'$ denoting normalized measure on the unit sphere. The scalar variable μ denotes the cosine of the angle made by the vector Ω and the positively directed τ-axis. The direction vector Ω_0 enters pareamtrically and is the direction of the incident field. Other incident fields distributed on the hemisphere will be obtained by integration.

The phase function p is taken as a nonnegative function of the scalar variable $\cos(\Theta) = \Omega \cdot \Omega'$, where Θ is the scattering angle. It is therefore a symmetric function of Ω and Ω', and

$$\int p(\Omega \cdot \Omega') d\omega' = \varpi_0 \leq 1. \tag{2.3}$$

We assume that p is a continuous function of $\cos(\Theta)$. This condition could be relaxed in the following analysis. An assumption necessary for the following analysis is that p is independent of τ, that is, the medium is homogeneous when measured in terms of optical depth. Sometimes for clarity we write $p(\Omega \cdot \Omega')$ as $p(\Omega, \Omega')$.

A standard procedure is to assume that p has an N-term expansion in Legendre polynomials. This can be used to replace equation (2.1) by N uncoupled equations (Chandrasekhar 1950). We do not follow this procedure for two reasons. First, such an expansion loses the property of positivity of the function p, which is helpful in certain spectral questions, and was used forcefully by Hopf (1934). Second, scattering functions for strong forward scattering require very large numbers of terms in an expansion, and we hope to derive results that may be useful in numerical computations which do not utilize this expansion. We will find that much of the structure in Chandrasekhar's work can be carried over to the analysis of equation (2.1) if scalar-valued functions are replaced by operator-valued functions.

We take an integral equation approach to equation (2.1) by using boundary conditions on the diffuse field that

$$I(0, \Omega, \Omega_0) = 0 \quad \text{for } 0 \leq \mu \leq 1,$$
$$I(\tau_1, \Omega, \Omega_0) = 0 \quad \text{for } -1 \leq \mu \leq 0. \tag{2.4}$$

In a standard way we obtain the equation for the source function:

$$J(\tau, \Omega, \Omega_0) = p(\Omega, \Omega_0)\mu_0^{-1} \exp\left(-\tau\mu_0^{-1}\right)$$

$$+ \int_{\mu' \geq 0} p(\Omega, \Omega') \int_0^\tau (\mu')^{-1} \exp\left(-(\tau - \tau')/\mu'\right) J(\tau', \Omega', \Omega_0) d\tau' d\omega'$$

$$+ \int_{\mu' \leq 0} p(\Omega, \Omega') \int_\tau^{\tau_1} |\mu'|^{-1} \exp\left(-(\tau - \tau')/\mu'\right) J(\tau', \Omega', \Omega_0) d\tau' d\omega' \tag{2.5}$$

for $\mu_0 > 0$.

The homogeneous version of this equation for $\tau = \infty$ and $\varpi_0 = 1$ was studied by Hopf (1934) and Maslennikov (1968), and was shown to have a solution unbounded for increasing τ. We are interested in solutions to equation (2.5) that are bounded and absolutely integrable functions of τ and Ω. Such problems are also studied by Maslennikov (1968).

We denote the integral operator L_{τ_1} in equation (2.5) by

$$L_{\tau_1}[f](\tau, \Omega) = \int \int_0^{\tau_1} k(\tau - \tau', \Omega, \Omega') f(\tau', \Omega') d\tau' d\omega', \tag{2.6}$$

where the kernel is defined by

$$k(\tau, \Omega, \Omega') = p(\Omega, \Omega') \exp(-\tau/\mu')/\mu' \qquad \text{for } 0 < \mu' \leq 1, 0 \leq \tau$$
$$= 0 \qquad \text{for } 0 < \mu' < 1, \tau \leq 0$$
$$= p(\Omega, \Omega') \exp(-\tau/\mu')/|\mu'| \qquad \text{for } -1 \leq \mu' < 0, \tau \leq 0$$
$$= 0 \qquad \text{for } -1 \leq \mu' < 0, \tau > 0. \quad (2.7)$$

The kernel is undefined for $\mu' = 0$ and unbounded for $\tau = 0$ as μ' tends to zero. Because L_{τ_1} is of difference-kernel type in the τ-variable, we can use Fourier transform methods of analysis (Hopf 1934; Krein 1962; Gohberg and Feldman 1974).

We consider first the more general problem than equation (2.5) of solving, with given functions g_i, the integral equations for f_i:

$$(E - L_{\tau_1})f_1 = g_1$$

and

$$(E - L^*_{\tau_1})f_2 = g_2,$$

where $L^*_{\tau_1}$ denotes the integral operator with transposed kernel and E denotes the identity operator.

We pose these problems in the Banach spaces $\mathcal{L}_q\{[0, \tau_1] \times S^2\}$ for all q, $1 \leq q \leq \infty$.

Theorem 1. *For $\tau_1 < \infty$ and $\varpi_0 \leq 1$, or $\tau_1 = \infty$ and $\varpi_0 < 1$, the operators* $(E - L_{\tau_1})$ *and* $(E - L^*_{\tau_1})$ *have bounded inverses in all $\mathcal{L}_q$ spaces.*
Proof. An elementary calculation shows that

$$\|L_{\tau_1}\| \leq \varpi_0[1 - \exp(-\tau_1/2)]$$

in both $\mathcal{L}_1$ and $\mathcal{L}_\infty$. The same is true of $L^*_{\tau_1}$. By the Riesz convexity theorem (Dunford and Schwartz 1967), this is valid in all spaces, and the inverse is expressed by a Neumann series.

We will need also to make use of the integral operator P on $C(S^2)$, the space of continuous functions on the unit sphere, defined by

$$P[f](\Omega) = \int p(\Omega \cdot \Omega')f(\Omega')d\omega'. \quad (2.8)$$

Theorem 2. *The operator P is compact, has only real eigenvalues, has simple eigenvalue ϖ_0 with eigenfunction $f(\Omega) \equiv 1$, and $\varpi_0 > |\lambda|$ for all other eigenvalues λ.*
Proof. As an operator on $\mathcal{L}_2(S^2)$, P is self-adjoint and compact, so has real eigenvalues only. Other properties follow because P leaves invariant the cone of nonnegative functions in $C(S^2)$ and theorems of Frobenius-Perron-Jentscz are applicable (Krein and Rutman 1951). These theorems require that some power of the operator P have positive kernel. This is obvious

because p is positive on some open subset of the sphere. If this is a proper subset of the sphere, then P^2 has a kernel which is positive on a larger subset of the sphere.

H-Operators and Factorizations

We consider in the next three sections a homogeneous, absorbing medium which fills a half-space, that is, $\tau_1 = \infty$ and $\varpi < 1$. Then equation (2.5) for J, with parameter Ω_0, has a unique solution in all spaces $\mathcal{L}_q$. In $\mathcal{L}_1$ this function, in its dependence on τ, Ω, and Ω_0, can be related to a boundary value of the resolvent kernel for the operator $(E - L_\infty)^{-1}L_\infty$. We do not prove the existence of a resolvent kernel, but it can be done as in simpler cases (Krein 1962).

We do, however, need a boundary value of the adjoint resolvent kernel as well, so we define J_r and J_l as unique solutions in $\mathcal{L}_1\{[0, \infty) \times S^2\}$ to

$$J_r(\tau, \Omega, \Omega_0) = k(\tau, \Omega, \Omega_0)$$

$$+ \int\!\!\int_0^\infty k(\tau - \tau', \Omega, \Omega')J_r(\tau', \Omega', \Omega_0)d\tau'd\omega' \qquad (3.1)$$

and

$$J_l(\tau, \Omega, \Omega_0) = k(-\tau, \Omega, \Omega_0)$$

$$+ \int\!\!\int_0^\infty k(\tau' - \tau, \Omega', \Omega_0)J_l(\tau', \Omega, \Omega')d\tau'd\omega' . \qquad (3.2)$$

In the first of these equations Ω_0 is a parameter, and in the second Ω is a parameter.

Given these functions, we define operators $\mathcal{K}_r(z)$ and $\mathcal{K}_l(z)$ on the space $C(S^2)$ for complex z, $\mathrm{Im}(z) \geq 0$, by

$$\mathcal{K}_r(z)[f](\Omega) = f(\Omega) + \int\!\!\int_0^\infty f(\Omega_0) \exp(iz\tau)J_r(\tau, \Omega, \Omega_0)d\tau d\omega_0 , \qquad (3.3)$$

and

$$\mathcal{K}_l(z)[f](\Omega) = f(\Omega) + \int\!\!\int_0^\infty f(\Omega_0) \exp(iz\tau)J_l(\tau, \Omega, \Omega_0)d\tau d\omega_0 . \qquad (3.4)$$

These analogs of Chandrasekhar's H-function play a fundamental role in our analysis.

With k^- denoting the kernel $k(-\tau, \Omega, \Omega_0)$, $0 \leq \tau < \infty$, we have $J_l = (E - L^*_\infty)^{-1}k^-$. If we interpret the integral in the definition of $\mathcal{K}_l$ as a linear functional of J_l with parameter Ω, we obtain the representation

$$\mathcal{K}_l(z)[f](\Omega) = f(\Omega)$$

$$+ \int_{\mu_0 \leq 0}\int_0^\infty p(\Omega, \Omega_0)j_r(\tau, \Omega_0, z) \exp(-\tau/|\mu_0|)d\tau \, \frac{d\omega_0}{|\mu_0|} . \qquad (3.5)$$

Here the function j_r is the unique solution in $\mathcal{L}_\infty$ for $\mathrm{Im}(z) \geq 0$ to the equation

$$j_r(\tau, \Omega_0, z) = f(\Omega_0) \exp{(i\tau z)}$$
$$+ \int\int\int_0^\infty k(\tau - \tau', \Omega_0, \Omega')j_r(\tau', \Omega', z)d\tau'd\omega' . \qquad (3.6)$$

The representation of $\mathcal{K}_l$ in equation (3.5) has several applications, the first being

Theorem 3. *As operators on $C(S^2)$, $\mathcal{K}_r(z)$ and $\mathcal{K}_l(z)$ are the sums of the identity and compact operators. They have bounded norms in $\mathrm{Im}(z) \geq 0$, and for $z \in [i, i\infty)$ the compact operators are integral operators with nonnegative kernels.*

Proof. It follows readily from the Neumann series solution to equations (3.1) and (3.2) and the definitions of $\mathcal{K}_r$ and $\mathcal{K}_l$ that $\mathcal{K}_r(z) - E$ and $\mathcal{K}_l(z) - E$ for z in $[i, i\infty)$ are integral operators with nonnegative kernels.

That $\mathcal{K}_l(z) - E$ is a compact integral operator on $C(S^2)$ for $\mathrm{Im}(z) > 0$ follows from equations (3.5) and (3.6) which display $\mathcal{K}_l(z) - E$ as the product of a bounded operator, represented by the transform of j_r, and a compact integral operator with kernel $p(\Omega, \Omega_0)$. That $\mathcal{K}_r(z) - E$ is a compact integral operator is most easily seen from the representation obtained in the next section in equation (4.9).

With our analysis of equations (3.1) and (3.2) based on Fourier transform methods, we need properties of the transform kernel

$$\int_{-\infty}^\infty \exp{(i\tau z)}k(\tau, \Omega, \Omega')d\tau = p(\Omega, \Omega')/(1 - iz\mu') . \qquad (3.7)$$

For complex z outside the intervals $[i, i\infty)$ and $[-i, -i\infty)$, we use this continuous kernel to define an operator $K(z)$ on $C(S^2)$ by

$$K(z)[f](\Omega) = \int \frac{P(\Omega, \Omega')f(\Omega')}{1 - iz\mu'} d\omega' \qquad (3.8)$$

It follows that if $M = \max p(\Omega, \Omega')$,

$$\|K(z)\| \leq \tfrac{1}{2}M \int_{-1}^1 \frac{d\mu'}{|1 - iz\mu'|} . \qquad (3.9)$$

In the following, when we speak of the adjoint $K^*(z)$, it will be appropriate to consider it as the integral operator with transposed kernel again defined on $C(S^2)$, rather than on the full dual space to $C(S^2)$.

Our first main result below follows from more general results of Gohberg (1964). For completeness, however, we give a simpler proof for this special case similar to that for N-group neutron transport in Mullikin (1973).

Theorem 4. *For $\varpi_0 < 1$, the operators $\mathcal{K}_l(z)$ and $\mathcal{K}_r(z)$ are operator-valued functions of z, analytic for $\mathrm{Im}(z) > 0$ and continuous for $\mathrm{Im}(z) \geq 0$. They* *are invertible for* $\text{Im}(z) \geq 0$ *and for* $\text{Im}(z) = 0$ *give the operator factorization*

$$[E - K(z)]\mathcal{H}_r(z)\mathcal{H}_l(-z) = E . \tag{3.10}$$

Proof. Analyticity of the operators $\mathcal{H}_r$ and $\mathcal{H}_l$ in $\text{Im}(z) > 0$ follows from definitions (3.3) and (3.4). This and other results for $\mathcal{H}_l$ also follow from properties of j_r in equation (3.6).

For f in $C(S^2)$ and $\text{Im}(z) \geq 0$, equation (3.6) for j_r has a unique solution in $\mathcal{L}_\infty\{[0, \infty) \times S^2\}$. It follows then from the right-hand side of equation (3.6) that j_r is a continuous function of τ and Ω. With $\text{Im}(z_0) > 0$ and j_{rn} defined by

$$(E - L_\infty)j_{rn} = f(\Omega)\frac{(i\tau)^n \exp (iz_0\tau)}{n!}$$

it follows from equation (3.6) and an easy estimate that

$$j_r(\tau, \Omega, z) = \sum_{n=0}^{\infty} j_{rn}(\tau, \Omega)(z - z_0)^n ,$$

with convergence in $\mathcal{L}_\infty$ for $|z - z_0| < \text{Im}(z_0)$. For each τ, j_{rn} is the image of the function f in $C(S^2)$ under the action of a bounded linear operator which depends continuously on τ. The above series shows that $j_r(\tau, z)$ is defined by action of a bounded operator on $C(S^2)$ which depends analytically on z for $\text{Im}(z) > 0$ and continuously on τ for $0 \leq \tau < \infty$.

From equations (3.2) and (3.6) we compute that

$$\int\int_0^\infty J_l(\tau, \Omega_0, \Omega)j_r(\tau, \Omega, z)d\tau d\omega = \int\int_0^\infty f(\Omega) \exp (i\tau z)J_l(\tau, \Omega_0, \Omega)d\tau d\omega$$

$$+ \int\int_0^\infty j_r(\tau', \Omega', z)\left[\int\int_0^\infty k(\tau - \tau', \Omega, \Omega')J_l(\tau, \Omega_0, \Omega)d\tau d\omega\right]d\tau' d\omega'$$

$$= \int\int_0^\infty f(\Omega) \exp (i\tau z)J_l(\tau, \Omega_0, \Omega)d\tau d\omega$$

$$+ \int\int_0^\infty j_r(\tau', \Omega', z)[J_l(\tau', \Omega_0, \Omega') - k(-\tau, \Omega_0, \Omega')]d\tau' d\omega' .$$

If we cancel the common term from the first and last expressions and use (3.6) for $\tau = 0^+$, we see that

$$0 = \int f(\Omega)d\omega \int_0^\infty J_l(\tau, \Omega_0, \Omega) \exp (i\tau z)d\tau + f(\Omega_0) - j_r(0^+, \Omega_0, z) ,$$

and conclude from equation (3.4) that

$$j_r(0^+, \Omega_0, z) = \mathcal{H}_l(z)[f](\Omega_0) . \tag{3.11}$$

For $\text{Im}(z) > 0$, we determine from equation (3.6) that

$$\frac{1}{h}[j_r(\tau + h, \Omega, z) - j_r(\tau, \Omega, z)] = \frac{\exp (ihz) - 1}{h}j_r(\tau, \Omega, z)$$

$$+ (E - L_\infty)^{-1}\left[\int \frac{1}{h} \int_{-h}^0 k(\cdot - s, \cdot, \Omega')j_r(s + h, \Omega', z)ds d\omega'\right](\tau, \Omega) .$$

Continuity in $\mathcal{L}_1$ of $(E - L_\infty)^{-1}$ and of the averaging operator in this last expression permit passage to the limit as h goes to zero, and proves, by equation (3.1), that

$$\frac{\partial}{\partial \tau} j_r(\tau, \Omega, z) - iz j_r(\tau, \Omega, z) = \int J_r(\tau, \Omega, \Omega') j_r(0^+, \Omega', z) d\omega' . \qquad (3.12)$$

By equations (3.11), this last equation is equivalent to

$$j_r(\tau, \Omega, z) \exp(-i\tau z) = \mathfrak{K}_l(z)[f](\Omega)$$

$$+ \int_0^\tau \int \exp(-iz\tau') J_r(\tau', \Omega, \Omega') \mathfrak{K}_l(z)[f](\Omega') d\omega' d\tau' . \qquad (3.13)$$

By continuity this extends to $\operatorname{Im}(z) \geq 0$.

Since $\mathfrak{K}_l(z)$ is the sum of the identity and a compact operator, it can fail to have a bounded inverse only by mapping some nontrivial function f in $C(S^2)$ into the zero function θ. Suppose that for some z_0, $\operatorname{Im}(z_0) \geq 0$, we have $\mathfrak{K}_l(z_0)f = \theta$. It follows from equation (3.13) that $j_r(\tau, \Omega, z_0) = 0$ for all $\tau, 0 \leq \tau < \infty$. But then from equation (3.6) we conclude that $f = \theta$. So $\mathfrak{K}_l(z)$ has a bounded inverse for $\operatorname{Im}(z) \geq 0$.

If we use equation (3.12) to express the innermost integral in the representation (3.5) for $\mathfrak{K}_l(z)$, we find for $\operatorname{Im}(z) \geq 0$

$$\mathfrak{K}_l(z)[f](\Omega) = f(\Omega)$$

$$+ \int_{\mu_0 \leq 0} p(\Omega \cdot \Omega_0) \mathfrak{K}_r(i/|\mu_0|)[\mathfrak{K}_l(z)[f]](\Omega_0) \frac{d\omega_0}{1 - |\mu_0| zi} . \qquad (3.14)$$

This demonstrates directly that $\mathfrak{K}_l(z)$ has a left inverse; but since $\mathfrak{K}_l(z)$ is the sum of the identity and a compact operator, it has an inverse. A similar representation for $\mathfrak{K}_r$ follows from the fact that J_r is simply related to j_r and is given in the following section in equation (4.6).

From equations (3.13) and (3.3) we conclude that for $\operatorname{Im}(z) = 0$

$$\lim_{\tau \to \infty} j_r(\tau, \Omega, z) \exp(-i\tau z) = \mathfrak{K}_r(-z)[\mathfrak{K}_l(z)[f]](\Omega) . \qquad (3.15)$$

Now combining equations (3.6) and (3.13), we get, with a change of integration variable, the representation

$$j_r(\tau, \Omega, z) \exp(-i\tau z) = f(\Omega)$$

$$+ \int d\omega' \int_{-\infty}^\tau \exp(-i\sigma z) k(\sigma, \Omega, \Omega') \Big\{ \{\mathfrak{K}_l(z)[f](\Omega')$$

$$+ \int_0^{\tau - \sigma} \exp(-i\sigma' z) \int J_r(\sigma', \Omega', \Omega'') \mathfrak{K}_l(z)[f](\Omega'') d\omega'' d\sigma' \Big\} d\sigma . \qquad (3.16)$$

The integrand in the last integral is bounded by the integrable function

$$k(\sigma, \Omega, \Omega') \Big[1 + \int\int_0^\infty J_r(\sigma', \Omega', \Omega'') d\omega'' d\sigma' \Big] \|\mathfrak{K}_l\| \, \|f\| .$$

The Lebesgue-dominated convergence theorem gives, with equation (3.3),

$$\mathcal{H}_r(-z)\mathcal{H}_l(z)[f] = f + K(-z)\mathcal{H}_r(-z)\mathcal{H}_l(z)[f]$$

for all f in $C(S^2)$. This completes the proof.

For z in any compact subset of $\mathrm{Im}(z) > 0$ which excludes the interval $[i, i\infty)$, the operator $\mathcal{H}_l^{-1}(z)$ is analytic and the operator $[E - K(z)]^{-1}$ is meromorphic. By equation (3.10) it then follows that the operator $H_l(z)$ can be extended into $\mathrm{Im}(z) < 0$ as an operator-valued function meromorphic off the branch-cut $[-i, -i\infty)$. A similar argument applies to $\mathcal{H}_r(z)$.

We have information about singularities in

Theorem 5. *The operator $[E - K(z)]^{-1}$ is meromorphic off the branch-cuts $[i, i\infty)$ and $[-i, -i\infty)$. Poles, $\pm z_j$, are restricted to the imaginary axis with the only possible accumulation point being at $\pm i$.*

Proof. It is a theorem of functional analysis (Gohberg 1951) that $[E - K(z)]^{-1}$ is meromorphic in the domain of analyticity of the compact operator-valued function $K(z)$. Poles can accumulate only on the boundary of analyticity. We have assumed that $\|K(0)\| = \varpi_0 < 1$, so 0 is not a pole of $[E - K(z)]^{-1}$. Suppose $z_j(\neq 0)$ is such a pole. Then we have a solution in $C(S^2)$ to

$$f(\Omega) = \int \frac{p(\Omega \cdot \Omega')}{1 - i\mu' z_j} f(\Omega')d\omega' . \tag{3.17}$$

Define F by

$$F(\Omega) = f(\Omega)/(1 - i\mu z_j) ,$$

and obtain

$$(1 - i\mu z_j)F(\Omega) = P[F](\Omega) .$$

The operator P is self-adjoint in $\mathcal{L}_2(S^2)$ with spectral radius ϖ_0 and hence

$$\int \bar{F}(\Omega)P[F](\Omega)d\omega \leq \varpi_0 \int |F(\Omega)|^2 d\omega .$$

It then follows that

$$(1 - \varpi_0)\|F\|_2^2 \leq iz_j \int \mu |F(\Omega)|^2 d\omega \leq (\varpi_0 + 1)\|F\|_2^2 ,$$

and that iz_j is a real number. Clearly $f(-\Omega)$ satisfies equation (3.17) if z_j is replaced by $-z_j$, so poles occur in pairs, $\pm z_j$.

Of particular interest are the poles $\pm z_0$ of $[E - K(z)]^{-1}$ nearest to $z = 0$. Properties of functions annihilated by $[E - K(z)]$ and its adjoint have been studied by several researchers (cf. Maslennikov 1968; Kuscer and Vidav 1969; Inönü 1970; Case 1974).

Theorem 6. *There is a pole z_0 of $[E - K(z)]^{-1}$ satisfying $|z_0| = \min|z_j|$ and $0 < |z_0| < 1$. A positive solution exists to*

$$u(\Omega) = \int \frac{p(\Omega, \Omega')}{1 + \mu' |z_0|} u(\Omega')d\omega' \tag{3.18}$$

*u is independent of the azimuth angle and unique up to scalar multiplication.
If also $p(\cos \Theta) = p(-\cos \Theta)$, then $u(\Omega) = u(-\Omega)$.*

We have independence of the azimuth angle for u_j a solution to

$$[E - K(z_j)]u_j = 0$$

provided solutions are unique up to multiplication by a scalar. This follows
from invariance of the equation under the group of rotations about the polar
axis. More details are given by Maslennikov.

The operator $[E - K(z)]^{-1}$ has in a neighborhood of each pole z_j and $-z_j$ a
Laurent expansion. We refer to Maslennikov (1968, Theorem 16) for a rather
long proof of

Theorem 7. *The poles of $[E - K(z)]^{-1}$ are simple and the residues*

$$R_j^{\pm} = \lim_{z \to z_j} (z \pm z_j)[E - K(\pm z)]^{-1} \tag{3.19}$$

have finite dimensional ranges.

The results of this section are particularly simple in case the function $p(\Omega \cdot \Omega')$
is a constant. It follows readily in this case that the operators $\mathcal{K}_r$ and $\mathcal{K}_l$ leave
invariant the subspace of $C(S^2)$ of constant functions. When restricted to this
subspace, they become multiplication operators which are usually viewed
merely as scalar-valued functions, rather than operators, and it also follows
that $\mathcal{K}_r = \mathcal{K}_l$. The result in equation (3.10) becomes

$$(1 - K(z))\mathcal{K}(z)\mathcal{K}(-z) = 1 \,,$$

with K a complex-valued function with $K(z) = K(-z)$. This is the familiar
Wiener-Hopf factorization (Hopf 1934) found in Chandrasekhar (1950) by
means of invariance principles and resulting analysis of nonlinear equations,
whose extensions we derive in the next section.

The Scattering Operator and *H*-Operator Equations

The specific intensity of radiation exiting at $\tau = 0$ in the direction $\Omega(\mu \leq 0)$
due to the incident field in direction $\Omega_0(\mu_0 \geq 0)$ is given by equations (2.1) and
(2.2) as

$$\mu_0|\mu|I(0, \Omega, \Omega_0) = \mu_0 \int_0^{\infty} \exp(-\tau/|\mu|)J_r(\tau, \Omega, \Omega_0)d\tau \,. \tag{4.1}$$

We define a scattering (or reflection) kernel by

$$s(\Omega, \Omega_0) = \mu_0|\mu|I(0, \Omega, \Omega_0) \,, \quad \mu \leq 0 \quad \text{and} \quad \mu_0 \geq 0$$

$$= 0 \,, \qquad\qquad \text{otherwise} \,. \tag{4.2}$$

A scattering operator is defined on $C(S^2)$ by

$$S[f](\Omega) = \int s(\Omega, \Omega')f(\Omega')d\omega' \tag{4.3}$$

Because J_r equals j_r if $z = i/\mu_0$ and $f(\Omega) = p(\Omega \cdot \Omega_0)/\mu_0$ ($\mu_0 \geq 0$), it follows from equations (3.12) and (4.2) that

$$s(\Omega, \Omega_0) = \frac{|\mu|\mu_0}{|\mu| + \mu_0} \; \mathfrak{K}_r(i/|\mu|)[\mathfrak{K}_l(i/\mu_0)[p(\cdot, \Omega_0)]](\Omega) . \tag{4.4}$$

This important kernel is determined if we can determine the $\mathfrak{K}$-operators on $[i, i\infty)$. The definition of these operators by equations (3.3) and (3.4) depends on having solutions to equations (3.1) and (3.2). We seek equations for these operators which do not require the solution of equations (3.1) and (3.2). One such set of equations was derived by Ambartsumian (1952) and Chandrasekhar (1950) by use of principles of invariance.

For simplicity of notation we define H-operators by

$$H_r(\nu) = \mathfrak{K}_r(i/\nu) \quad \text{and}$$

$$H_l(\nu) = \mathfrak{K}_l(i/\nu) , \quad \text{for} \quad 0 \leq \nu \leq 1 . \tag{4.5}$$

Theorem 8. *Necessary conditions on the H-operators are*

$$H_r(\nu)[f](\Omega) = f(\Omega) + \nu \int_{\mu' \geq 0} f(\Omega') H_r(\nu)[[H_l(\mu')[p(\cdot, \Omega')]](\Omega) \frac{d\omega'}{\nu + \mu'} \tag{4.6}$$

and

$$H_l(\nu)[f](\Omega) = f(\Omega) + \nu \int_{\mu' \leq 0} p(\Omega \cdot \Omega') H_r(|\mu'|)[H_l(\nu)[f]](\Omega') \frac{d\omega'}{\nu + |\mu'|} \tag{4.7}$$

for all F in $C(S^2)$.

Proof. The function J_r of equation (3.1) is a special case of the function j_r of equation (3.6) since $J_r(\tau, \Omega, \Omega) = j_r(\tau, \Omega, i/\mu_0)$ for $f(\Omega) = p(\Omega \cdot \Omega)/\mu_0$ and $\mu_0 > 0$. From equation (3.12) we compute that

$$\int_0^\infty \exp(iw\tau) j_r(\tau, \Omega, z) = \frac{-1}{i(w + z)} \; \mathfrak{K}_r(w)[\mathfrak{K}_l(z)[f]](\Omega) . \tag{4.8}$$

Returning to equation (3.3) and using the last result, we find

$$\mathfrak{K}_r(w)[f](\Omega) = f(\Omega) + \int_{\mu_0 \geq 0} f(\Omega_0) \mathfrak{K}_r(w)[H_l(\mu_0)[p(\cdot, \Omega_0)]](\Omega) \frac{d\omega_0}{1 - i\mu_0 w} . \tag{4.9}$$

If we set $w = i/\nu$, we obtain equation (4.6). If in equation (3.14) we set $z = i/\nu$, we obtain equation (4.7) and complete the proof.

The nonlinear equations of the last theorem seem to offer a method for determining the $\mathfrak{K}$-operators without solving the J-equations. Unfortunately, these nonlinear equations are necessary but not sufficient conditions for determining the $\mathfrak{K}$-operators of physical interest. This is well known in the simple case of a constant phase function for which there is the nonlinear H-equation.

$$H(\mu) = 1 + \varpi_0 \mu \int_0^1 \frac{H(\mu)H(\nu)}{\mu + \nu} \, d\nu .$$

For the case of phase functions with N-term Legendre polynomial expansion, Chandrasekhar (1950) obtained systems of coupled nonlinear equations for ψ_l^m-functions. Existence and uniqueness of solutions to these equations were studied by Mullikin (1964c).

We have not related the above nonlinear H-operator equations to Chandrasekhar's ψ_l^m-functions. This should be possible by restricting the H-operators to the subspace of $C(S^2)$ spanned by the set of spherical harmonics used in the representation of a phase function with N-term Legendre polynomial expansion. Instead of exploring this question, we proceed to obtain additional necessary conditions on the general operator equations and to study their independence of the above nonlinear equations.

Linear singular equations have been used successfully in the study of simple phase functions (cf. Sobolev 1956; Mullikin 1962, 1964a, b). This idea is generalized in

Theorem 9. *Suppose that the phase function p is Hölder continuous. For z outside $[i, i\infty)$ and $\mathrm{Im}(z) > 0$, the $\mathcal{K}$-operators are expressed in terms of the H-operators by the linear relations*

$$(E - K(z))[\mathcal{K}_r(z)[f]](\Omega) = f(\Omega) - \int_{\mu' \leq 0} \frac{p(\Omega, \Omega')H_r(|\mu'|)[f](\Omega')}{1 - i\mu'z}\, d\omega' \quad (4.10)$$

and

$$\mathcal{K}_l(z)[(E - K(-z))[f]](\Omega) = f(\Omega)$$
$$- \int_{\mu' \geq 0} \frac{H_l(\mu')[p(\cdot, \Omega')](\Omega)f(\Omega')}{1 + i\mu'z}\, d\omega' . \quad (4.11)$$

Proof. By definition (3.3) we have for $\mathrm{Im}(z) > 0$

$$\mathcal{K}_r(z)[f](\Omega) = f(\Omega) + \int_0^\infty \exp{(iz\tau)} \int J_r(\tau, \Omega, \Omega_0)f(\Omega_0)d\Omega_0 d\tau .$$

We interpret the τ-integral for fixed f and Ω as the $\mathcal{L}_2$-inner product of

$$h(\tau) = \exp{(iz\tau)} , \quad 0 < \tau , \ \mathrm{Im}(z) > 0$$
$$= 0 , \quad \tau < 0 ,$$

and the function obtained from equation (3.1) by extension to $(-\infty, \infty)$

$$g(\tau, \Omega) = \int J_r(\tau, \Omega, \Omega_0)f(\Omega_0)d\Omega_0 , \qquad \text{for } 0 < \tau$$

$$= \int k(\tau, \Omega, \Omega_0)f(\Omega_0)d\Omega_0 + \int\int_0^\infty k(\tau - \tau', \Omega, \Omega')$$

$$\times \int J_r(\tau', \Omega', \Omega_0)f(\Omega_0)d\omega_0 d\tau' d\omega' , \qquad \text{for } \tau < 0.$$

It follows from equation (3.1) that, as a function of τ, g is square integrable, as is h, so the Parseval formula gives

$$\mathfrak{IC}_r(z)[f](\Omega) = f(\Omega) + \frac{1}{2\pi i} \int_{-\infty}^{\infty} \frac{1}{t-z} G(t)dt \,, \qquad (4.12)$$

where G denotes the Fourier transform of g. From the definition of g in terms of a convolution integral in τ and definitions (3.8) and (3.3) of K and $\mathfrak{IC}_r$, it follows that

$$G(t, \Omega) = K(t)[\mathfrak{IC}_r(t)[f]](\Omega) \,. \qquad (4.13)$$

Combining the last two equations, we obtain the representation for $\text{Im}(z) > 0$:

$$\mathfrak{IC}_r(z)[f](\Omega) = f(\Omega) + \frac{1}{2\pi i} \int_{-\infty}^{\infty} \frac{1}{t-z} K(t)[\mathfrak{IC}_r(t)[f]](\Omega)dt \,. \qquad (4.14)$$

Since $\mathfrak{IC}_r(t)$ is analytic in $\text{Im}(t) > 0$, $K(t)$ is analytic in $\text{Im}(t) > 0$ except on the interval $[i, i\infty)$, $\|\mathfrak{IC}_r(t)\|$ is bounded for $\text{Im}(t) \geq 0$, and $\|K(t)\|$ satisfies equation (3.9), we obtain the representation (4.10) by moving the contour of integration in equation (4.14) to the branch-cut $[i, i\infty)$ of $K(t)[\mathfrak{IC}_r(t)[f](\Omega)]$ as a function of t for fixed Ω. Hölder continuity of the phase function p permits use of the Plemelj formulae (Muskhelishvili 1963). A similar argument gives equation (4.11) and completes the proof.

If in equations (4.10) and (4.11) limits are taken as z approaches a point in $[i, i\infty)$, jump conditions lead to linear singular integral equations. These have been used successfully in numerical computation for phase functions with finite expansions in Legendre polynomials (cf. Sobolev 1956; Carlstedt and Mullikin 1966; Sweigert 1970). We will not pursue this here.

Nonlinear equations as in Theorem 8 have been derived by use of heuristic principles of invariance (Ambartsumian 1952; Chandrasekhar 1950). The following necessary conditions do not seem to be derivable from such heuristic arguments. One reason, perhaps, is that the conditions result from analytic continuation in the variable $\mu = \cos(\theta)$ outside the interval of physical significance.

Since the operators $\mathfrak{IC}_r$ and $\mathfrak{IC}_l$ are analytic in $\text{Im}(z) > 0$, we obtain from the last theorem the

Corollary 1. Necessary linear conditions on the H-operators are that for each pole z_j of $[E - K(z)]^{-1}$

$$\theta = R_j^+\left[f - \int_{\mu' \leq 0} \frac{P(\cdot, \Omega')H_r(|\mu'|)[f](\Omega')}{1 + \mu'|z_j|} d\omega' \right] \qquad (4.15)$$

and

$$\theta = R_j^- \mathfrak{IC}_l^{-1}(z_j)\left[f - \int_{\mu' \geq 0}^{0} \frac{H_l(\mu')[p(\cdot, \Omega')]f(\Omega')}{1 - \mu'|z_j|} d_{\omega'} \right] \qquad (4.16)$$

for all functions f in $C(S^2)$.

We have not done an investigation of all solutions to the nonlinear equations

(4.6) and (4.7). Using a construction due to Victory (1974) in a model of neutron transport theory, we have some partial results concerning independence of the necessary conditions in Theorem 9 and in Corollary 1.

For each pole z_j of $[E - K(z)]^{-1}$ we construct the operator S_j by use of solutions to

$$[E - K(z_j)]u_j = 0 \quad \text{and} \quad [E - K^*(-z_j)]v_j = 0 . \qquad (4.17)$$

We define the number c_j by

$$c_j \int v_j(\Omega) \mathcal{K}_l^{-1}(z_j)[\mathcal{K}_r^{-1}(z_j)[u_j]](\Omega) d\omega = 1 . \qquad (4.18)$$

If c_j is finite for some choice of u_j and v_j, we define the operator S_j by

$$S_j[f](\Omega) = c_j \mathcal{K}_r^{-1}(z_j)[u_j](\Omega) \int v_j(\Omega') \mathcal{K}_l^{-1}(z_j)[f](\Omega') d\omega' , \qquad (4.19)$$

and note that

$$S_j^2 = S_j . \qquad (4.20)$$

The case $j = 0$ deserves special attention.

Lemma 1. *If $p(\Omega \cdot \Omega_0) = p(-\Omega \cdot \Omega_0)$, the operator S_0 is well defined and maps positive functions onto a positive function.*
Proof. From equation (4.6) we find for $f = u$ that

$$\mathcal{K}_r^{-1}(z_0)[u](\Omega) = u(\Omega) - \int_{\mu' \geq 0} u(\Omega') H_l(\mu')[p(\cdot, \Omega')](\Omega) \frac{d\omega'}{1 + \mu' |z_0|} .$$

From equation (4.11) we find for $f(\Omega) = u(-\Omega)$, which by hypothesis and Theorem 6 is the same as $f(\Omega) = u(\Omega)$,

$$\theta = u(\Omega) - \int_{\mu' \geq 0} u(\Omega') H_l(\mu')[p(\cdot, \Omega')](\Omega) \frac{d\omega'}{1 - \mu' |z_0|} .$$

Subtracting these two equations shows that $\mathcal{K}_r^{-1}(z_0)[u]$ is a positive function. A similar argument shows that $\mathcal{K}_l^{*-1}(z_0)[v]$ is a positive function. Thus c_0 in equation (4.18) is finite.

This lemma may be true without the hypothesis about the phase function p. We have been unable to prove it, however.

The mathematical motivation for the construction in the following theorem is that operators defined by

$$\mathcal{K}_{rj}(z) = \mathcal{K}_r(z) \Big[E + \frac{2z_j}{z - z_j} S_j \Big] \qquad (4.21)$$

and

$$\mathcal{K}_{lj}(z) = \Big[E + \frac{2z_j}{z - z_j} S_j \Big] \mathcal{K}_l(z) \qquad (4.22)$$

provide a factorization, for $\text{Im}(z) = 0$,

$$[E - K(z)] \mathcal{K}_{rj}(z) \mathcal{K}_{lj}(-z) = E .$$

This follows from equation (3.10), and the fact that $S_j^2 = S_j$ gives the result

$$\left[E + \frac{2z_j}{z - z_j} S_j\right]\left[E - \frac{2z_j}{z + z_j} S_j\right] = E \,. \tag{4.23}$$

Theorem 10. *In addition to the solutions of physical significance to the non-linear equations (4.6) and (4.7), there are the solutions*

$$H_{rj}(\nu) = H_r(\nu)\left[E + \frac{2|z_j|\nu}{1 - \nu|z_j|} S_j\right] \tag{4.24}$$

and

$$H_{lj}(\nu) = \left[E + \frac{2|z_j|\nu}{1 - \nu|z_j|} S_j\right]H_l(\nu) \tag{4.25}$$

for each z_j for which S_j is defined by equation (4.19). This is true for any phase function satisfying $p(\Omega \cdot \Omega') = p(-\Omega \cdot \Omega')$ for the pole z_0, and the operators H_{r0} and H_{l0} do not satisfy the linear constraints (4.15) and (4.16).
Proof. We treat S_j as an unknown operator which satisfies $S_j^2 = S_j$ and substitute H_{rj} and H_{lj} into equations (4.6) and (4.7). Use of the fact that H_r and H_l satisfy these equations together with elementary algebraic manipulations leads to the conditions

$$\theta = S_j\left[f - \int_{\mu' \geq 0} \frac{H_l(\mu')[p(\cdot, \Omega')]f(\Omega')}{1 - \mu'|z_j|} d\omega'\right] \tag{4.26}$$

and

$$\theta = S_j[f](\Omega) - \int_{\mu' \leq 0} \frac{p(\Omega, \Omega')H_r(|\mu'|)[S_j f](\Omega)}{1 + \mu'|z_j|} d\omega' \tag{4.27}$$

From equations (4.10) and (4.11) it follows that S_j can be defined by

$$S_j[f] = c_j \mathcal{K}_r^{-1}(z_j)[u_j]$$

and

$$S_j[f](\Omega) = c_j \mathcal{K}_r^{-1}(z_j)[u_j](\Omega) \int \mathcal{K}_l^{*-1}(z_0)[v](\Omega')f(\Omega') d\omega' \,,$$

provided c_j can be so chosen that $S_j^2 = S_j$. This agrees with definition (4.19).

When the phase function satisfies $p(\Omega \cdot \Omega') = p(-\Omega \cdot \Omega')$, the operator S_0 is well defined. Also, by Theorem 7, the pole z_0 is of order one. Substitution of H_{r0} and H_{l0} into the linear equations (4.15) and (4.16) reduces these to

$$\theta = R_0^+\left[\int_{\mu' \leq 0} \frac{p(\cdot, \Omega')H_r(|\mu'|)[S_0[f]](\Omega')}{1 - (\mu'|z_0|)^2} d\omega'\right],$$

and

$$\theta = R_0^- \mathcal{K}_l^{-1}(z_0)S_0\left[f - \int_{\mu \geq 0} \frac{H_l(\mu')[p(\cdot, \Omega')]f(\Omega')}{1 - \mu'|z_0|} d\omega'\right].$$

The first of these is not satisfied for the choice $f \equiv 1$, for this gives an integrand which is positive, and R_0^+ does not annihilate positive functions. This completes the proof.

An unresolved question is the sufficiency of the nonlinear equations (4.6) and (4.7) together with the linear constraints (4.15) and (4.16) for general phase functions. Independence of these necessary conditions is also of importance because it is difficult to incorporate the linear constraints in any numerical scheme that might be devised for using the nonlinear equations to approximate the kernels in the integral operators which appear in the H-operators.

The integral operator terms in the H-operator act on functions supported on hemispheres. It might be that nonuniqueness of solutions to the nonlinear equations is "worst" when the phase function is symmetric with respect to forward and backward scattering. This conjecture is suggested in a negative way by the fact that in the last theorem we have been unable to prove non-uniqueness except for such symmetric phase functions.

In the simple case of $p(\Omega \cdot \Omega') \equiv \varpi_0$, the nonlinear equations and linear constraints are independent and both necessary and sufficient for determination of the H-function of physical importance (cf. Busbridge 1960; Mullikin 1962). A clever combination of these equations has been found by Pahor (1968) which gives an equation rapidly convergent under iteration even when ϖ_0 is near 1. Iterative methods do not work well for the nonlinear H-equation when ϖ_0 is near 1, because there are two solutions near each other which have to be separated by use of the linear constraint.

In these last two sections we have restricted attention to the case $\varpi_0 < 1$. It is possible to extend many of the above results to the case $\varpi_0 = 1$ by a limiting procedure. A complication arises because the operator $[E - K(z)]^{-1}$ then definitely has a pole of higher order than one at $z = 0$. We will not develop here the results analogous to those given by Mullikin (1968a) and Mullikin and Victory (1975), which are based on bifurcation analysis of the nonlinear equations.

The Source Function

The source function of equation (3.1) has been used in equation (3.3) to define the operator $\mathfrak{K}_r$, related to the scattering kernel. The source function itself is of interest as a function of τ, and this in turn can be expressed in terms of the $\mathfrak{K}$-operators. We still impose the condition that absorption is present, $\varpi_0 < 1$.

Since J_r is related to j_r of equation (3.1), we first consider equation (3.13) and interpret integration with respect to τ as an $\mathfrak{L}_2(-\infty, \infty)$-inner product to which we apply the Parseval relation. If we replace f in equation (4.13) by $\mathcal{K}_l(z)[f]$, we obtain

Theorem 11. *Given the $\mathcal{K}$-operators, the function j_r is given for* $\mathrm{Im}(z) > 0$ *by*

$$j_r(\tau, \Omega, z) = \exp(i\tau z)\mathcal{K}_l(z)[f](\Omega)$$

$$-\frac{1}{2\pi i} \int_{-\infty}^{\infty} K(-t)[\mathcal{K}_r(-t)[\mathcal{K}_l(z)[f]]](\Omega) \frac{\exp(i\tau t) - \exp(i\tau z)}{t - z} dt. \quad (5.1)$$

We can relate j_r to the H-operators by moving the contour of integration in this last expression to the branch-cut $[i, i\infty)$. It is best to first use equation (3.10) to replace $\mathcal{K}_r(-t)$ in the integrand by

$$\mathcal{K}_r(-t) = [E - K(-t)]^{-1}\mathcal{K}_l^{-1}(t).$$

In moving the contour in equation (5.1) to the branch-cut, we obtain contributions from the poles of $[E - K(-t)]^{-1}$. This gives complicated results unless the poles are finite in number. (See Case 1974 for conditions on p sufficient for this to be true.) For some special cases, results are given by Carlstedt and Mullikin (1966).

We obtain asymptotic formulae for large τ by moving the contour past the pole at z_0 (cf. Sobolev 1956; Maslennikov 1968).

Corollary 2. The source function deep in a semi-infinite atmosphere is given by

$$J_r(\tau, \Omega, \Omega_0) = O(\exp(-\tau|z_1|))$$

$$+ \frac{1}{i} R_0^{-}\mathcal{K}_l^{-1}(z_0) H_l(\mu_0)[p(\cdot, \Omega_0)](\Omega) \frac{\exp(-\tau|z_0|)}{1 - \mu_0|z_0|}. \quad (5.2)$$

This approximates J_r by a positive multiple of the positive function $u_0(\mu) \exp(-\tau|z_0|)$, which is independent of azimuth angle.

From equations (2.1) and (2.4) we obtain

Corollary 3. The radiation field due to a unidirectional uniform field at $\tau = 0$ is given for $-1 \leq \mu \leq 0$ by the asymptotic formula

$$I(\tau, \Omega, \Omega_0) = O[\exp(-\tau|z_1|)]$$

$$+ \frac{1}{i} R_0^{-}\mathcal{K}_l^{-1}(\mu_0)[p(\cdot, \Omega_0)](\Omega) \frac{\exp(-\tau|z_0|)}{(1 - \mu_0|z_0|)(1 + |\mu z_0|)}. \quad (5.3)$$

The dominant term is independent of azimuth angle.

The source function can also be used to obtain interesting probabilistic results (cf. Sobolev 1966, 1967; Mullikin 1968a, 1969; Abu-Shumays 1967). We consider a photon which scatters at depth τ in a semi-infinite atmosphere from direction Ω into a cone of solid angle $d\omega_0$ and direction Ω_0. We define a random variable X which depends on the parameters τ, Ω, and Ω_0, and $X = n$ if the photon exists at $\tau = 0$ in the cone about Ω_0 after n scatterings. We let $P_n(\tau, \Omega, \Omega_0)$ be the probability that $X = n$, and define the probability-generating function F by

$$F(\tau, \Omega_0, \zeta) = \sum_{n=0}^{\infty} P_n(\tau, \Omega, \Omega_0)\zeta^n \qquad 0 \leq \zeta \leq 1 . \tag{5.4}$$

It is easy to see that F satisfies the integral equation

$$F(\tau, \Omega, \Omega_0, \zeta) = \mu_0\zeta k(\tau, -\Omega, -\Omega_0)d\omega_0$$

$$+ \int\int_0^{\infty} \zeta k(\tau - \tau', -\Omega, \Omega')F(\tau', \Omega', \Omega_0, \zeta)d\tau' d\omega' .$$

The solution, by equation (2.5), is expressed by

$$F(\tau, \Omega, \Omega_0, \zeta) = |\mu_0| J_r(\tau, -\Omega, -\Omega_0, \zeta)d\omega_0 . \tag{5.5}$$

The parameter ζ enters by the dependence of J_r on the phase function $\zeta p(\Omega, \Omega')$.

From this last result we can obtain several facts. A photon which enters the medium at $\tau = 0$ in the direction Ω will interact first in the interval $(\tau, \tau + d\tau)$ with probability $\exp(-\tau/\mu)d\tau/\mu$. The probability-generating function F_1 for the number of scatterings before exit in the cone with solid angle $d\omega$, and direction Ω_0, given that exit occurs, is given by equation (5.5) as

$$F_1(\Omega, \Omega_0, \zeta) = \left[\frac{d\omega_0|\mu_0|}{\mu} \int_0^{\infty} \exp(-\tau/\mu)J_r(\tau, -\Omega, -\Omega_0, \zeta)d\tau\right] \Bigg/$$

$$\left[\int_{\mu_0 \leq 0} \int_0^{\infty} \frac{|\mu_0|}{\mu} \exp(-\tau/\mu)J_r(\tau, -\Omega, -\Omega_0, 1)d\tau d\omega_0\right] . \tag{5.6}$$

By equation (4.1) this can also be expressed by

$$F_1(\Omega, \Omega_0, \zeta) = \frac{\mu^{-1}s(-\Omega, -\Omega_0, \zeta)d\omega_0}{\mu^{-1}\int s(-\Omega, -\Omega_0, 1)d\omega_0} . \tag{5.7}$$

The denominator gives the probability $P_e(\Omega)$ of exit from the half-space, and by equation (4.4) is, for $0 \leq \mu \leq 1$,

$$P_e(\Omega) = \int_{\mu_0 \geq 0} \left(1 - \frac{\mu}{\mu + \mu_0}\right) H_r(\mu)H_l(\mu_0)[P(\cdot, \Omega_0)](-\Omega)d\omega_0 . \tag{5.8}$$

Using equations (4.6) and (4.9), we can express P_e by

$$P_e(\Omega) = 1 - H_r(\mu)[1](-\Omega) + H_r(\mu)[(1 - \mathcal{K}_r^{-1}(0))[1]](\Omega)$$

$$= 1 - H_r(\mu)[\mathcal{K}_r^{-1}(0)[1]](-\Omega) .$$

From equation (3.10) it follows that

$$\mathcal{K}_r^{-1}(0)[1](-\Omega) = (1 - \varpi_0)\mathcal{K}_l(0)[1](-\Omega) ,$$

and finally that

$$P_e(\Omega) = 1 - (1 - \varpi_0)H_r(\mu)[\mathcal{K}_l(0)[1]](-\Omega) . \qquad (5.9)$$

In the simple case of constant phase function, we can use (3.10) to compute that

$$H_r^{-1}(0)[1](-\Omega) = \mathcal{K}_l^{-1}(0)[1](-\Omega) = (1 - \varpi_0)^{1/2} .$$

This gives the result (Mullikin 1968a)

$$P_e(\Omega) = 1 - (1 - \varpi_0)^{1/2}H_r(\mu) .$$

It is possible to use equation (5.7) in this case to obtain a formula for the expected number of scatterings before exit or absorption and asymptotic formulae for the probabilities of n scatterings, asymptotic for large n (cf. Mullikin 1968a for results and other references).

X and *Y* Operators

We consider now the case of finite atmospheres, $0 \leq \tau \leq \tau_1$, for $\varpi_0 \leq 1$. We will later in the section again impose the condition $\varpi_0 < 1$.

Of particular interest are scattering and transmission kernels. For simplicity we consider only the scattering kernel defined by

$$s(\Omega, \Omega_0) = \mu_0 \int_0^{\tau_1} \exp{(-\tau/|\mu|)}J(\tau, \Omega, \Omega_0)d\tau , \qquad \mu \leq 0 , \mu_0 \geq 0$$

$$= 0 , \qquad\qquad\qquad\qquad\qquad \text{otherwise} , \qquad (6.1)$$

where J satisfies equation (2.5) for $\tau_1 < \infty$.

In order to analyze this kernel, we define a one-parameter family of kernels for $\text{Im}(z) \geq 0$ by

$$s(\Omega, \Omega_0, z) = \mu_0 \int_0^{\tau_1} \exp{(i\tau z)}J(\tau, \Omega, \Omega_0)d\tau . \qquad (6.2)$$

For arbitrary g in $C(S)^2$ we compute

$$\int s(\Omega, \Omega_0, z)g(\Omega)d\omega = \mu_0 \int \int_0^{\tau_1} \exp{(i\tau z)}g(\Omega)J(\tau, \Omega, \Omega_0)d\tau d\omega ,$$

and use the fact that $J = (E - L_{\tau 1})^{-1}k$ to express this as

$$\int s(\Omega, \Omega_0, z)g(\Omega)d\omega = \int \int_0^{\tau_1} j_l(\tau, \Omega, z)p(\Omega, \Omega_0) \exp{(-\tau/\mu_0)}d\tau d\omega , \qquad (6.3)$$

where j_l is the unique solution in $\mathfrak{L}_\infty\{(0, \tau_1) \times S^2\}$ to

$$l(\tau, \Omega, \nu) = g(\Omega) \exp{(i\tau z)} + \int \int_0^{\tau_1} k(\tau' - \tau, \Omega', \Omega)j_l(\tau', \Omega', z)d\tau' d\omega' . \qquad (6.4)$$

In equation (6.3) we integrate by parts to obtain

$$\int s(\Omega, \Omega_0, z) g(\Omega) d\omega = \mu_0 \int p(\Omega, \Omega_0) [j_l(0, \Omega, z) - \exp(-\tau_1/\mu_0) j_l(\tau_1, \Omega, \nu)] d\omega$$

$$+ \mu_0 \int \int_0^{\tau_1} \frac{\partial j_l}{\partial \tau} (\tau, \Omega, z) p(\Omega, \Omega_0) \exp(-\tau/\mu_0) d\tau d\omega .$$

In a manner similar to that in equation (3.12) we find

$$\frac{\partial}{\partial \tau} j_l(\tau, \Omega, z) = iz j_l(\tau, \Omega, z)$$

$$+ (I - L_{\tau_1}^*)^{-1} \int [(k(-\cdot, \Omega', \cdot) j_l(0, \Omega', z)$$

$$- k(\tau_1 - \cdot, \Omega', \cdot) j_l(\tau_1, \Omega', z) d\omega'](\tau, \Omega) . \qquad (6.5)$$

Combining the last two equations, we have

$$\left(\frac{1}{\mu_0} - iz \right) \int s(\Omega, \Omega_0, z) g(\Omega) d\omega$$

$$= \int [j_l(0, \Omega, z) - \exp(-\tau_1/\mu_0) j_l(\tau_1, \Omega, z)] p(\Omega, \Omega_0) d\omega$$

$$+ \int \int_0^{\tau_1} J(\tau, \Omega, \Omega_0) \left[\int k(-\tau, \Omega', \Omega) j_l(0, \Omega', z) d\omega' \right.$$

$$\left. - \int k(\tau_1 - \tau, \Omega', \Omega) j_l(\tau_1, \Omega', z) d\omega' \right] d\tau d\omega .$$

It then follows from equation (2.5) that

$$\left(\frac{1}{\mu_0} - iz \right) \int s(\Omega, \Omega_0, z) g(\Omega) d\omega$$

$$= \int [j_l(0, \Omega, z) J(0, \Omega, \Omega_0) - j_l(\tau_1, \Omega, z) J(\tau_1, \Omega, \Omega_0)] d\omega . \qquad (6.6)$$

We define operators on $C(S^2)$ for $0 \leq \nu, \mu_0 \leq 1$, by

$$X_r^*(\nu)[g](\Omega) = j_l(0, \Omega, i/\nu) ,$$

$$X_l(\mu_0)[f](\Omega) = j_r(0, \Omega, i/\mu_0) ,$$

$$Y_r^*(\nu)[g](\Omega) = j_l(\tau_1, \Omega, i/\nu) ,$$

$$Y_l(\mu_0)[f](\Omega) = j_r(\tau_1, \Omega, i/\mu_0) , \qquad (6.7)$$

where now j_r satisfies the equation

$$j_r(\tau, \Omega, z) = f(\Omega) \exp(iz\tau) + \int \int_0^{\tau_1} k(\tau - \tau', \Omega, \Omega') j_r(\tau', \Omega', z) d\tau' d\omega' . \qquad (6.8)$$

Since J is a special case of j_r, and equation (6.6) holds for all g in $C(S^2)$, we obtain

$$s(\Omega, \Omega_0, i/\nu)$$

$$= \frac{\nu \mu_0}{\nu + \mu_0} [X_r(\nu) X_l(\mu_0) - Y_r(\nu) Y_l(\mu_0)][p(\cdot, \Omega_0)](\Omega) , \qquad 0 \leq \nu, \mu_0 \leq 1$$

$$= 0 , \qquad \qquad \qquad \qquad \qquad \text{otherwise} . \qquad (6.9)$$

The scattering kernel is a Fourier transform of the function J for parameters Ω and Ω_0. We can set $\nu = |\mu|$ in equation (6.9) to obtain

Theorem 12. *For a finite atmosphere, the scattering kernel is expressed by* X *and* Y *operators as*

$$s(\Omega, \Omega_0)$$

$$= \frac{|\mu|\mu_0}{|\mu| + \mu_0} [X_r(|\mu|)X_l(\mu_0) - Y_r(|\mu|)Y_l(\mu_0)]$$

$$\times [p(\cdot, \Omega_0)](\Omega), \quad 0 \le \mu_0, |\mu| \le 1$$

$$= 0, \qquad\qquad\qquad\qquad \text{otherwise}. \tag{6.10}$$

It is possible to derive nonlinear equations for X and Y operators analogous to those for H-operators in Theorem 8. The uniqueness question for such equations is quite complicated. We investigated this question for Chandrasekhar's ψ_l^m and ϕ_l^m equations which come from expansions for phase functions with N-term Legendre polynomial expansions (Mullikin 1964c).

An alternative to nonlinear equations is to treat X and Y operators for $\tau_1 < \infty$ as perturbations to the H-operators for $\tau_1 = \infty$. For relatively simple problems this approach has been used for numerical computations (Carlstedt and Mullikin 1966; Sekera and Kahle 1966; Kahle 1967; Sweigart 1970). The case $\varpi_0 = 1$ is exceptional, so with the assumption $\varpi_0 < 1$ we now develop some of the results in this approach.

For brevity, we consider only the $\mathcal{X}_r$ and $\mathcal{Y}_r$ operators defined by

$$\mathcal{X}_l(z)[f](\Omega) = j_r(0, \Omega, z)$$

$$\mathcal{Y}_l(z)[f](\Omega) = j_r(\tau_1, \Omega, z). \tag{6.11}$$

Theorem 13. *For* $\varpi_0 < 1$, $\mathcal{X}_l$ *and* $\mathcal{Y}_l$ *operators satisfy, for* $\mathrm{Im}(z) > 0$,

$$\mathcal{X}_l(z)[f](\Omega) = \mathcal{H}_l(z)[f](\Omega) + O(\exp(-\tau_1|z_0|)), \tag{6.12}$$

$$\mathcal{Y}_l(z)[f](\Omega) = O(\exp(i\tau_1 z)) + O(\exp(-\tau_1|z_0|)). \tag{6.13}$$

Proof. We rewrite equation (6.8) for $-\infty < \tau < \infty$ as

$$j_1(-\tau, \Omega, z) + j_r(\tau, \Omega, z) + j_2(\tau - \tau_1, \Omega, z)$$

$$= \int\int_0^{\tau_1} k(\tau - \tau', \Omega, \Omega') j_r(\tau, \Omega', z) d\tau' d\omega'$$

$$+ \begin{cases} f(\Omega) \exp(iz\tau), & \tau \ge 0 \\ 0, & \tau < 0, \end{cases} \tag{6.14}$$

where j_1, j_r, and j_2 have support on the interval $0 \le \tau < \infty$.

By a computation similar to that on page 121, we use J_l of equation (3.2) to find that

$$j_r(0, \Omega, z) = \mathcal{K}_l(z)[f](\Omega) - \int\!\int_{\tau_1}^{\infty} J_l(\tau, \Omega, \Omega') j_2(\tau - \tau_1, \Omega', z) d\tau d\omega' \,.$$

If we apply the Parseval formula to the τ-integration, we find

$$\mathcal{X}_l(z)[f](\Omega) \doteq \mathcal{K}_l(z)[f](\Omega)$$

$$- \frac{1}{2\pi} \int_{-\infty}^{\infty} \mathcal{K}_l(-\zeta)[K(\zeta)[\mathcal{J}_2(\zeta, \cdot, z)]](\Omega) \exp(i\tau_1\zeta) d\zeta \,, \qquad (6.15)$$

where $\mathcal{J}_2$ is the Fourier transform of j_2 and $\mathcal{K}_l(-\zeta)K(\zeta)$ is obtained for the Fourier transform of J_l in a manner similar to that in deriving equation (4.13).

If in this last equation we replace $\mathcal{K}_l(-\zeta)$ by $\mathcal{K}_r^{-1}(\zeta)[E - K(\zeta)]^{-1}$, we can move the contour of integration to the branch-cut $[i, i\infty)$ taking account of poles of $[E - K(\zeta)]^{-1}$. The details are complicated and require a determination of the function $\mathcal{J}_2$. The result, stated in equation (6.12), summarizes this in an asymptotic formula.

The definition of the $\mathcal{Y}_l$ operator and use of the Parseval formula gives the representation

$$\mathcal{Y}_l(z)[f](\Omega) = f(\Omega) \exp(i\tau_1 z)$$

$$+ \frac{1}{2\pi} \int_{-\infty}^{\infty} \exp(i\tau_1\zeta) K(-\zeta)[\mathcal{J}_r(-\zeta, \cdot, z)](\Omega) d\zeta \,. \qquad (6.16)$$

From equation (6.14) we have

$$\mathcal{J}_1(-\zeta, \Omega, z) + [E - K(\zeta)][\mathcal{J}_r(\zeta, \cdot, z)](\Omega)$$

$$+ \mathcal{J}_2(\zeta, \Omega, z) \exp(i\tau_1\zeta) = \frac{-f(\Omega)}{i(\zeta + z)} \,, \qquad (6.17)$$

and consequently

$$\mathcal{Y}_l(z)[f](\Omega) = f(\Omega) \exp(i\tau_1 z) - \frac{1}{2\pi} \int_{-\infty}^{\infty} K(\zeta)[\mathcal{J}_2(\zeta, \cdot, z)](\Omega) d\zeta$$

$$+ \frac{1}{2\pi} \int_{-\infty}^{\infty} K(-\zeta)(E - K(-\zeta))^{-1}\left[\frac{f(\cdot)}{i(\zeta - z)} - \mathcal{J}_1(\zeta, \cdot, z)\right] (\Omega)$$

$$\times \exp(i\tau_1\zeta) d\zeta \,. \qquad (6.18)$$

Because $\mathcal{J}_1(\zeta)$ and $\mathcal{J}_2(\zeta)$ are analytic in $\mathrm{Im}(\zeta) > 0$, we obtain the result in equation (6.13) by moving contours past the pole of $[E - K(\zeta)]^{-1}$ at z_0. This completes the proof.

The detailed structure of the $\mathcal{X}_l$ and $\mathcal{Y}_l$ operators is quite complicated, even for rather simple phase functions (cf. Carlstedt and Mullikin 1966). In addition to needing knowledge of the $\mathcal{K}$-operators, and Laurent expansions of the operator $[E - K(z)]^{-1}$, we need also the functions $\mathcal{g}_1$ and $\mathcal{g}_2$ which appear in equations (6.15) and (6.18).

The factorization

$$\mathcal{K}_r(\zeta)\mathcal{K}_l(-\zeta)[E - K(\zeta)] = E \qquad (6.19)$$

makes possible

Theorem 14. *The functions $\mathcal{g}_1$ and $\mathcal{g}_2$ satisfy, for $\mathrm{Im}(w) > 0$ and $\mathrm{Im}(z) > 0$, the equations*

$$\mathcal{g}_1(w, \Omega, z) = \frac{i}{z - w} [\mathcal{K}_l^{-1}(w)\mathcal{K}_l(z) - E][f](\Omega)$$

$$+ \frac{1}{2\pi i} \int_{-\infty}^{\infty} \frac{\mathcal{K}_l^{-1}(w)\mathcal{K}_r^{-1}(\zeta)[E - K(\zeta)]^{-1}\mathcal{g}_2(\zeta, \cdot, z)(\Omega) \exp{(i\tau\zeta)}}{\zeta + w} d\zeta ,$$

$$(6.20)$$

and

$$\mathcal{g}_2(w, -\Omega, z) = \frac{1}{2\pi i} \int_{-\infty}^{\infty} \mathcal{K}_l^{-1}(w)\mathcal{K}_r^{-1}(\zeta)[E - K(\zeta)]^{-1}$$

$$\times \left[\mathcal{g}_1(\zeta, -\cdot, z) - \frac{f(-\cdot)}{i(\zeta - z)} \right](\Omega) \frac{\exp{(i\tau\zeta)}}{\zeta + w} d\zeta . \quad (6.21)$$

Proof. We operate on equation (6.17) with $\mathcal{K}_l(-\zeta)$ and use the factorization (6.19) to write

$$\mathcal{K}_l(-\zeta)\mathcal{g}_1(-\zeta) + \mathcal{K}_r^{-1}(\zeta)\mathcal{g}_r(\zeta)$$

$$+ \mathcal{K}_l(-\zeta)\mathcal{g}_2(\zeta) \exp{(i\tau_1\zeta)} = - \frac{\mathcal{K}_l(-\zeta)[f]}{i(\zeta + z)}. \quad (6.22)$$

The term $\mathcal{K}_r^{-1}(\zeta)\mathcal{g}_r(\zeta)$ is analytic in $\mathrm{Im}(\zeta) > -\mathrm{Im}(z_0)$ and has norm vanishing exponentially as $\mathrm{Im}(\zeta) \to +\infty$. If we multiply this last equation by $d\zeta/(\zeta + w)2\pi i$, integrate, and use analyticity of $\mathcal{g}_1$ and $\mathcal{g}_2$ in the upper half-plane, we obtain equation (6.20).

If we denote by $\mathcal{g}_r^-$ the function

$$\mathcal{g}_r^-(\zeta, \Omega, z) = \mathcal{g}_r(-\zeta, -\Omega, z) \exp{(i\tau_1\zeta)} \qquad (6.23)$$

and replace ζ by $-\zeta$ and Ω by $-\Omega$ in equation (6.17), we obtain

$$\mathcal{g}_2(-\zeta, -\Omega, z) + [E - K(\zeta)][\mathcal{g}_r^-(\zeta, \cdot, z)](\Omega) = - \exp{(i\tau_1\zeta)}$$

$$\times \left[\mathcal{g}_1(\zeta, -\Omega, z) - \frac{f(-\Omega)}{i(\zeta - z)} \right]. \qquad (6.24)$$

We have made use of properties of the operator $K(\zeta)$ of equation (3.8) under changes of sign of ζ and of integration variables. The function $\mathcal{g}_r^-(\zeta)$ is analytic in $\mathrm{Im}(\zeta) > -\mathrm{Im}(z_0)$ and has norm vanishing exponentially as $\mathrm{Im}(\zeta) \to +\infty$. We again operate with $\mathfrak{IC}_l(-\zeta)$ and perform an integration to obtain equation (6.21), and complete the proof.

If the contour of integration in equations (6.20) and (6.21) is moved to the branch-cut $[i, i\infty)$, we obtain integral equations for $\mathcal{G}_1$ and $\mathcal{G}_2$ on $[i, i\infty)$ together with unknown values of $\mathcal{G}_1$ and $\mathcal{G}_2$ resulting from residue computations at poles of $[E - K(\zeta)]^{-1}$. The resulting integral operators, when expressed by kernels defined on the square $0 \le \mu, \nu \le 1$, will be of the form

$$\frac{\mu' \exp\left(-\tau_1/\mu\right) H_r^{-1}(\mu)\psi(\Omega', \Omega)}{\mu' + \mu} \, d\omega , \tag{6.25}$$

where ψ is determined by the jump across $[i, i\infty)$ of $[E - K(z)]^{-1}$.

For τ_1 sufficiently large, the resulting integral equations can be solved by an iterative procedure. In special cases we have shown that the integral equations can be solved by such a procedure for all values of τ_1 (Mullikin 1964a) and used for numerical computations (Carlstedt and Mullikin 1966).

In a study of anisotropic problems in neutron transport theory, Leonard and Mullikin (1964) were able to prove convergence of such iterative methods provided the kernel functions were nonnegative. It follows from equation (4.6) that $\mathfrak{IC}_r^{-1}(\zeta)$ is a positive operator for ζ in $[i, i\infty)$. It may be possible to prove positivity of ψ in equation (6.25) and to extend our previous results concerning convergence of iterative methods for all τ_1. We have not resolved this question for general phase functions p.

References

Abu-Shumays, I. K. 1967, *J. Math. and Appl.*, **18**, 453.
Ambartsumian, V. A. 1952, *Theoretical Astrophysics* (Moscow: Gosudstu. Isdat. Tehn.-Theor. Let.; London: Pergamon Press, 1968).
Bellman, R. E., Kalaba, R. E., and Prestrud, M. C. 1963, *Invariant Imbedding and Radiative Transfer in Slabs of Finite Thickness* (New York: American Elsevier).
Busbridge, I. W. 1960, *The Mathematics of Radiative Transfer* ("Cambridge Tracts," No. 50 [Cambridge University Press]).
Carlstedt, J. L., and Mullikin, T. W., 1966, *Ap. J. Suppl.*, **12**, No. 113, 449.
Case, K. M. 1974, *J. Math. Phys.*, **15**, 974.
Case, K. M., and Zweifel, P. F. 1967, *Linear Transport Theory* (Reading: Addison-Wesley).
Chandrasekhar, S. 1950, *Radiative Transfer* (London: Oxford University Press [reprinted New York: Dover, 1960]).
Dunford, N., and Schwartz, J. 1969, *Linear Operators, Part I: General Theory* (New York: Wiley Interscience).
Gohberg, I. C. 1964, *Izv. Akad. Nauk SSSR*, **28**, 1055 (AMS transl. [1966] **49**, 130).

Gohberg, I. C. 1951, *Dokl. Akad. Nauk SSSR*, **78**, 629.

Gohberg, I. C., and Feldman, I. A. 1971, *Convolution Equations and Projection Methods for their Solution* (Moscow: Naukal; Trans. Math. Monographs [1974], Vol. **41**, Am. Math. Soc., Providence).

Hille, E., and Phillips, R. S. 1957, *Functional Analysis and Semigroups* (Providence: Am. Math. Soc.).

Hopf, E. 1934, *Mathematical Problems of Radiative Equilibrium* ("Cambridge Tracts," No. 31 [Cambridge: Cambridge University Press]).

Hunt, G. E., ed. 1971, *Transport Theory* (*Atlas Symposium* No. 3. *J. Quant. Spectrosc. Rad. Transfer*, **11**, 511).

Inönü, E. 1970, *J. Math. Phys.*, **11**, 568.

Kahle, A. B. 1967, *Global radiation emerging from a Rayleigh scattering atmosphere of large optical thickness* (Santa Monica, Calif.: RAND Corp., RM-5343-PR).

Kourganoff, V., and Busbridge, I. W. 1952, *Basic Methods in Transfer Problems* (Oxford: Clarendon Press).

Krein, M. G. 1962, *Am. Math. Soc. Transl.*, **22**, 163.

Krein, M. G., and Rutman, M. A. 1951, *Am. Math. Soc. Transl.*, No. 26.

Kuriyan, J. G., ed. 1973, *The UCLA International Conference on Radiation and Remote Probing of the Atmosphere* (North Hollywood: Western Periodicals Company).

Kuscer, I., and Vidav, I. 1969, *J. Math. Appl.*, **25**, 80.

Lenoble, J. 1974, *Standard Procedures to compute atmospheric Radiative Transfer in a Scattering Atmosphere*, Vol. **1**, (Lille: University of Lille).

Leonard, A., and Mullikin, T. W. 1964, *J. Math. Phys.*, **5**, 399.

Maslennikov, M. V. 1968, *The Milne Problem with Anisotropic Scattering* (Moscow: Nauka [Providence: Am. Math. Soc., 1969]).

Mullikin, T. W. 1962, *Ap. J.*, **136**, 627.

——. 1964*a*, *Trans. Am. Math. Soc.*, **113**, 316.

——. 1964*b*, *Ap. J.*, **139**, 379.

——. 1964*c*, *Ap. J.*, **139**, 1267.

——. 1968*a*, *J. Appl. Prob.*, **5**, 357.

——. 1968*b*, "Neutron Branching Processes," in *Probabilistic Methods in Applied Mathematics*, Vol. **1**, ed. Bharucha-Reid (New York: Academic Press).

——. 1969, *Neutron Transport Theory Conference*, January, 1969 (Clearinghouse for Federal Scientific and Technical Information, National Bureau of Standards, U.S. Dept. of Commerce, Springfield, Va.).

——. 1973, *J. Transport Theory and Statistical Phys.*, **2**, 335.

Mullikin, T. W., and Victory, D. 1975, *J. Math. Anal. Appl.*, in press.

Muskhelishvili, N. I. 1963, *Singular Integral Equations* (Groningen: P. Noordhoff).

Pahor, S., 1968, *Nucl. Sci. Eng.*, **31**, 110.

Sekera, Z., and Kahle, A. B. 1966, Scattering Functions for Rayleigh Atmospheres of Arbitrary Thickness (The RAND Corp., R-452-PP, Santa Monica, Calif.).

Siewert, C. E. 1972, *J. Quant. Spectrosc. Rad. Transfer*, **12**, 683.

Sobolev, V. V. 1956, *A Treatise on Radiative Transfer* (Moscow [Princeton: Van Nostrand 1963]).

———. 1966, *Akad. Nauk Armenian SSR* **2**, 135.

———. 1967, *Akad. Nauk Armenian SSR*, **3**, 6.

Sweigert, A. V. 1970, *Ap. J. Suppl.*, No. 182, **22**, 1.

Victory, D. 1974, *Multigroup Critical Problems for Slabs in Neutron Transport Theory* (unpublished Ph.D. thesis, Purdue University).

7

Relativistic Astrophysics and Observation   L. Woltjer

Introduction

During the last decade extensive observational studies have been made of several classes of objects which are believed to be connected to general relativity in the sense that strong gravitational fields appear to be involved. Prominent among these are quasars, pulsars, and compact X-ray sources. Also the universe itself may be regarded as belonging to this class.

Newtonian gravitational theory was derived from observations in the solar system. However, its applicability elsewhere has been demonstrated in several cases. In particular the rather close fit of main-sequence stellar structure theory and observation should be mentioned in this context. Even the approximate constancy of the gravitational constant G through our Galaxy, in other galaxies, and through much of the history of the universe may be demonstrated, starting out from the fact that in stars like the Sun, the luminosity is proportional to G^7. Variations of a factor of 2 in G would already have very noticeable effects.

In the case of general relativity the situation is different. The theory was developed without immediate reference to astronomical observations. Very soon, however, it was found that the theory provided a striking explanation of the long standing problem of the perihelion precession of Mercury; also, subsequent studies of the bending of the path of electromagnetic waves near the Sun seemed to be fully in accordance with prediction. Recent radio astronomical measurements confirm this to an accuracy of 5 percent or better.

In the solar system one deals with extremely weak gravitational fields; the Schwarzschild radius for one solar mass is only of the order of a kilometer, while the radius of the Sun is more than 10^5 times as large. The recent discovery of a binary pulsar may allow one to study perihelion precession and related effects in conditions of stronger—but still weak—gravitational forces. While the present data seem to be within the range of possible predictions, the uncertainties in the masses of the objects involved combined with the possibility of precession due to nonrelativistic shape effects (possibly due to tidal interaction) preclude a valid conclusion at the present time.

Strong gravitational fields undoubtedly occur in several objects. As we shall show in the following, however, the complexity of the relevant astrophysical

L. Woltjer is at the European Southern Observatory, c⁄o CERN, in Geneva, Switzerland.

systems is so great that at present the connection with relativity is at best very tenuous. The most one can say is that from the point of view of a consistent theory not only of the present state but also of the past and future evolution of these systems, some theory of strong gravitational fields is required. However, empirical evidence relating to such a theory is still completely lacking, as will become apparent from the discussion of the various types of objects which follows.

Quasars

The quasars were discovered more than 10 years ago, but their nature still is controversial. They are characterized by a starlike optical appearance—with sometimes some surrounding fuzz—and by spectra which show emission lines with substantial redshifts, amounting to $z = (\lambda_{obs} - \lambda_0)/\lambda_0 \approx 3.5$ in the most extreme cases. The optical continuous spectrum underlying the emission lines tends to be blue and variable; numerous narrow absorption lines are frequently seen, especially at large redshifts. Radio emission is found in a few percent of all quasars, sometimes with a morphology that is indistinguishable from that of the radio galaxies. The so-called BL Lacertae objects—strongly variable radio and optical continuum sources without emission lines—may be a related class of objects.

Every possible interpretation of the redshift has been explored. Gravitational redshifts appear to be rather definitely excluded in some objects, because the volume within which the gravitational potential is large is insufficient to produce the emission-line strengths with the spectroscopically determined densities. Kinematical redshifts appear to be excluded because in a brightness-limited sample of randomly moving objects blueshifts should predominate, and none have been found. Interactions between the emission-line photons and the photons from the general radiation field in the quasar have been postulated, but no very convincing physical picture has been presented. As a result most authors have adopted a cosmological interpretation in which the redshifts are due to the expansion of the universe and the quasars therefore are at very large distances. Others, however, have claimed nonrandom spatial coincidences between quasars and objects at much smaller distances and have inferred thereby the existence of a completely new and as yet unknown redshift mechanism.

If the quasars are, in fact, at cosmological distances, then the most plausible picture is one in which they represent extremely active nuclei in galaxies. Many galaxies are known in which there is much nonthermal emission from the nucleus at radio and optical wavelengths. Examples are the Seyfert galaxies—mainly spirals—in which the luminosity from the nucleus may be as high as that from the rest of the galaxy, and also certain elliptical galaxies which are strong radio galaxies. If one imagines cases in which the activity in the nucleus would be a factor of 10–100 stronger, then objects resembling quasars should result.

The total energies emitted by quasars would be quite large. The associated radio sources indicate lifetimes of at least 10^5–10^6 years, and as a result the total emitted radiation should amount to 10^{59}–10^{60} ergs in typical cases and possibly more in the most extreme ones. Because of its spectrum and polarization, much of the radiation must be nonthermal. The situation is no different in the stronger radio galaxies. These are frequently associated with giant elliptical galaxies for which there is no doubt about the interpretation of the redshifts and the resulting distances. To explain the (synchrotron) radio emission, a reservoir of relativistic electrons and magnetic fields with a total energy of up to 10^{60}–10^{61} ergs is required, that is, an energy equivalent of 10^6–10^7 solar masses. The radio structures appear to indicate that much of this energy is released in a rather small volume, probably less than a few light-years across.

Theories for the energy generation in these objects tend to involve total masses in the range of perhaps 10^8–10^{10} solar masses. In some models all of this mass is put in one large rotating disk with a magnetic field, and energy is extracted with a pulsar-like mechanism. In other models it is subdivided into a large number of starlike objects from which supernovae and pulsars result. In either case it is unnecessary to assign to the overall configuration relativistic parameters. In view of the rather tenuous connection between these models and actual observations, it is clear that we are still far away from being able to make inferences about relativity effects. It is true, however, that it appears probable that these configurations ultimately will evolve into denser systems and that therefore a relativistic theory may be required if a full theoretical description is to be given.

Pulsars

Within a year following the discovery of pulsars, their identification with rotating, magnetic neutron stars had been universally accepted. The main reasons for this were the lack of a plausible alternative and the fact that neutron stars would be expected to rotate and to have magnetic fields simply as a result of the great compression involved in their formation. The most quantitative information comes from the Crab Nebula, the remnant of the supernova of 1054. First of all, this confirmed the identification of a pulsar with the leftover object from a supernova explosion—i.e., a neutron star. In addition, information about the moment of inertia of the pulsar could be obtained.

Much of the radiation from the Crab Nebula is synchrotron radiation from relativistic electrons moving in the magnetic field of the nebula. The lifetime due to radiative losses of the electrons that emit the bulk of the energy is less than the age of the nebula, and consequently they must be replenished continuously. The energy required for this is estimated as about $(1-2) \times 10^{38}$ ergs per second. From the period of the pulsar and its moment of inertia, its rotational energy can be evaluated. When this is combined with the time rate of change of the period, the loss of rotational energy follows. With the moment of inertia from a standard neutron star model with a mass equal to that of the Sun, this loss is found to be about 4×10^{38} ergs per second. The close similarity between the energy gain of the electrons in the Crab Nebula and the energy loss of the pulsar, suggests that the two are related. In fact, semiquantitative models for the acceleration of the particles in the pulsar magnetic field may easily be constructed which demonstrate that this relation is plausible. But even if it is granted that all pulsar rotational energy is converted into relativistic particle and magnetic field energy, the conclusion that the moment of inertia of a neutron star is in accord with prediction remains highly uncertain. In fact, the uncertainties in the parameters of the Crab Nebula, including the lack of knowledge about accelerated protons and about the mass of the neutron star, are so large that an error of a factor of 3 in the predicted moment of inertia would have gone unnoticed. As a result, the confirmation of neutron star theory is marginal at best. Since, moreover, neutron star models depend sensitively on the adopted equation of state for nuclear matter, no conclusions can be reached about the comparatively small general-relativistic effects (redshift on surface ≈ 0.10).

Compact X-ray Sources

About 150 X-ray sources are now known. Identifications for these include clusters of galaxies, radio galaxies, quasars, supernova remnants, and starlike objects in our Galaxy. Several of the latter turn out to be double stars composed of a fairly normal rather massive (20 solar masses) star and a compact companion. Mass loss from the star to the compact component appears to be responsible for the X-ray emission—the matter falling onto the compact component becomes quite hot as a result of the dissipation of the kinetic energy acquired. In order to achieve the temperatures indicated by the observations, infall onto a neutron star or possibly a white dwarf appears to be required, and several fairly detailed models have been constructed involving such objects.

The case of Cygnus X-1 appears to be particularly interesting. The mass of the normal star may be estimated spectroscopically. From radial velocity measurements its orbit may be found (apart from a projection factor). As a result, the mass of the compact body may be estimated—at least if no additional objects exist in the system. The data indicate that this mass can hardly be less than 5 times that of the Sun. According to theoretical models the masses of white dwarfs and neutron stars should be less than about two solar masses, and it is therefore believed that in Cygnus X-1 a black hole may have been found.

While it certainly would have some interest if a black hole really has been found, it is good to realize that all one is saying is that a compact object has been discovered which *theory* suggests must be a black hole. For the sake of argument, suppose one were convinced of the validity of Newtonian theory. It then would be found that the mass limit of configurations of degenerate fermions is about 6 (mass of proton)/(mean mass per degenerate fermion) times the mass of the Sun, or six solar masses for a neutron star. There is nothing in present-day observations to distinguish such a neutron star from a black hole.

Of course, it is not at all impossible that future observations will establish some phenomenon which would lead to a more positive identification. In particular, it has been suggested that X-ray observations with high time resolution might do this. However, it will not be easy to reach unambiguous conclusions. It has been mentioned that since the period of the last stable orbit around a black hole of five solar masses is of the order of a millisecond, variations on that time scale should be looked for. It appears to us optimistic to believe that the free particle orbit periodicities would be recognizable in the kind of hydro-dynamical accretion flow that is involved. In addition, instabilities in such a flow may well lead to more rapid variations than would be surmised from the periods of the orbits around a neutron star. As a consequence, the eventual identification of a black hole—if at all possible—would have to be based on much more sophisticated arguments.

The Universe

Cosmology is perhaps the subject most closely tied in with general relativity. Large numbers of cosmological models have been constructed, but clear observational conclusions have eluded astronomers until now. The main ingredients for an analysis of our Universe are the following:

The Hubble constant $H_0 = \dot{R}/R$.
The deceleration parameter $q_0 = -R\ddot{R}/\dot{R}^2$.
The blackbody radiation field characterized by a temperature of about 3 K.
The mean mass density ρ_0.
The abundances of isotopes like ^2H, ^3He, ^4He.
The age of the oldest objects in our Galaxy (globular clusters).

In the absence of a cosmological constant, a simple relation should connect q_0, H_0, and ρ_0, namely,

$$q_0 = \frac{4\pi G \rho_0}{3H_0^2},$$

which is simply stating that the deceleration of the expansion of the universe is due to the gravitational interactions in its mass distribution. The connection between theory and observation in cosmology is still sufficiently vague that even the correctness of the sign in this relation cannot really be demonstrated.

The Hubble constant, which measures the present expansion rate of nearby portions of the universe, is determined by measuring the recession velocity (redshift) of objects of independently determined distance. To eliminate the effects of local fluctuations, objects more distant than the Virgo cluster have to be used. The distance determination is done in several steps. For example, one could determine the distance of the Andromeda Galaxy on the assumption that its Cepheids are intrinsically similar to those in our own Galaxy. Then one can investigate the intrinsic properties of H II regions in the Andromeda Galaxy and, on the assumption that these are the same in more remote galaxies, obtain the distances for these. Undoubtedly there is much possibility for accumulating errors at each of these steps. In addition, the very first step—namely, the determination of the distance to the nearest stellar aggregate, the Hyades—is already uncertain by as much as 5–10 percent. Taking into account the steps which follow, we would estimate an uncertainty of at least 25 percent in the most recent determinations, not taking into account the effects of smaller scale flows in the universal expansion and the very real possibility that some major systematic error is still lurking somewhere. The fact that the Hubble constant has decreased by a factor of exactly 10 in the last 25 years should serve as a warning in this latter respect.

The latest determination of Sandage and Tammann yields $H_0 = 55$ km s^{-1} per megaparsec, corresponding to an age of the universe of $H_0^{-1} = 19 \times 10^9$ years if the expansion has been at a constant rate.

Recently, early suggestions of using supernovae for the determination of H_0 have again been taken up. In principle, from a supernova spectrum the effective temperature may be obtained as well as the expansion velocity of the matter from which the radiating area may be found. Hence the intrinsic luminosity may be calculated and the distance found without any intermediate steps. While the idea is quite interesting and simple, a reliable determination will have to be made on the basis of a good model. For example, the surface of unit optical depth from which most radiation comes need not at all expand with the spectroscopically inferred velocity, and only a detailed model can connect the two.

The deceleration parameter q_0 may be determined from the curvature of the redshift-magnitude (or angular diameter) relation at redshifts of order unity. In practice many difficulties enter, in particular in connection with corrections for evolutionary changes in the intrinsic luminosities and luminosity distributions of galaxies. At present even the sign of q_0 is not observationally established.

The 3 K radiation field appears now to be relatively well understood. All recent measurements seem to be consistent with a blackbody spectrum. This fact, and also the absence of any convincing models in which special discrete sources are responsible for the emission, give support to the belief that this is radiation left over from an early hot phase in the evolution of the universe. The isotropy of the radiation field has been well established at centimeter wavelengths. It provides the strongest evidence for an overall isotropy of the universe.

The mean density of matter ρ_0 still is as uncertain as ever. If we denote as ρ^* the density that would "close" the Universe ($q_0 = 0.5$), then the observed value of the average luminosity density due to galaxies combined with mass-to-light ratios, inferred from the measured galactic velocity fields, would lead to about $\rho_0 = 0.03\rho^*$. However, "invisible" matter may well be present. In particular, observations of the dynamics of clusters of galaxies seem to indicate the presence of additional mass, possibly increasing the quoted value by a factor of order 10. Such mass might be in the form of low-mass stars possibly in galactic halos or distributed through the clusters.

The age T of the oldest globular clusters follows from the evolutionary turnoff of stars in the color-magnitude diagrams. The stellar structure theory that enters here appears to be comparatively well established; however, some uncertainties remain, and also the adopted helium abundance affects the results. The latter might give rise to a 10 percent uncertainty, while the observational determination of the turnoff should have an uncertainty at least as large as that. As a result, the expansion age of the universe for $q_0 = 0$ (19×10^9 years) and for $q_0 = 0.5$ (12×10^9 years) both are compatible with the globular cluster data.

It is believed that most of the helium in the universe was made in an early dense and hot phase. Detailed models allow one to calculate the cosmological abundances of elements on the basis of a knowledge of the blackbody radiation field and ρ_0. Results appear to be roughly compatible with observation for $0.05\rho^* < \rho < \rho^*$. The main source of uncertainty is that the (positive or negative) contribution of element production in galaxies is hard to evaluate, especially for a sensitive isotope like deuterium.

In summary, at present we are still extremely uncertain as to the basic parameters of the Universe. All we can say is that the data are compatible with an extremely simple model of a universe in which expansion is the only motion and in which everything was determined by the initial explosive event. But we are still a long way from being able to ascertain if a theory based on general relativity gives a proper description.

Prospects for future improvements in observational cosmology certainly exist. The technology for observing faint objects is making rapid progress. For example, recently the redshift of a galaxy was determined at Palomar with an SIT vidicon in less than half an hour. For the same object it had taken Humason 78 hours of photographic exposure. Even more important than the speed of the modern detectors is the possibility of sky background subtraction which allows accurate measurements of objects whose surface brightness is below that of the sky. With the new instruments now becoming available it should be possible to extend the Hubble diagram (magnitude-redshift) into the interesting redshift range where different cosmological models lead to quite different predictions. Unless the evolutionary effects turn out to be prohibitive, this at least might yield a determination of q_0. In addition, if we ever learn to understand quasars, they may still become a useful tool. Systematic studies of the diameters of galaxy clusters and of radio sources may also make a contribution.

Conclusion

From the foregoing it should be clear that there is far less connection between general relativity and the world of observation than one might have hoped for. Much further observational work will have to be done before the situation improves. Of course, this is no reason to abandon work in relativity. As Chandrasekhar has so frequently emphasized, the lack of possibilities for experimentation in astronomy necessitates detailed theoretical extrapolation to provide a framework for the analysis of our data. To quote from his recent article on the "Development of General Relativity" (*Nature*, **252**, 15, 1974): "In my judgement, theory has a double role to play in astronomy: the common one of providing interpretations for observed phenomena; and the uncommon one of providing for astronomy the kind of basis which experiments provide for physics. The latter role is largely unrecognised and largely not practised." Nevertheless, especially in the case of relativity a stronger observational foundation would be extremely desirable. The extent to which this may be obtained from the present types of astrophysical observation remains unclear.

8

General-Relativistic Astrophysics

Kip S. Thorne

I. A Brief Overview of General-Relativistic Astrophysics

In 1964 S. Chandrasekhar and I both entered the field of general-relativistic astrophysics—he, in one of those big changes of research direction that he makes roughly once per decade; I, as a very green graduate student at Princeton. Chandra and I met soon thereafter. Since then, as our field has grown and blossomed, Chandra has shared a close friendship with me, my contemporaries, and our students, which makes us regard him as one of us rather than as a distinguished scientist of a previous generation.

During this period our field has developed a large body of knowledge about the ways in which general relativity should influence astrophysical systems. Later in this chapter I shall review in detail a small portion of that knowledge; but first I want to comment on the overall relevance of general relativity to astrophysics.

The discovery of quasars (1963), cosmic X-ray sources (1963), and the cosmic microwave radiation (1964) suggested in 1964 that contact between astronomical observation and general relativity theory might not be far off. Unfortunately, that contact has been slow in materializing—as L. Woltjer argues, perhaps overly strongly, elsewhere in this volume. On the other hand, we have come a long way since 1966 when one of my more hard-nosed colleagues at Caltech asked me, "How many angels can dance on the head of a neutron star?" and when another told me that "Somehow all stars probably find a way to avoid catastrophic gravitational collapse [to a black-hole state]." Nowadays it is generally accepted dogma that neutron stars, black holes, and gravitational waves exist and play important roles in the universe. Although we have no completely definitive observational case of a black hole, X-ray astronomy is close to providing one or more (Cygnus X-1; globular-cluster X-ray sources; see, e.g.,

Kip S. Thorne is at the W. K. Kellogg Radiation Laboratory, California Institute of Technology, Pasadena, California.

This chapter is an expanded version of a lecture presented at the (Chandrasekhar) Symposium on Theoretical Principles in Astrophysics and Relativity, held in Chicago on 27–29 May 1975. The work described was supported in part by the National Science Foundation [AST 75-01398 A01] and by the National Aeronautics and Space Administration [NGR 05-002-256].

Bolton 1975; Shapiro, Lightman, and Eardley 1976; Bahcall and Ostriker 1975; Silk and Arons 1975).[1] And although we have no definitive proof that general relativity is at work in the observed neutron stars, evidence for its role in the binary pulsar is mounting (Taylor *et al.* 1976). And although gravitational waves have probably not been detected yet, prospects for the future are fairly bright (see § IIb below).

Actually, it is unreasonable to ask for much observational evidence about general relativity from astronomical data. The physics of matter is so complex that an astrophysicist has trouble enough figuring out what is happening in a given object by blind-faith application of all the standard laws of physics. Only very rarely can he manage to solve simultaneously for the nature of the object and for some of the physical laws that govern it. What definitive proof do we have that convection inside stars occurs in rough accordance with standard mixing-length theory? What definitive proof that the Lorentz force law correctly describes the motion of charged particles in magnetic fields outside the solar system?

Thus it is that, despite the lack of any definitive proof of its relevance, general relativity has by now become a standard working tool of the astrophysicist. For example, it plays important roles in: computations of nucleosynthesis in the big bang (Peebles 1966; Wagoner *et al.* 1967; Wagoner 1973); the theory of galaxy formation (Lifshitz 1946; Doroshkevich *et al.* 1974; Peebles 1974; Jones 1976); deductions of constraints on the equation of state of nuclear matter from recently measured neutron-star masses (Rappaport *et al.* 1976; Middleditch and Nelson 1976); computations of neutron-star moments of inertia (Hartle 1967, 1970, 1973; Cohen 1970; Arnett and Bowers 1977)—which are key elements in the starquake theory of pulsar glitches (Ruderman 1969; Baym and Pines 1971; Pines *et al.* 1974; Ruderman 1976); the theory of the innermost regions of accretion disks around black holes in binary X-ray systems and in galactic nuclei (Novikov and Thorne 1973; Page and Thorne 1974; Bardeen and Petterson 1975; Cunningham 1976); computations of gamma-ray production by the evaporation of primordial black holes (Hawking 1974, 1975; Page 1976a, b); and computations of gravitational-wave generation by astrophysical systems (see §§ IIa and IIc below).

In this chapter I shall largely ignore cosmology. I shall focus attention primarily on finite-sized astrophysical systems (stars, globular clusters, galactic nuclei, primordial black holes).

One can distinguish three stages in the evolution of finite-sized, relativistic,

---

[1] Here and below no attempt at completeness of references is made. I usually cite the earliest significant papers on a topic, and one or two recent papers—preferably review articles—which can be used as stepping stones into the previous literature.

systems: a Newtonian stage, which is slow; a dynamical relativistic stage, which is fast; and a quasi-static relativistic stage, which is slow. Very occasionally the quasi-static relativistic stage may be interrupted by renewed dynamical activity.

For a normal (but massive) star the Newtonian stage is the subject of the standard theory of stellar evolution (see the lecture by Schwarzschild in this volume). The dynamical relativistic stage is the catastrophic collapse to a neutron star or black hole, which terminates the Newtonian evolution (see § IIa below). The quasi-static relativistic stage is the subsequent evolution of the neutron star or black hole (neutron-star cooling [Tsuruta and Cameron 1966a; Tsuruta et al. 1972]; pulsar emission [Ginzburg and Zheleznyakov 1975]; accretion of gas from a companion star to produce X-ray emission [Pringle and Rees 1972; Shakura and Sunyaev 1973; Davidsen and Ostriker 1973; Lamb et al. 1973; Novikov and Thorne 1973; Lamb 1975]; accretion of interstellar gas onto an isolated black hole to produce variable optical emission [Shwartzman 1971; Shapiro 1973a, b]). The occasional renewed dynamical activity includes corequakes in neutron stars followed by rapidly damped, gravitational-wave–emitting pulsations (§ IIa[vii] below); and coalescence of two relativistic objects (neutron stars and/or black holes) that previously lived together as a binary pair.

For a globular cluster or galactic nucleus the Newtonian stage is characterized by a slowly growing central condensation with the most massive stars sinking to the center first, followed by an era of star-star collisions and perhaps formation of supermassive stars (Spitzer 1971; Spitzer and Shull 1975; Saslaw 1975). From time to time subregions presumably go through a dynamical relativistic stage including collapse to form black holes, and/or hole-hole collisions (§§ IIa [iii] and IIa[iv] below). In the quasi-static relativistic stage one or more black holes accrete stars and gas from the surrounding medium (Lynden-Bell 1969; Hills 1975; Frank and Rees 1976; Lightman and Shapiro 1976), perhaps producing globular-cluster X-rays (Bahcall and Ostriker 1975; Silk and Arons 1975) and galactic-nucleus optical, infrared, and radio emission (Lynden-Bell 1969; Lynden-Bell and Rees 1971; Pringle et al. 1973; Norman and ter Haar 1973); perhaps powering quasar and Seyfert-nucleus activity (Hills 1975; Frank and Rees 1976); and perhaps also powering extended double radio sources around galaxies (Blandford 1976; Lovelace 1976). Renewed dynamical activity will occur when two black holes collide and coalesce (Hawking 1971b; Gibbons and Schutz 1972; Tipler 1975; Smarr et al. 1976).

For a primordial black hole there is no Newtonian stage. The hole may form in a dynamical relativistic stage either by direct collapse of an overdense region of the early universe (Hawking 1971a; Carr 1975), or by transformation from a white-hole state (region of delayed expansion) into a black-hole state (Eardley

1974; Zel'dovich *et al.* 1974). The subsequent quasi-static relativistic stage is characterized by accretion from the surrounding universe (Zel'dovich and Novikov 1966; Carr and Hawking 1974)—and, later, if the black hole has a mass $M \lesssim 4 \times 10^{25}$ g, by a quantum-mechanical evaporative emission of particles into the surrounding universe (Hawking 1974, 1975). For initial masses $M \lesssim 6 \times 10^{14}$ g, the black hole ultimately destroys itself by explosion (Hawking 1974; Page 1976a, b). Gamma-rays and other particles from such evaporations and explosions may be detectable at Earth (Page and Hawking 1976; Carr 1976). From observed limits on such gamma-rays one can infer interesting constraints on the inhomogeneity of the early universe.

For the observer who wishes to study the effects of relativistic gravity in astrophysical systems, different tools are appropriate for the dynamical and the quasi-static stages.

The dynamical relativistic stage is very brief: $\tau \sim (10^{-3}\text{ s})(M/M_\odot)$. During this stage the innermost region of the system, where gravity is relativistically strong, will be surrounded by enormous amounts of obscuring gas—so much, in fact, that typically there is no hope of receiving any electromagnetic radiation directly from the relativistic region. In some cases one might hope to receive neutrinos or antineutrinos (e.g., Bludman and Ruderman 1975); but typically gravitational waves will be the only type of radiation that emerges, unimpeded, in profuse amounts. Gravitational radiation is potentially a powerful tool for studying the dynamical relativistic stage because it carries off direct information about the bulk motion of the matter in the relativistic region, and about the time evolution of the relativistic gravitational fields (see § IIc below).

In the quasi-static stage gravitational radiation is a very *un*promising observational tool. When an object of mass m accretes quasi-radially onto a black hole of mass $M \gg m$, the energy it emits as gravitational radiation is only $\sim 0.01\ (m/M)mc^2$ (Zel'dovich and Novikov 1967; Davis *et al.* 1971). If the accretion is along a spiral orbit, the energy emitted per circuit around the hole is only $\sim (m/M)mc^2$ (Detweiler 1977). In realistic accretion scenarios, where $m \lll M$, this energy output is hopelessly small. However, the electromagnetic energy emitted in accretion can easily exceed $0.01\ mc^2$; and during the quasi-static stage the black hole or neutron star may be sufficiently devoid of surrounding matter, along our line of sight, that profuse numbers of electromagnetic quanta fly unimpeded from the relativistic regime to Earth. Thus, electromagnetic quanta appear to be the best tool for observing quasi-static, relativistic objects. Examples include pulsar emission (radio, optical, and X-ray) from rotating neutron stars; X-rays and radio waves from accretion disks around black holes; and optical emission from optically thin, magnetized interstellar gas accreting quasi-radially onto a black hole. For references see above.

II. Gravitational-Wave Astronomy

Having completed these general remarks, I shall focus attention in the rest of this written version of my lecture on one special topic—a topic chosen because (i) I happen to be particularly interested in it at the moment; (ii) it involves theoretical results that should have a long useful lifetime; and (iii) I expect it to become increasingly important during the next decade. This topic is "gravitational-wave astronomy."

I shall divide my discussion into three parts: *section a*, which is my guess as to the characteristics of the strongest gravitational waves that bathe the Earth; *section b*, which describes present obstacles and future prospects for detection of the predicted waves; and *section c*, which reviews the theory of small perturbations of relativistic stars and black holes, and the gravitational waves they generate. These three topics are not independent; rather, they are intimately intertwined. Results, or hoped-for results, from each one have strong influence on subsequent research in the other two. However, the research methods used in the three are totally different: phenomenological arguments and order-of-magnitude analyses; sophisticated experimental technique; and elegant mathematical physics.

a) Estimates of the Strongest Waves That Bathe the Earth

To guess what kinds of strong gravitational-wave sources actually occur in our universe, and where and how often, is a very dangerous business. Nevertheless, it is necessary to guess so that experimenters can have guidelines in the design of gravitational-wave detectors. In this section I shall offer my best current guess about the strongest sources in the universe, and about the resulting gravitational waves that bathe the Earth in the frequency band of interest to experimenters: $10^4 \text{ Hz} \gtrsim \nu \gtrsim 10^{-5} \text{ Hz}$. (I shall argue that strong waves are unlikely to exist above 10^4 Hz and below 10^{-5} Hz. Also, detectors of high sensitivity, operating below 10^{-5} Hz, are probably not feasible technologically; cf. Table 1 of Press and Thorne 1972.) My attention will focus primarily on transient sources of waves, since they should be much stronger and easier to detect than the overall time-averaged wave background. Estimates of the background have been given by Rosi and Zimmerman (1976).

Experience dictates that one not take my estimates of waves bathing Earth too seriously: (i) Estimates have changed somewhat from year to year due to changing astronomical knowledge, changing astrophysical fashion, and improving mathematical analyses of possible sources. (Compare my estimates below with previous estimates by Ruffini and Wheeler 1971, Press and Thorne 1972, and Rees 1974.) (ii) Theorists did a lousy job of estimating, in advance, the

X-rays bathing Earth from outside the solar system; the X-rays discovered in 1963 were much stronger than theorists had predicted. (See the review in the Introduction to Giacconi and Gursky 1974.)

i) *Characteristics of Waves from Black-Hole Events*

The very strongest sources of gravitational waves in the universe are probably black holes in collision, and black holes in their birth throes. Nobody has yet performed a reliable computation of the gravitational waves from such sources. However, analyses of weak perturbations of holes (Fig. 1 and § IIc below) suggest strongly that the radiation will consist of a short, initial broad-band burst, followed by longer, damped, narrower-band ringing. The initial burst is produced rather directly by the relative motions of various parts of the source (imploding matter or colliding holes); the subsequent ringing is due to normal-mode pulsations of the final black hole or holes.

The best guide we have today to the details of the initial burst is a perturbation-theory computation of the radiation emitted by a test particle falling radially into a nonrotating hole (Davis *et al.* 1971; Davis *et al.* 1972; Chung 1973; Fig. 1 of this paper). Figure 1*b* suggests that the initial burst and the subsequent ringing may both contain substantial fractions of the radiated energy.

In establishing goals for gravitational-wave searches, I shall focus attention on the initial burst: The burst's instantaneous wave amplitudes are likely to exceed those of the subsequent ringing; the burst will be spread over a larger frequency band than the ringing, so that a resonant detector with predetermined frequency will be more likely to see it; and even at the ringing frequency the total energy per bandwidth in the burst may be comparable to that in the ringing, unless the black hole is rapidly rotating (see Fig. 1*a* and § IIIc below).

I shall characterize the initial burst by the following parameters:

$$\nu \equiv (\text{frequency at which the spectrum peaks}) ; \tag{1a}$$

$$\tau \equiv \frac{1}{2\pi\nu} \equiv \begin{pmatrix} \text{characteristic time scale for changes in the} \\ \text{gravitational-wave field as it passes Earth} \end{pmatrix} ; \tag{1b}$$

$$\hat{\tau} \equiv (\text{duration of the initial burst}) ; \tag{1c}$$

$$\hat{\psi} \equiv \begin{pmatrix} \text{dimensionless amplitude of variation of a typical component} \\ \text{of the gravitational-wave field as it passes Earth} \end{pmatrix} ; \tag{1d}$$

$$\mathscr{F}_\nu \equiv \begin{pmatrix} \text{``burst''} \\ \text{intensity''} \end{pmatrix} \equiv \begin{pmatrix} \text{energy per unit frequency per unit area carried} \\ \text{past Earth by the burst, at the peak frequency } \nu \end{pmatrix} \tag{1e}$$

$$\text{units: ``Gravitational Pulse Units''} \equiv \text{GPU} \equiv 10^5 \text{ ergs cm}^{-2} \text{ Hz}^{-1} ;$$

cf. Misner (1974). The dimensionless amplitude $\hat{\psi}$ is defined more precisely as

$$\hat{\psi} = \tfrac{1}{2}[\tfrac{1}{4}\, \Sigma_{j,k}\, (h^{TT}_{jk\,\text{max}} - h^{TT}_{jk\,\text{min}})^2]^{1/2}\,, \qquad (1d')$$

where the sum is over the four transverse-traceless components of the metric perturbation associated with the wave, and $h^{TT}_{jk\,\text{max}}$ and $h^{TT}_{jk\,\text{min}}$ are the maximum and minimum values of those components during the passage of the wave. (For further details of notation and definitions of basic concepts see, e.g., chapters 35–37 of Misner, Thorne, and Wheeler 1973—cited henceforth as MTW.) The amplitude of the Riemann curvature associated with the wave is

$$\hat{R} = \tfrac{1}{2}\hat{\psi}/\tau^2\,. \qquad (2)$$

The burst intensity $\mathcal{F}_\nu$ will be roughly constant for frequencies between $0.5\,\nu$ and $1.5\,\nu$, but may die out very rapidly above $1.5\,\nu$ and somewhat more slowly below $0.5\,\nu$; cf. Figure 1a. The intensity is related to the amplitude by

$$\mathcal{F}_\nu = (\text{flux of energy}) \times \frac{(\text{duration of burst})}{(\text{peak frequency})} = \left[\frac{c^3}{G}\frac{1}{8\pi}\left(\frac{\hat{\psi}}{\tau}\right)^2\right] \times \frac{\hat{\tau}}{\nu}$$

$$= \frac{c^3}{G}\left(\frac{\hat{\psi}}{2}\right)^2\frac{\hat{\tau}}{\tau} = \left(\frac{\hat{\psi}}{3\times 10^{-17}}\right)^2\frac{\hat{\tau}}{\tau}\ \text{GPU} \qquad (3)$$

(cf. eq. [4.1] of Isaacson 1968 or § 35.7 of MTW for justification of above formula for the flux of energy in the wave). It is useful to remember that

$$c^3/G = 4.04 \times 10^{33}\ \text{GPU}\,. \qquad (4)$$

The above characteristics of the initial wave burst from a generic black-hole event will be determined by the following properties of the source and the universe:

$$M \equiv (\text{total mass in strong-gravity region of source}), \qquad (5a)$$

$$\varepsilon \equiv (\text{"efficiency factor"})$$

$$\equiv \frac{1}{Mc^2} \times \binom{\text{total energy carried off by burst,}}{\text{as measured near source}}, \qquad (5b)$$

$$r \equiv \binom{\text{distance between source and Earth, if source}}{\text{is not at a cosmological distance}}, \qquad (5c)$$

$$z \equiv (\text{cosmological redshift of source, for distant sources}), \qquad (5d)$$

$$H_o \equiv (\text{Hubble expansion rate of universe}), \qquad (5e)$$

$$q_o \equiv (\text{deceleration parameter of universe}). \qquad (5f)$$

The characteristic time scale $\tau = 1/(2\pi\nu)$ of the wave will be of order the light-travel time across the strong-gravity region, which means (several) $\times$ GM/c^3. The precise value of the coefficient will depend on the details of the col-

Figure 1. The gravitational radiation produced when a particle of mass m falls radially into a nonrotating black hole of mass M, in the limit $m \ll M$ and (size of particle) $\ll GM/c^2$.

Figure 1a shows the spectrum of the emitted radiation as computed independently by Davis and Ruffini and by Press and Price, and published jointly in Davis *et al.* (1971). Plotted horizontally is angular frequency $\sigma = 2\pi\nu$; plotted vertically is the total energy emitted per unit angular frequency $dE/d\sigma$—which is related to the burst intensity $\mathcal{F}_\nu$ measured at distance r, averaged over all directions, by

$$(\mathcal{F}_\nu)_{\text{avg}} = (1/2r^2)dE/d\sigma \ .$$

The curve marked TOTAL is the total spectrum; the curves marked $l = 2$, $l = 3$, and $l = 4$ are the spectra of the radiation contained in spherical harmonics of indices 2, 3, and 4. The total energy radiated, $E = \int(dE/d\sigma)d\sigma$, is $0.0104(m/M)mc^2$. The black hole has two quadrupole ($l = 2$) normal modes, three octupole ($l = 3$) normal modes, and three hexadecapole ($l = 4$) normal modes according to calculations by Chandrasekhar and Detweiler (1975a); see Table 1. Radiation emitted in these normal modes has a time dependence $\exp(-i\omega t)$, where $\omega = \sigma - i/\tau_d$ is the complex eigenfrequency of the normal mode. The quadrupole eigenfrequencies, as computed by Chandrasekhar and Detweiler, are given in Figure 1a; and the real parts of the octupole eigenfrequencies are indicated by vertical arrows.

Figure 1b shows the gravitational-wave field (transverse-traceless part of the metric perturbation) as a function of retarded time, as computed by Davis, Ruffini, and Tiomno (1972)—and also, using a completely different technique, by Chung (1973). Spherical polar coordinates are used, with the polar axis ($\theta = 0$) oriented along the particle's infall trajectory. Plotted vertically is the physical component $- h_{\theta\theta}^{TT} = h_{\varphi\varphi}^{TT}$ of the $l = 2$ part of the gravitational-wave field. The $l = 3$ and higher-order contributions are much much smaller, as one can see from Figure 1a; and the other component $h_{\theta\varphi}^{TT}$ of the field vanishes. Plotted horizontally is retarded time.

Figure 1c shows the "power" dE/dt_{ret} emitted in $l = 2$ radiation as a function of retarded time t_{ret}, as computed by Davis, Ruffini, and Tiomno (1972). This "power" is defined by

$$\frac{dE}{dt_{\text{ret}}} = \frac{c^5}{G}\int \frac{1}{16\pi}\left(\frac{dh_{\theta\theta}^{TT}}{dt_{\text{ret}}}\right)^2 r^2 \sin\theta\, d\theta d\varphi \ .$$

(The word "power" is placed in quotations because the instantaneous power output is not really a well-defined concept; the above definition is arbitrary. See the discussion of averaging in § 35.7 of Misner, Thorne, and Wheeler [1973].)

Perusal of these figures suggests the following interpretation of the radiation: The infalling particle emits an initial burst of gravitational waves; and it also excites normal-mode oscillations of the black hole which radiate ("ringing" waves). A clean separation of burst and ringing is not possible; the radiation switches over gradually from one to the other during the time interval $-10 \lesssim t_{\text{ret}}/(GM/c^3) \lesssim +10$. At late times, $t_{\text{ret}} \gtrsim 30 GM/c^3$, the radiation is due primarily to ringing of the most slowly damped quadrupole mode, which has $\sigma = 0.37367 (GM/c^3)^{-1}$. I suspect that numerical error may cause the computed radiation to damp out more slowly than it should at late times. (The amplitude e-folding time should be $\tau_d = 11.24 GM/c^3$ if the black hole is really vibrating in its most slowly damped quadrupole mode.)

The displacement of the peak of the spectrum (Fig. 1a) leftward from the normal-mode frequencies is due, presumably, to the lower-frequency radiation from the initial burst.

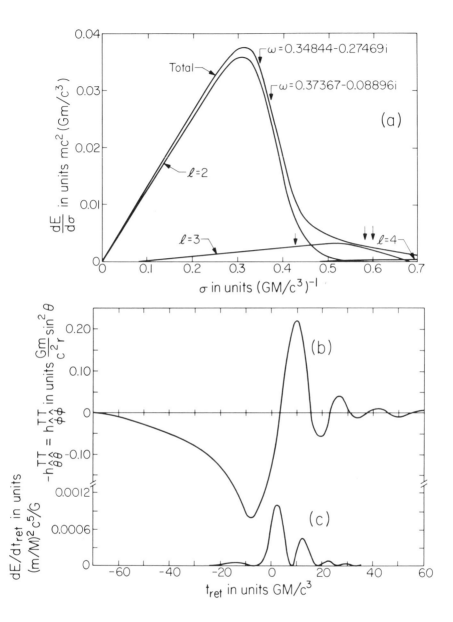

lapse or collision. Presumably the most rapid burst will be that produced by quasi-radial free-fall collapse. In other sources centrifugal forces may slow down the burst some—but not much, since the burst is at its peak when internal gravity is so strong that it can probably overwhelm the centrifugal forces. Thus, the quasi-radial problem is probably a good guide to τ and $\hat{\tau}$—and waves from a particle falling radially into a hole are probably a good guide to quasi-radial collapse. For the particle infall problem, Figure 1 augmented by the cosmological redshift reveals

$$\tau = 1/2\pi\nu \approx (3GM/c^3)(1 + z) = (1.5 \times 10^{-5}\text{ s})(M/M_\odot)(1 + z) \,, \quad (6a)$$

$$\nu \approx (6\pi GM/c^3)^{-1}(1 + z)^{-1} = (11{,}000\text{ Hz})(M/M_\odot)^{-1}(1 + z)^{-1} \,, \quad (6b)$$

$$\hat{\tau} \approx (10GM/c^3)(1 + z) \approx 3\tau = (5 \times 10^{-5}\text{ s})(M/M_\odot)(1 + z) \,. \quad (6c)$$

These values of τ, ν, $\hat{\tau}$ are probably correct to within a factor 2 for the initial burst from generic black-hole events (the birth throes of a hole; collisions of holes).

Because nuclear forces probably prevent black-hole formation with $M < 2\ M_\odot$ (except in the early stages of the big bang), and because the radiation from a hole probably cuts off sharply above the frequency 1.5 ν, I would expect all strong gravitational-wave bursts hitting Earth to have frequencies lower than $1.5 \times (11{,}000\text{ Hz})/2 \approx 10{,}000$ Hz. (Bursts from neutron stars will also be at lower frequencies than 10,000 Hz; see Table 2 in § IIc and Detweiler 1975a.)

There are two types of sources that might violate my suggested limit of $\nu_{max} = 10{,}000$ Hz: (i) The normal-mode oscillations of a very rapidly rotating black hole, for which $\nu_{max} \approx (16{,}000\text{ Hz})(M/2\ M_\odot)^{-1}\ (l/2)$, where $l \geq 2$ is the spherical-harmonic order of the pulsations; see Figure 10 in § IIIc. I doubt that generic collapses or collisions can strongly excite normal modes of such high frequency; and if I am wrong, then probably only the quadrupole mode will be excited substantially (cf. Fig. 1a), and it does not violate $\nu_{max} = 10{,}000$ Hz by much. (ii) The ergotoroid instability of a very rapidly rotating, highly relativistic neutron star (§ IIIc[iv]), which can produce radiation of very high frequency. Nothing is known at present about the strength of the radiation from such a source or about whether the ergotoroid instability actually occurs in realistic neutron-star models. Thus, at present ergotoroid radiation creates nothing more than nagging uncertainty in the limit $\nu_{max} = 10{,}000$ Hz.

At the other end of the spectrum, it seems likely to me that black-hole events involving more than $10^9\ M_\odot$ will occur in the universe much less often than once per year. Thus, I expect all strong-wave bursts hitting Earth more often than once per year to have $\nu > 11{,}000\text{ Hz}/10^9 \approx 10^{-5}$ Hz.

The correct value of the efficiency factor ε (eq. [5b]) is highly uncertain. Rough order-of-magnitude calculations, using the "quadrupole-moment formalism" for wave generation (which is surely invalid for black-hole events), give $\varepsilon \sim 1$; see § 36.5 of MTW. However, I do not believe that *any* calculation of ε

for black-hole events using the quadrupole-moment formalism can be trusted
to within a factor 100, because one is ignorant of how to correctly mock-up the
redshift effects that cut off the radiation just when it is getting strongest, at
collapse radii $r = $ (several) $\times (GM/c^2)$. More reliable guides to ε include (i) the
second law of black-hole mechanics which gives, for the collision and coalescence
of two holes (Hawking 1971b; Gibbons and Schutz 1972)

$$\dot{\varepsilon} < 1 - 1/\sqrt{2} = 0.29 \quad \text{if holes are nonrotating,}$$

$$\varepsilon < \tfrac{1}{2} = 0.50 \quad \text{if holes are rapidly rotating;}$$

(ii) solutions of the strong-gravity initial-value equations by Eardley (1975),
which suggest but do not prove $\varepsilon \lesssim 0.034$; and (iii) the Davis et $al.$ (1971) calculation of $\varepsilon = 0.010\ m/M$ for waves from a test particle of mass m falling into a
hole of mass M (Fig. 1). It is conceivable that the strongest sources have $\varepsilon \approx$
0.01; but my own half-educated guess is $\varepsilon \approx 0.1$. Ultimately computer calculations of black-hole collisions and nonlinear stellar collapse will give us a more
reliable estimate of ε. Much progress on such calculations has been made by
Smarr et $al.$ (1976).

The amplitude $\hat{\psi}$ and pulse intensity $\mathfrak{F}_\nu$ can be calculated in terms of the
efficiency factor ε and other properties of the source by equating the energy
radiated, εMc^2, to the redshift-corrected energy in the wave today

$$\varepsilon Mc^2 = 4\pi r^2\, \mathfrak{F}_\nu \nu \quad \text{for noncosmological sources}$$

$$= (4\pi R^2\, \mathfrak{F}_\nu \nu)(1 + z) \quad \text{for cosmological sources.} \tag{7}$$

Here R is the "circumference function" of Friedmann cosmology, given by

$$\frac{H_o R}{c} = \frac{1 - q_o + q_o z - (1 - q_o)(2q_o z + 1)^{1/2}}{q_o^2(1 + z)}$$

$$= z/(1 + z) \quad \text{if } q_o = 1$$

$$= z(1 + \tfrac{1}{2}z)/(1 + z) \quad \text{if } q_o \ll 1 \quad \text{and} \quad z \ll q_o^{-1}$$

$$= q_o^{-1} \quad \text{if } q_o \ll 1 \quad \text{and} \quad z \gg q_o^{-1}; \tag{8}$$

cf. eq. (29.33) of MTW. The result for $\hat{\psi}$ and $\mathfrak{F}_\nu$, after combining with equations
(6) and (3) and imposing a Hubble constant of $H_o = 55$ km s^{-1} Mpc^{-1}, is

$$\hat{\psi} = (2\varepsilon)^{1/2}\frac{GM/c^2}{r}$$

$$= 2.1 \times 10^{-20}\left(\frac{\varepsilon}{0.1}\right)^{1/2}\frac{(M/M_\odot)}{(r/1\ \text{Mpc})} \quad \text{(noncosmological)}$$

$$= (2\varepsilon)^{1/2}\frac{GM/c^2}{R}$$

$$= 3.9 \times 10^{-24}\left(\frac{\varepsilon}{0.1}\right)^{1/2}\frac{M/M_\odot}{H_o R/c} \quad \text{(cosmological)}, \tag{9a}$$

$$\mathfrak{F}_\nu = \frac{3\varepsilon}{2} \frac{G}{c} \frac{M^2}{r^2}$$

$$= (1.4 \times 10^{-6} \text{ GPU}) \left(\frac{\varepsilon}{0.1}\right) \frac{(M/M_\odot)^2}{(r/1 \text{ Mpc})^2} \quad \text{(noncosmological)}$$

$$= \frac{3\varepsilon}{2} \frac{G}{c} \frac{M^2}{R^2}$$

$$= (5 \times 10^{-14} \text{ GPU}) \left(\frac{\varepsilon}{0.1}\right) \frac{(M/M_\odot)^2}{(H_o R/c)^2} \quad \text{(cosmological)} . \quad (9b)$$

There are a number of places in the universe where violent black-hole events are likely to occur. It seems to me that the four places most promising for gravitational-wave astronomy are pregalactic condensations, the nuclei of quasars and galaxies, the cores of globular clusters, and the deaths of normal stars in nearby galaxies.

ii) *Pregalactic Black-Hole Events*

Current theories of galaxy formation suggest that the first generation of objects to condense out of the primordial plasma may have had masses $M \approx 10^5$ to $10^6 M_\odot$, and may have condensed out at a redshift $z \approx 30$ to 100 when the age of the universe was ~ 10 million to 100 million years (Doroshkevich *et al.* 1967; Peebles and Dicke 1968). Some of these objects may have become globular clusters, while others may have become supermassive stars. A large fraction of the supermassive stars may have died rather quickly ($\lesssim 10^6$ years after birth) by gravitational collapse to form massive black holes (see Fowler 1966; Fricke 1973, 1974, and references therein). And many of the original condensations may have fragmented to form smaller objects ($M \sim 10$ to $10^5 M_\odot$) which subsequently collapsed to black-hole states (cf. Hartquist and Cameron 1977). It is not entirely unreasonable to think that in this way more than 50 percent of the mass of the universe went into holes with masses $M \lesssim 3 \times 10^5 M_\odot$ at a cosmological epoch $z \approx 30$ to 100. Paczyński (1975) has suggested that those holes today could reside in the outer regions of galaxies, where their collective effects would have the form of a massive, gravitating galactic halo. There is much astronomical evidence for the existence of such halos (Ostriker *et al.* 1974); and this pregalactic black-hole hypothesis is not an entirely unreasonable explanation.

If the above scenario is correct, then there were many gravitational-wave bursts produced by the births of black holes with $M \lesssim 3 \times 10^5 M_\odot$ at a redshift $z \sim 60$. The amplitude, intensity, time scale, and duration of each burst as it passes Earth would be

$$\hat{\psi} \sim 1 \times 10^{-18} \left(\frac{M}{3 \times 10^5 M_\odot}\right), \quad \mathfrak{F}_\nu \sim 4 \times 10^{-3} \text{ GPU} \left(\frac{M}{3 \times 10^5 M_\odot}\right)^2,$$

$$\tau \sim 5 \text{ minutes} \left(\frac{M}{3 \times 10^5 M_\odot}\right), \quad \hat{\tau} \sim 15 \text{ minutes} \left(\frac{M}{3 \times 10^5 M_\odot}\right) \quad (10a)$$

(cf. eqs. [9] and [6] where I assume $q_o \approx 1$). It is straightforward to show that the time interval between bursts arriving at Earth today, Δt, and the mean density of burst centers (pregalactic holes) in the universe today, n_o, are related by

$$\Delta t = \frac{1}{4\pi R^2 n_o c}. \tag{10b}$$

If 10 percent of the mass of the universe is in such holes, then

$$n_o \sim (10^{-30} \text{ g cm}^{-3})/M \sim 5 \times 10^4 \text{ Mpc}^{-3} (M/3 \times 10^5 M_\odot)^{-1} \tag{10c}$$

and (cf. eq. [8]) $R \sim 5000$ Mpc, so

$$\Delta t \sim 7 \text{ seconds } (M/3 \times 10^5 M_\odot). \tag{10d}$$

Thus, roughly 100 bursts ($= \hat{\tau}/\Delta t$) are passing Earth at once. Their amplitudes will add stochastically, producing a total amplitude and intensity

$$\hat{\psi}_{\text{total}} \approx (100)^{1/2}\hat{\psi} \sim 1 \times 10^{-17} (M/3 \times 10^5 M_\odot),$$

$$(\mathfrak{F}_\nu)_{\text{total during time } \tau} \approx 100 \, \mathfrak{F}_\nu \approx 0.4 \text{ GPU } (M/3 \times 10^5 M_\odot)^2, \tag{11}$$

which fluctuates on the same time scale, $\tau \sim (5 \text{ minutes}) (M/3 \times 10^5 M_\odot)$, as the individual sources.

The above estimates are made assuming $q_o \sim 1$. If, instead, $q_o \ll 1$, $\hat{\psi}$ may be smaller by an order of magnitude, Δt may be smaller by two orders of magnitude, and $\hat{\psi}_{\text{total}}$ will be roughly unchanged. On the other hand, if far less than 10 percent of the mass of the universe went into such holes, then $\hat{\psi}$ and $\hat{\psi}_{\text{total}}$ will be far less than the above; and Δt will be far larger.

Note added in proof: Since the above paragraphs were written, Ipser and Price (1977) have estimated the electromagnetic radiation produced by accretion today onto pregalactic black holes, as they plunge through the gas of our galaxy's disk. Ipser and Price predict very strong emission at far ultraviolet, visual, and far infrared wavelengths. By comparing their predictions with observations they conclude that "the number of holes with masses $3 \times 10^4 M_\odot \lesssim M \lesssim 3 \times 10^6 M_\odot$ is too small by a factor $\gtrsim 100$ to provide a galactic halo that would explain the suggested linear mass-radius relation for spiral galaxies, stabilize galactic disks, or close the universe." It is not yet clear whether similar limits can be placed on holes with masses $M < 3 \times 10^4 M_\odot$.

iii) *Black-Hole Events in Galactic Nuclei and Quasars*

When building models for the cores of quasars, of strong radio sources, and of active galactic nuclei, theorists typically conclude that, whatever may be in the core, it is likely to generate one or more supermassive black holes ($M \sim 10^6$ to $10^{10} M_\odot$) in a time short compared to the age of the universe. See, e.g., Lynden-Bell (1969), Wolfe and Burbidge (1970), Spitzer (1971), and Saslaw (1975).

A typical scenario involves the collapse of supermassive stars, gas clouds, or star clusters to form holes, and perhaps subsequent collisions between holes.

There are so few quasars and galaxies in the universe ($\sim 10^{10}$), and each one can produce so few strong gravitational-wave bursts (perhaps $\lesssim 10$), that to detect several bursts per year, one must search for them over the entire volume of the observable universe. Most such bursts will probably come from a redshift $z \approx 2.5$, when the universe was about 2 billion years old, since quasar activity seems to have peaked very sharply at that time. The characteristics of such bursts, according to equations (6), (8), and (9), are

$$\left.\begin{aligned}
\hat{\psi} &\approx 5 \times 10^{-18} \ (M/10^6 M_\odot) \\
\mathcal{F}_\nu &\approx 0.1 \ \mathrm{GPU} \ (M/10^6 M_\odot)^2 \\
\tau &\approx 50 \ \mathrm{s} \ (M/10^6 M_\odot) \\
\hat{\tau} &\approx 3 \ \mathrm{minutes} \ (M/10^6 M_\odot)
\end{aligned}\right\} M \approx 10^5 \ \text{to} \ 10^8 M_\odot \ \text{is reasonable} . \quad (12a)$$

Thorne and Braginsky (1976) have discussed such wave bursts in some detail. From observed data on quasars and galaxies, and from equation (10b), they have estimated that the mean time between arrivals of such bursts at Earth lies in the range

$$1 \ \text{week} < \Delta t < 300 \ \text{years} . \quad (12b)$$

The great uncertainty in Δt is a measure of our ignorance about quasars and galactic nuclei.

iv) *Black-Hole Events in Globular Clusters*

Many globular clusters in our own Galaxy have such short relaxation times ($< 10^9$ years) that it is reasonable to believe that all objects more massive than $\sim 2 \ M_\odot$ have sunk to the center by now (Spitzer and Shull 1975; Bahcall and Ostriker 1975). One does not know with any certainty how the growing core of such a cluster will evolve; but motivated by the discovery of X-ray sources in globular clusters, Bahcall and Ostriker (1975) and Silk and Arons (1975) have suggested that the evolution may produce one or more black holes of mass $M \sim 10^2$ to $10^4 \ M_\odot$. If such holes grow gradually, by swallowing gas, stars, and small-mass holes one after another, then the gravitational waves they produce will be relatively weak ($\varepsilon \approx 0.01 \ m/M$). But it is conceivable that sometimes two or more big holes may be created in one globular cluster, and may subsequently collide and coalesce, producing a large gravitational-wave burst ($\varepsilon \sim 0.1$). Moreover, dynamical friction may drag several globular clusters into the nucleus of the galaxy (Tremaine *et al.* 1975; Tremaine 1976), and their central holes may then collide and coalesce.

From a galaxy like ours, with ~ 300 globular clusters, it is not unreasonable to imagine ~ 300 black-hole events in the age of the universe—each event involving a mass $M \sim 100$ to $10^4 \, M_\odot$ and an efficiency factor $\varepsilon \sim 0.1$. (It is also not unreasonable to imagine less than one such event per galaxy.) In order to see one such event per month, it is then necessary to look out to a distance $r \approx 1000$ Mpc—i.e., one-third the distance to the edge of the universe. The wave bursts at this distance will have

$$\hat{\psi} \approx 2 \times 10^{-20} \, (M/10^3 \, M_\odot) \,,$$

$$\mathfrak{F}_\nu \approx 1 \times 10^{-6} \, \mathrm{GPU} \, (M/10^3 \, M_\odot)^2$$

$$\tau \approx 0.015 \, \mathrm{s} \, (M/10^3 \, M_\odot) \,,$$

$$\nu \approx 11 \, \mathrm{Hz} \, (M/10^3 \, M_\odot)^{-1} \,,$$

$$\hat{\tau} \approx 0.05 \, \mathrm{s} \, (M/10^3 \, M_\odot) \,. \tag{13}$$

v) *The Collapse of Normal Stars to Form Black Holes*

It is generally presumed that stars less massive than the Chandrasekhar (1931) limit terminate their normal evolution by contraction into a white-dwarf state; that stars between $M_\mathrm{Chandrasekhar} \approx 1.4 \, M_\odot$ and a limit $M_\mathrm{n-h} \sim 5$ to $20 \, M_\odot$ undergo core collapse to form a neutron star and simultaneously eject their envelopes in a supernova outburst; and that stars above the limit $M_\mathrm{n-h}$ collapse to form black holes. See, e.g., § 3 of Novikov and Thorne (1973). Of course, this is a great oversimplification: The ultimate fate of a star must depend not only on its mass, but also on its angular momentum and on the amount of mass loss during the late stages of its evolution. Also, there is no firm reason why core collapse to a black-hole state should *fail* to trigger a supernova outburst, nor is it obvious that collapse to a neutron star must *always* trigger a supernova.

The most one can claim with confidence is that the total number of collapses to neutron star or black hole exceeds the number of observed supernovae (~ 1 per 30 years in galaxies of our type; Tammann 1974), and is less than the number of star births with mass above $M_\mathrm{Chandrasekhar}$ (~ 1 per year in galaxies of our type; page 465 of Zel'dovich and Novikov 1971). Nowadays it is fashionable to believe in much mass loss in the late stages of stellar evolution, and to thence conclude that the number of collapses is near the number of supernovae.

The supernova rate as a function of distance from Earth has been computed by Talbot (1976) using Tammann's (1974) data. His results, as shown in Figure 2, require that an observer look out a distance of 10 Mpc (M101 group) if he wants one supernova per year. Beyond this distance the supernova rate is

$$\mathfrak{R}_\mathrm{SN}(\text{distance} < r) \approx (1 \text{ per year})(r/10 \text{ Mpc})^3 \,. \tag{14}$$

Equations (6) and (9) suggest the following characteristics for black-hole events involving $M \sim 2$ to $10\ M_\odot$ at these distances:

$$\nu \approx (5000\ \mathrm{Hz})(M/2\ M_\odot)^{-1},$$

$$\hat{\psi} \approx (4 \times 10^{-21})(M/2\ M_\odot)(r/10\ \mathrm{Mpc})^{-1},$$

$$\mathcal{F}_\nu \approx (0.5 \times 10^{-7}\ \mathrm{GPU})(M/2\ M_\odot)^2(r/10\ \mathrm{Mpc})^{-2}. \qquad (15)$$

vi) *The Collapse of Normal Stars to Form Neutron Stars*

The gravitational waves from collapse to form a neutron star are likely to resemble those from collapse to black hole: an initial burst (produced by mass infall) followed by ringing (produced by neutron-star pulsations and/or rapid, nonaxisymmetric rotation; cf. §§ IIc[iii] and [iv]). However, because a neutron star is larger than a hole for given mass, its waves will have lower frequency, its emission efficiency will be lower, the damping time for the ringing will be longer —and, consequently, most of the energy might be carried off in the ringing rather than in the initial burst.

Misner (1974) suggests for pulsation-produced ringing a total energy output of $\sim 0.05\ M_\odot c^2$, most of it within a bandwidth $\Delta\nu \sim 100$ Hz around the lowest

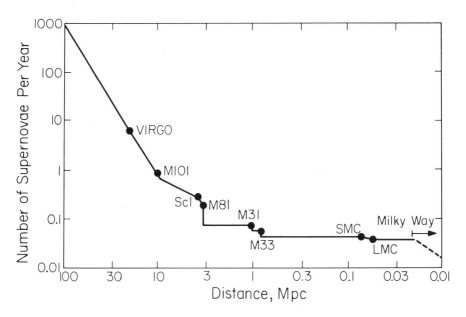

Figure 2. Number of supernovae per year out to a given distance from Earth, as estimated by Talbot (1976) using supernova rates per galaxy due to Tammann (1974). The distances corresponding to several well-known galaxies and clusters of galaxies are indicated by large dots.

quadrupole eigenfrequency of the star $\nu \approx 1000$ to 3000 Hz, and with e-folding time for the ringing $\hat{\tau} \approx 0.1$ to 1 s (cf. Table 2, below). The corresponding wave amplitude and burst intensity are

$$\hat{\psi} \approx (1 \times 10^{-22}) \left(\frac{\varepsilon M c^2}{0.05\, M_\odot c^2}\right)^{1/2} \left(\frac{\nu}{2000\ \text{Hz}}\right)^{-1} \left(\frac{\hat{\tau}}{0.3\ \text{s}}\right)^{-1/2} \left(\frac{r}{10\ \text{Mpc}}\right)^{-1},$$

$$\mathfrak{F}_\nu \approx (0.8 \times 10^{-6}\ \text{GPU}) \left(\frac{\varepsilon M c^2}{0.05\, M_\odot c^2}\right) \left(\frac{\Delta \nu}{100\ \text{Hz}}\right)^{-1} \left(\frac{r}{10\ \text{Mpc}}\right)^{-2}. \qquad (16a)$$

For ringing due to rotation the waves should be roughly similar (Ruffini and Wheeler 1971, p. 146). It is conceivable that a large amount of energy (~ 0.01 $M_\odot c^2$) might come off in the initial broad-band burst. Such a burst would have

$$\nu \approx 1000\ \text{Hz},$$

$$\hat{\psi} \approx (2 \times 10^{-21}) \left(\frac{\varepsilon M c^2}{0.01\, M_\odot c^2}\right)^{1/2} \left(\frac{r}{10\ \text{Mpc}}\right)^{-1},$$

$$\mathfrak{F}_\nu \approx (2 \times 10^{-8}\ \text{GPU}) \left(\frac{\varepsilon M c^2}{0.01\, M_\odot c^2}\right) \left(\frac{r}{10\ \text{Mpc}}\right)^{-2}. \qquad (16b)$$

vii) *Corequakes in Neutron Stars*

Neutron stars more massive than $\sim 0.7\ M_\odot$ are suspected to form crystalline cores with crystal structure governed by nuclear forces rather than electrostatic forces (Anderson and Palmer 1971; Clark and Chao 1972; Canuto and Chitre 1974). Pines *et al.* (1972) have suggested that deformations $\varepsilon_0 \sim 0.01$ could freeze into the core of such a neutron star when it is rapidly rotating, and be released in a series of subsequent corequakes occurring once every few years. They suggest a strain release $\Delta\varepsilon \sim 10^{-6}$ in each corequake so that the total number of quakes over the lifetime of the star is $\sim 10^4$. If the rate of birth of neutron stars more massive than $0.7\ M_\odot$ is the same as the supernova rate (1 per 30 years in our Galaxy), there would then be ~ 300 corequakes in our Galaxy each year, and ~ 1 per month within a distance of ~ 3 kpc. If the recently measured neutron-star masses ($1.3\ M_\odot$ and $1.6\ M_\odot$; Rappaport *et al.* 1976; Middleditch and Nelson 1976) are a valid indicator, then neutron stars tend to be massive; and the energy release in each corequake will be

$$\Delta E \approx (1 \times 10^{53}\ \text{ergs})\varepsilon_0 \Delta\varepsilon \approx 1 \times 10^{45}\ \text{ergs} \qquad (17a)$$

(Pines *et al.* 1972). Dyson (1972) has suggested that a large fraction of this energy may be deposited in torsional oscillations of the neutron star, and thence be radiated as "current-quadrupole" gravitational waves. Such waves would have frequencies and damping times

$$\nu \approx 3000\ \text{Hz}, \qquad \hat{\tau} \approx 1\ \text{s} \qquad (17b)$$

(Thorne and Żytkow 1977), and would be very monochromatic ($\Delta \nu \ll \nu$). The burst intensity and wave amplitude at Earth would be

$$\mathcal{F}_\nu \approx (0.8 \times 10^{-6} \text{ GPU}) \left(\frac{\Delta E}{10^{45} \text{ ergs}}\right) \left(\frac{\Delta \nu}{10 \text{ Hz}}\right)^{-1} \left(\frac{r}{3 \text{ kpc}}\right)^{-2},$$

$$\hat{\psi} \approx (1 \times 10^{-23}) \left(\frac{\Delta E}{10^{45} \text{ ergs}}\right)^{1/2} \left(\frac{\nu}{3000 \text{ Hz}}\right)^{-1} \left(\frac{\hat{\tau}}{1 \text{ s}}\right)^{-1/2} \left(\frac{r}{3 \text{ kpc}}\right)^{-1}. \quad (17c)$$

Note that these waves from nearby neutron-star corequakes are very similar to waves from neutron-star births at the distance of M101 (eqs. [16]).

viii) *Summary of Strongest Waves Bathing Earth*

Figure 3 summarizes the above guesses as to the strongest gravitational waves that bathe the Earth. Each entry in that figure is perfectly reasonable in the light of current astrophysical knowledge and prejudices. However, none of the entries is particularly compelling, except (to within an order of magnitude in $\mathcal{F}_\nu$) the neutron-star-birth entries.

b) Remarks on Gravitational-Wave Detection

Of the many conceivable designs for gravitational-wave antennas (see, e.g., Press and Thorne 1972), three are now being pursued seriously: Weber-type resonant bars, Michelson interferometers using laser beams that bounce off pendulum-suspended mirrors, and Doppler tracking of interplanetary space-craft. In this lecture I shall describe briefly the state of the art and the ultimate goals of each of these systems. For more detailed reviews see, e.g., the lectures by Weber, by Braginsky, by Fairbank, by Hamilton, by Kafka, and by Tyson in various conference proceedings including Bertotti (1974), DeWitt-Morette (1974), and Weber (1977); also the Weber-Drever-Kafka-Tyson panel discussion in Rosen and Shaviv (1975); also a book by Braginsky and Manukin (1974).

i) *Weber-Type Detectors*

By now it appears to be rather well established that frequent bursts of gravitational waves have *not* been detected using Weber-type resonant bars—though Drever *et al.* (1973) have detected one impressive event, and the origin of Weber's (1969, 1970) coincidences is not yet adequately understood; see Lee *et al.* (1976).

In view of the failure of other groups to reproduce his coincidences, in March 1975 Joseph Weber convened a small meeting of gravitational-wave experimenters from all over the world in Erice, Sicily; see Weber (1977). At that meeting the future of gravitational-wave astronomy was discussed in depth. The experimenters (and a few kibitzing theorists including me) emerged from the meeting agreeing on the following: (i) The most promising detectors in the kilohertz band for the next few years are antennas of the type Weber (1960, 1961) pioneered: massive, high-Q bars with sensors to measure the amplitude and phase of the bar oscillations. (ii) Enormous improvements in detector sensitivity are possible. (iii) By very hard work one can hope to build detectors sensitive enough for detecting strong-wave bursts from nearby groups of galaxies. The required improvement over 1975 detectors is at least 3 orders of magnitude in amplitude sensitivity.

I shall describe briefly the noise problems faced by a Weber-type detector, and the steps that are being taken to surmount them. My description of the noise is a much abbreviated version of the analysis by Braginsky (1975)—which in turn is an adaptation of previous analyses by Braginsky (1970, 1974a, b) and by Braginsky and Manukin (1974).

Consider a normal-mode oscillation of a bar ("antenna"), with angular eigenfrequency ω_m and dimensionless eigenfunction $\xi(x)$ normalized so that

$$\int |\xi|^2 d \text{ volume} = \text{(volume of antenna)} . \tag{18a}$$

(The subscript m means "mass" and refers always to properties of the bar's normal-mode oscillation.) When the bar is vibrating in this normal mode, its internal displacements are

$$\delta x = \text{Real} [X \exp(-i\omega_m t) \xi(x)] , \tag{18b}$$

where X is the complex amplitude of excitation. Because of the normalization (18a), the absolute value of X is the volume-averaged amplitude of internal motion of the bar. Actually, X is not constant. It changes with time due to external forces acting on the bar (gravitational waves, and the back-action force of the sensor) and also due to internal frictional forces by which this normal mode couples to the bar's other normal modes ("Nyquist forces"; "Brownian-motion forces").

The sensor measures the average value of $X(t)$ during some averaging time τ_a—which we assume is longer than typical gravitational-wave bursts $\hat{\tau}$, but shorter than the damping time of the normal-mode amplitude $\tau_d = 2Q_m/\omega_m$.

If a gravitational-wave burst arrives between measurements of X, it produces a change of X, from one measurement to the next, of magnitude

$$|\delta X_{\text{GW}}| = [(\tfrac{1}{2}M_m\omega_m^2)^{-1}(\int \sigma d\nu)\mathcal{F}_\nu(\nu = \omega_m/2\pi)]^{1/2} \tag{19a}$$

Here M_m is the mass of the bar, $\mathcal{F}_\nu(\nu = \omega_m/2\pi)$ is the burst intensity at the bar's eigenfrequency, and $\int \sigma d\nu$ is the bar's cross section for gravitational waves, integrated in frequency over the resonance associated with its normal

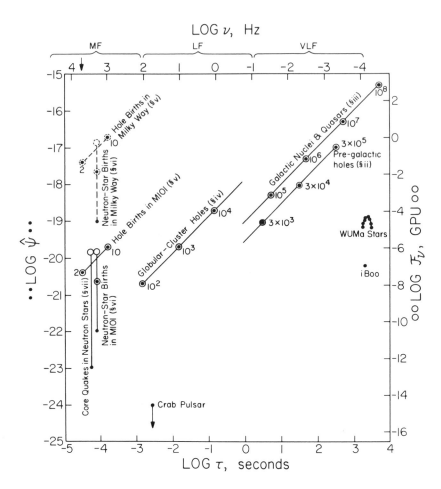

Figure 3. Summary of half-educated guess as to the strongest gravitational waves bathing Earth. Plotted horizontally at the top is frequency ν; and at the bottom, the time scale $\tau = 1/2\pi\nu$ on which the waves change. The frequency scale is divided into a very-low-frequency band (VLF), a low-frequency band (LF), and a medium-frequency band (MF); cf. Table 1 of Press and Thorne (1972). Plotted vertically on the left is the dimensionless amplitude of the waves (eq. [1d']); and on the right, the burst intensity (eq. [1e]). Inside the figure, dots refer to amplitude (*left scale*); open circles, to intensity (*right scale*). Dots surrounded by circles indicate broad-band bursts (*both scales*) with a frequency spread roughly from 0.5 ν to 1.5 ν, and with duration $\hat{\tau} \sim 1/2\nu$ (cf. § i). A dot standing alone refers to everlasting monochromatic waves (*left scale*). A dot connected to an open circle by a vertical line refers to a damped, ringing wave with amplitude as indicated by the dot and total intensity as indicated by the open circle. The distances to all transient sources are chosen such that one can reasonably hope for several events per year—except in the case of hole births and neutron-star births in the Milky Way (*upper left-hand corner*), which are shown dashed because they presumably occur somewhat less than once per year.

All of the transient sources are discussed in the text in the subsections (§ ii to § vii) mode:

$$\int \sigma d\nu = \begin{pmatrix} \text{factor of order unity that depends} \\ \text{on shape of bar and normal mode} \end{pmatrix} \frac{1}{10} \frac{G}{c^3} M_m \omega_m^2 \left(\frac{L}{n+1}\right)^2$$

$$= \begin{pmatrix} \text{factor of} \\ \text{order unity} \end{pmatrix} \frac{G}{c^3} M_m v_s^2$$

if bar is excited in a mode with $n = 0$, $\pi v_s/\omega_m \approx L$. (19b)

See, e.g., equation (8b) of Box 37.4 and equation (37.45) of the text of MTW. Here L is the length of the bar, v_s is the speed of sound in the bar, and n is the number of nodes in the eigenfunction along the length of the bar. By combining equations (19a, b) we see that

$$|\delta X_{\text{GW}}| = \alpha \frac{L}{n+1} \left[\frac{G}{5c^3} \mathcal{F}_\nu \left(\nu = \frac{\omega_m}{2\pi}\right) \right]^{1/2}$$

$$= \left(\alpha \frac{L}{n+1}\right) (7 \times 10^{-18}) \left[\frac{\mathcal{F}_\nu(\nu = \omega_m/2\pi)}{1 \text{ GPU}} \right]^{1/2}$$

$$= \alpha' \left(\frac{v_s}{\omega_m}\right) \left[\frac{G}{c^3} \mathcal{F}_\nu \left(\nu = \frac{\omega_m}{2\pi}\right) \right]^{1/2} \quad \text{if} \quad n = 0, \quad \pi v_s/\omega_m \approx L$$

$$= (7 \times 10^{-16} \text{ cm}) \alpha' \left(\frac{v_s}{3 \text{ km s}^{-1}}\right) \left(\frac{2\pi \times 10^3 \text{ Hz}}{\omega_m}\right) \left[\frac{\mathcal{F}_\nu(\nu = \omega_m/2\pi)}{1 \text{ GPU}} \right]^{1/2},$$

(20)

where α and α' are geometrically determined factors of order unity. Currently operating detectors can detect bursts of intensity $\sim$10 GPU, corresponding to wave-induced displacements of $\sim$2 $\times$ 10^{-15} cm (Kafka 1975; Douglass et al. 1975). An improvement of only 3 to 10 in amplitude sensitivity [$|\delta X| \sim$ (6 to 2) $\times$ 10^{-16} cm; $\mathcal{F}_\nu \sim$ 0.1 to 1 GPU] may be adequate to permit detection of wave bursts from the births of black holes and neutron stars throughout our

---

indicated on the figure. In the case of black-hole events, the mass of the hole involved is indicated beside the circled dot. The indicated properties of the transient events are surely no more accurate than a factor $\sim$3 in ν, $\sim$3 in ψ, and $\sim$10 in $\mathcal{F}_\nu$.

The monochromatic sources (dots not associated with circles; left-hand scale applicable but not right-hand) are not discussed in the text. They include the most favorable known binary star, i Boo (Braginsky 1965); the total gravitational-wave emission from all W Ursae Majoris binary-star systems in the Milky Way Galaxy, as estimated by Mironovskii (1965); and the Crab pulsar. The dot for the Crab pulsar is an upper limit based on the demand that the slowdown rate due to gravitational-radiation reaction not exceed the observed slowdown rate. The actual amplitude of the waves from the Crab could be as large as this limit if the core of the neutron star is crystalline and has a nonaxial ellipticity $\epsilon \sim 10^{-3}$ (cf. eq. [21] of Press and Thorne 1972; also Pines, Shaham, and Ruderman 1972). If the nonaxial deformation is primarily associated with the star's mantle, then $\epsilon \lesssim 10^{-5}$ to 10^{-7} is more likely and the wave amplitude should be $< 10^{-27}$ (cf. Press and Thorne 1972).

Galaxy—bursts that presumably come much less often than once per year. (See Fig. 3.) To detect the several-per-year bursts expected from corequakes in neutron stars at distances $\sim$3 kpc, and from the births of neutron stars and black holes at the distance $\sim$10 Mpc of the M101 group, will require a 3000- to 10,000-fold improvement in amplitude sensitivity [$|\delta X| \sim$ (6 to 2) $\times$ 10^{-19} cm, $\mathcal{F}_\nu \sim (10^{-6}$ to $10^{-7})$ GPU]. This 3000- to 10,000-fold improvement marks the threshold for viable gravitational-wave astronomy, if the estimates of § IIa make any sense. Thereafter the number of observable stellar-collapse sources should increase as

Number $\propto$ (volume accessible)

$$\propto \text{(radius accessible)}^3 \propto \text{(amplitude sensitivity)}^{-3};$$

see Figure 2. A factor 3 increase in sensitivity should improve the event rate by a factor 27.

To be measurable, the wave-induced amplitude change in the antenna's oscillations (eq. [20]) must exceed the expected change in oscillation amplitude, $\delta X_{\text{antenna}}$, due to the antenna's internal frictional forces ("Nyquist forces"). This "antenna noise" is

$$|\delta X_{\text{antenna}}| \approx \left(\begin{array}{c}\text{typical Brownian-}\\\text{motion amplitude}\end{array}\right) \times \left(\frac{\text{averaging time}}{\text{damping time}}\right)^{1/2}$$

$$= \left[\frac{2kT_m}{M_m\omega_m{}^2}\frac{\omega_m\tau_a}{2Q_m}\right]^{1/2}. \tag{21}$$

Here M_m, T_m, and Q_m are the antenna mass, temperature, and Q factor.

Any sensor for monitoring the antenna amplitude $X(t)$ can make a more accurate measurement if one gives it a longer time τ_a over which to measure.[2] This is because it can then work in a narrower bandwidth $\Delta\omega \approx 1/\tau_a$. The energy noise it faces is proportional to $\Delta\omega$, so the amplitude noise is proportional to $(\Delta\omega)^{1/2}$:

$$|\delta X_{\text{sensor}}| = S(\Delta\omega)^{1/2} = S/\tau_a{}^{1/2}. \tag{22}$$

Here $|\delta X_{\text{sensor}}|$ is the noise which the sensor imposes on the measurement of δX; and S—the "sensitivity" of the sensor—is a constant depending on the sensor's design and temperature.

The best sensors are probably parametric-amplifier devices. An example, being developed by Braginsky (1975), is a microwave cavity with one wall a face of the antenna, so that the cavity shape and eigenfrequency are modulated

---

[2] Here and below I assume that the measurement bandwidth $\Delta\omega = 1/\tau_a$ is small compared to the natural bandwidth of the sensor—which, in the case of the microwave cavity of Figure 4, is $\Delta\omega_{\text{sensor}} \equiv 1/(\text{damping time for electromagnetic oscillations of the cavity})$. I also assume that $\Delta\omega_{\text{sensor}} < \omega_m$. When $\Delta\omega > \Delta\omega_{\text{sensor}}$ and/or $\omega_m < \Delta\omega_{\text{sensor}}$, the details of the discussion are different, but its qualitative features are the same.

by the antenna vibrations. See Figure 4. Such a cavity might be excited near its eigenfrequency $\omega_e \approx 2\pi \times 10^{10}$ Hz by an external driving voltage of constant amplitude; and the antenna vibrations, by modulating the cavity shape at frequency $\omega_m \approx 2\pi \times 10^3$ Hz, would modulate the response of the cavity to the driving voltage. Some sort of sophisticated voltmeter would monitor the cavity response.

Such a cavity (or any other parametric-amplifier sensor) will have randomly fluctuating excitations at frequency ω_e due to its finite temperature. These noise excitations are superposed on the cavity's driven oscillations; and they give rise

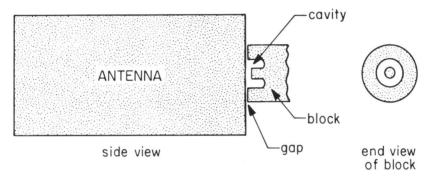

Figure 4. Schematic diagram of the microwave cavity which Braginsky (1975) proposes to use as a sensor for the oscillations of a Weber-type antenna. One wall of the cavity is the flat end face of the oscillating antenna. The other walls are machined out of a block of adjacent material. There is no physical contact between the antenna face and the machined block. By appropriate shaping of the walls one can focus the field in the cavity so that leakage of energy through the block-antenna gap is negligible.

to the $|\delta X_{\text{sensor}}|$ of equation (22). The larger the driven oscillations, the smaller will be the sensor noise

$$|\delta X_{\text{sensor}}| = \text{const.} \times N_{\text{sensor}}^{-1/2}\tau_a^{-1/2}. \qquad (22')$$

Here N_{sensor} is the total number of quanta in the driven oscillations of the cavity; and the constant depends on the sensor design and temperature, but not on N_{sensor}, τ_a, or the antenna properties (except its eigenfrequency ω_m).

The thermal oscillations of the cavity not only give rise to the sensor noise (22'); they also combine with the driven oscillations of the cavity to produce a randomly fluctuating back-action force on the antenna, at the antenna's eigenfrequency ω_m (Braginsky 1970, 1975; Braginsky and Manukin 1974). More specifically: the driving voltage which has frequency $\omega_e - \omega_m$, because it acts on a cavity with walls that pulsate at frequency ω_m, produces a driven voltage inside the cavity with frequencies ω_e and $\omega_e - \omega_m$. (The superposition of these frequencies gives a beating which is the response of the cavity to the antenna oscillations.) The driven voltage at the frequency $\omega_e - \omega_m$ superposes itself on the cavity's noise voltage, with frequency ω_e, to produce a force on the cavity walls at the frequency ω_m. Since one of the cavity walls is a face of the antenna (Fig. 4), the antenna experiences a back-action force from the sensor at its own eigenfrequency ω_m with a randomly fluctuating, unpredictable amplitude. This force produces antenna amplitude changes, during time τ_a, of magnitude

$$|\delta X_{\text{back action}}| = \text{const.} \times N_{\text{sensor}}^{1/2}\tau_a^{1/2} . \tag{23}$$

The constant depends on the construction of the sensor, but not on N_{sensor}, τ_a, or the properties of the antenna (except ω_m).

One important source of noise remains: the "voltmeter" which monitors the excitations of the cavity. We shall return to it in a moment. But first let us examine the dependence of the above noise sources on the measurement time τ_a (Fig. 5).

Detectors currently in operation have the relative noise magnitudes indicated in Figure 5a. The optimal averaging time τ_a (typically $\sim$0.1 s) is determined by competition between sensor noise and antenna noise. However, a revolution in antenna construction is now under way. This revolution includes lowering the antenna temperature by as much as five orders of magnitude (Fairbank *et al.*

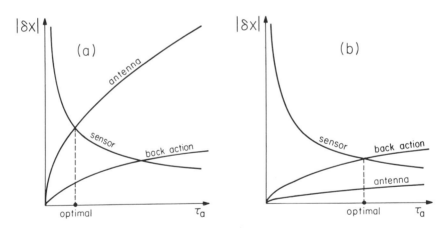

Figure 5. The relative magnitudes of three noise sources in Weber-type detectors, as functions of the averaging time τ_a over which one makes a single measurement of the bar's amplitude X. The three noise sources are $|\delta X_{\text{antenna}}| \propto \tau_a^{1/2}$ (eq. [21]), $|\delta X_{\text{sensor}}| \propto \tau_a^{-1/2}$ (eqs. [22] and [22']), and $|\delta X_{\text{back action}}| \propto \tau_a^{1/2}$ (eq. [23]). In currently operating detectors (Fig. 5a) back-action noise is totally negligible compared to sensor noise and antenna noise, so the optimal averaging time is determined by a competition between sensor and antenna. After the revolution in antenna technology, antenna noise will be reduced so much (Fig. 5b) that the experimenter will be faced by a competition between sensor noise and back-action noise.

1974; Boughn *et al.* 1974), and replacing the old aluminum antennas ($M_m \approx$ several $\times 10^3$ kg, $Q \approx 10^5$) by antennas made from materials with higher Q's—e.g., niobium with $M_m \approx 10^3$ kg (Hamilton 1975) which may achieve $Q \approx 10^9$ at millidegree temperatures; and monocrystal sapphire with $M_m \approx 10$ to 100 kg (Braginsky 1974*b*, 1975; Douglass 1976), which should achieve $Q \approx 10^{11}$ to 10^{14} at millidegree temperatures. The technology of this revolution is highly nontrivial. However, the experimenters expect with some confidence to succeed in lowering $|\delta X_{\text{antenna}}|$ (eq. [21]) by factors $\geq 10^3$ within the next few years; and an ultimate improvement of $\sim 10^6$ looks feasible. If this were matched by comparable improvements in the sensor and back-action noises, one could expect to detect bursts from collapsing stars in the M101 group in a few years, and ultimately bursts from stellar collapses at the Hubble distance.

Unfortunately, sensor technology does not look as promising as antenna technology. Given the above antenna improvements, one faces a competition between sensor noise and back-action noise (Fig. 5*b*). In that competition one can adjust the measurement time τ_a and also the excitation level of the sensor, N_{sensor} (eqs. [22'] and [23]). The optimal point of operation is where

$$\tau_a N_{\text{sensor}} = (\text{value which makes } |\delta X_{\text{back action}}| = |\delta X_{\text{sensor}}|) \ .$$

When one takes account of the facts that (i) τ_a must exceed the duration $\hat{\tau}$ of the wave burst, and (ii) existing techniques for probing a cavity cannot measure the number of electromagnetic quanta N_e at a given frequency more accurately than $\sqrt{N_e}$, then one finds the following remarkable result: Even with optimal operation and design, the sensor noise plus back-action noise produce uncertainty in the measurement of the antenna's oscillations at the level

$$\delta N_m \geq \sqrt{N_m} \ ; \quad N_m \equiv \left(\begin{array}{c} \text{number of quanta} \\ \text{of mechanical oscillation} \end{array} \right) = \left(\frac{M \omega_m^2 |X|^2}{2\hbar\omega_m} \right)^{1/2} . \quad (24)$$

This corresponds to an uncertainty in the amplitude of the bar of

$$|\delta X_{\text{sensor}}| + |\delta X_{\text{back action}}| \geq \left(\frac{2\hbar\omega_m}{M\omega_m^2} \right)^{1/2}$$

$$\approx (1 \times 10^{-18} \text{ cm})(M/100 \text{ kg})^{-1/2}(\omega_m/2\pi \times 5000 \text{ Hz})^{1/2} . \quad (24')$$

This uncertainty is roughly a factor 3 too great to permit detection of the predicted gravitational waves from the M101 cluster of galaxies.

This fundamental limit on gravitational-wave detectors was first derived by Braginsky (1970) for the case where the sensor is a Fabry-Perot resonator rather than a microwave cavity; see his equation (3.17). Later Braginsky and colleagues showed that the same limit holds for other specific types of sensors, such as microwave cavities; and recently Giffard (1976) has given a very general and elegant derivation for the case of any sensor which functions as a linear amplifier. The grave problems for gravity-wave detection posed by the limit (24′) have been pointed out and emphasized with vigor by Braginsky (1975) and by Giffard (1976).

Fortunately, the limit (24′) is not absolute. It applies to all sensors available today; but if one could measure the number of electromagnetic quanta N_e in a cavity or other oscillator more accurately than $\sqrt{N_e}$, then one would be able to beat the limit (24′). The ultimate in sensitivity would occur if one could measure the quantum state N_e of an oscillator with perfect precision. Braginsky has called such a measurement "quantum non-demolition" because it must not perturb or "demolish" the oscillator's quantum state in the process of measurement—or, if it *does* perturb the state, it must (i) give precise information about the original state, and (ii) leave the oscillator in a new, completely known state with which to compare the result of a subsequent measurement.

If one could perform a quantum non-demolition measurement, then the only fundamental limit on measurements of changes of antenna amplitudes would be

$$|\delta X| = \left(\begin{array}{c} \text{amplitude change associated} \\ \text{with a unit change in the number} \\ \text{of quanta in the antenna} \end{array} \right) = \frac{1}{\sqrt{N_m}} \left(\frac{2\hbar\omega_m}{M\omega_m{}^2} \right)^{1/2}$$

$$= \left(\frac{\hbar\omega_m}{kT_0} \right)^{1/2} \left(\frac{2\hbar\omega_m}{M\omega_m{}^2} \right)^{1/2}$$

$$\approx (1 \times 10^{-20}\ \text{cm}) \left(\frac{M}{100\ \text{kg}} \right)^{-1/2} \left(\frac{T_0}{10^{-3}\ \text{K}} \right)^{-1/2}. \tag{25}$$

Here $kT_0 \equiv \frac{1}{2}M\omega_m{}^2|X|^2$ is the energy of oscillation of the antenna before the gravitational wave hits it. Clearly a quantum non-demolition sensor would be a fantastic boon to gravitational-wave astronomy!

Braginsky and his research group have proposed a specific design for a quantum non-demolition sensor using electron-microscope–type technology. (See Braginsky and Vorontsov 1974; Braginsky, Vorontsov, and Krivchenkov 1975.) However, very recently a fatal flaw was discovered in the design of that sensor (Unruh 1977; Braginsky, private communication)—a flaw which makes it "demolition" ($\delta N_e \approx \sqrt{N_e}$) rather than non-demolition. Braginsky reports, on the other hand, that he has found a new non-demolition design; and I think it likely that other non-demolition sensors will be devised in the future.

Because of these sensor difficulties, I would be surprised if Weber-type detectors reached the sensitivities of strong-wave bursts from the M101 group (Fig. 3) before 1980. On the other hand, since all the foreseeable problems look surmountable, it seems likely that within a decade Weber-type detectors will be reaching beyond the M101 group and will be teaching us much about the guts of supernovae and other forms of stellar collapse.

ii) *Broad-Band, Free-Mass Detectors*

Weber-type bars, instrumented as described above, are narrow-band detectors. They can measure the intensity of a burst at discrete points on its spectrum; but they cannot study the details of the burst's waveform. A modified, wide-band version of a Weber-type bar, capable of studying waveforms, has been built and operated by Drever *et al.* (1973). However, a totally different type of broad-band detector may prove superior. This is the free-mass detector. (For the theory of free-mass detectors see, e.g., Press and Thorne 1972.)

Laboratory prototypes of the free-mass detector were constructed in 1971 by Moss *et al.* (1971), and are now being constructed at MIT by Ray Weiss, in Glasgow, Scotland by Ronald Drever and his colleagues, and in Munich, Germany by H. Billing, W. Winkler, and their colleagues. These devices consist of a multipass Michelson interferometer in which laser light bounces back and forth between mirrors that are suspended as pendula from an overhead support (Fig. 6). The pendulum eigenfrequencies are far below the characteristic frequencies of the searched-for waves; so under the action of the waves, the mirrors move like free masses. The wave-induced mirror motions produce time-changing lengths in the two arms of the interferometer, and thence produce a changing interference pattern. With the prototype antenna now under construction (arm lengths of 1 meter; 100 passes for the laser beam in each arm) Weiss expects a laser-noise-limited sensitivity of $\hat{\psi} \sim 10^{-17} \, (\Delta\nu/\mathrm{Hz})^{1/2}$ at frequencies $\nu \sim 30$ to 2000 Hz, where $\Delta\nu$ is the bandwidth. With this sensitivity Weiss could detect bursts of intensity $\mathcal{F}_\nu \approx 1$ GPU, which is better than current Weber-type bars but not so good as the bars now under construction in Moscow, Stanford, Rome, and Louisiana State University.[3] Future generations of such antennas may be superior to future bars at low frequencies ($\nu \lesssim 100$ Hz), but perhaps not in the middle-frequency band ($\nu = 100$ to 10^5 Hz).

In the very-low-frequency band ($\nu \approx 0.1$ to 10^{-5} Hz), where supermassive holes should radiate, the most promising detector is a free-mass system involving the Earth and one or more interplanetary spacecraft. The sensor that monitors the Earth-spacecraft relative motion—and searches for gravitational-wave effects therein—will be the Doppler tracking signal which is routinely used in the navigation of unmanned spacecraft. Such a detection system was first discussed in detail by Davies (1974); its response to gravitational waves has been analyzed by Estabrook and Wahlquist (1975); and its use to search for gravitational waves from massive black-hole events has been proposed by Thorne and Braginsky (1976). Early attempts to search for gravitational waves in this way were made by A. J. Anderson (1971). Detailed feasibility studies and

---

[3] I thank Professor Weiss for permission to give this brief description of his experiment in advance of any publication.

noise studies for future attempts are now beginning at the Caltech Jet Propulsion Laboratory.

One can understand the operation of this detector as follows: A tracking station on Earth transmits monochromatic electromagnetic waves with their frequency locked onto a highly stable clock (master oscillator). The spacecraft receives the waves, amplifies them, and transmits them back to Earth (i.e., it "transponds" them). The tracking antenna at Earth receives the transponded waves, compares them with the transmitted waves by counting the number of cycles and fractions thereof during an averaging time τ_a, and derives a frequency shift from the comparison. (For further details see, e.g., J. D. Anderson 1974.) The measured frequency shift is due primarily to the velocity of the spacecraft relative to Earth (Doppler shift), but it also contains contributions

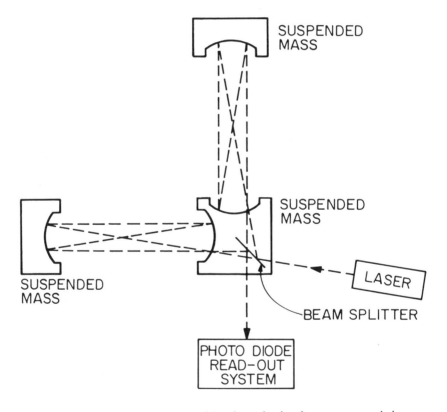

Figure 6. Schematic diagram of a broad-band gravitational-wave antenna being constructed independently by Ray Weiss at MIT, Ronald Drever *et al.* in Glasgow, and H. Billing *et al.* in Munich. The antenna is, essentially, a multipass Michelson interferometer with arm lengths that are perturbed by passing gravitational waves.

("noise") from frequency fluctuations and drift of the master oscillator, from time-varying dispersion of the electromagnetic waves in the Earth's ionosphere and troposphere and in the interplanetary medium, and from passing gravitational waves.

The gravitational-wave contribution as computed by Estabrook and Wahlquist (1975) can be described as follows (Fig. 7). Orient the spatial axes of a Euclidean coordinate system so the gravitational waves travel in the z direction, and the Earth-spacecraft line of sight has length l and lies in the x-z plane at an angle θ relative to the z axis. Then the component of the gravitational waves that influences the Doppler signal is $\psi(t - z) \equiv h_{xx}^{TT}(t - z)$. The influence of ψ on the Doppler signal is given by

$$\frac{\Delta \nu_e}{\nu_e} = -\left(\frac{1 - \cos \theta}{2}\right) \psi_R - \cos \theta \psi_T + \left(\frac{1 + \cos \theta}{2}\right) \psi_E \qquad (26a)$$

(Estabrook and Wahlquist 1975). Here $\Delta \nu_e / \nu_e$ is the frequency shift of a specific short piece of the electromagnetic wave train—a piece received at Earth at time t_R; and ψ_E, ψ_T, ψ_R are the values of ψ encountered by that same piece of wave

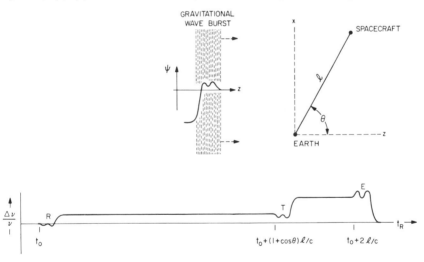

Figure 7. The response of an Earth-spacecraft Doppler signal to a passing gravitational-wave burst, as computed by Estabrook and Wahlquist (1975). *Top*, the gravitational-wave burst encounters the Earth-spacecraft system. The waveform $h_{xx}^{TT} = \psi(t - z)$ is plotted as a function of z at fixed time t. *Bottom*, the Doppler shift $\Delta \nu / \nu$ of the radio signal received from the spacecraft at time t_R shows the influence of the gravitational waves. Here t_0 is the time of arrival of the leading edge of the gravitational waveform at Earth. One can think of the bumps "R" as due to interaction of the gravitational wave with the receiver; the bumps "T," to interaction with the transponder; and the bumps "E," to interaction with the emitter of the radio waves.

train at its moments of emission, transponding, and reception:

$$\psi_E \equiv \psi(t_R - 2l/c) , \qquad \psi_T \equiv \psi(t_R - l[1 + \cos\theta]/c) , \qquad \psi_R \equiv \psi(t_R) . \quad (26b)$$

Two features of equation (24) are important: *First*, the magnitude of the gravitational-wave–induced frequency shift is $\Delta\nu_e/\nu_e \approx \psi$; second (as Estabrook and Wahlquist emphasize), the gravitational waveform is repeated three times in the Doppler signal, with relative amplitudes and time separations that depend on only one unknown parameter: the angle θ. This second feature may allow one to extract gravitational-wave effects in the presence of much larger Doppler noise, and may also allow one to determine the angle θ. By simultaneous tracking of several spacecraft one could dig even deeper into the noise, and/or one could determine the two-dimensional direction of the source.

In recent years, when using single-frequency radio waves ($\lambda_e \approx 14$ cm), the NASA deep-space network has tracked spacecraft under favorable conditions (no planet or Sun or high-density plasma cloud in the radio beam) with rms Doppler residuals of

$$\delta(\Delta\nu_e/\nu_e) \approx 3 \times 10^{-13} \qquad \text{for averaging times } \tau_a \gtrsim 7 \text{ minutes} .$$

This precision is far short of that required for detection of the VLF wave bursts shown in Figure 3 [e.g., $\delta(\Delta\nu_e/\nu_e) \sim \hat\psi \sim 1 \times 10^{-16}$ for $\tau_a < \tau \sim 15$ minutes]. However, three major improvements in NASA's best Doppler system have been made during the past year: (i) single-frequency tracking has given way to dual-frequency tracking ($\lambda_e \approx 14$ cm from Earth to spacecraft; $\lambda_e \approx 14$ cm and 3 cm from spacecraft back to Earth); (ii) rubidium clocks (stability $\sim 3 \times 10^{-13}$) have been replaced as the master oscillators by hydrogen-maser clocks (stability $\sim 2 \times 10^{-15}$); and (iii) a new "Doppler extractor," accurate to better than 10^{-3} cycles of the radio wave, has become available. With these improvements the system residuals could drop as low as $\delta(\Delta\nu_e/\nu_e) \sim 2 \times 10^{-15}$ for $\tau_a \gtrsim 15$ minutes—though dispersion of the signal, even with dual-frequency monitoring, may make the overall residuals somewhat worse than this. By using the triplet structure of the gravitational-wave signal to dig into the noise, and by transponding off spacecraft more distant than $c\tau \approx 2$ AU $\times (\tau/15$ min), one might be able to detect waves with $\hat\psi \sim 1 \times 10^{-15}$ and $\tau \gtrsim 15$ minutes. Davies (1974) and Thorne and Braginsky (1976) describe further possible Doppler improvements, which might ultimately permit detection of waves with $\hat\psi \sim 1 \times 10^{-17}$ and $\tau \gtrsim 1$ minute.

c) Black Holes and Relativistic Stars:
   Stability, Pulsations, and Gravitational-Wave Emission

I turn attention now from the experimental side of gravitational-wave astronomy to the mathematical side. Here I shall focus attention on the mathematical theory of the strong-wave sources which I invoked in § II*a:* dynamical black holes and stars.

Someday one may understand the full nonlinear, nonspherical dynamics of newborn black holes and neutron stars, and of highly relativistic supermassive stars. But no vaguely reliable analyses of the nonlinear regime have yet been completed; so at present for our best insight we must rely on the mathematical theory of small, linear perturbations of holes and stars. In this section I shall review that theory.

It is impossible for a hole or star to acquire enough electric charge that electromagnetic forces compete with gravitational forces in determining its pulsation properties and the gravitational waves it emits (see, e.g., Gibbons 1975, and § 5.2 of Eardley and Press 1975). For this reason I shall confine my attention to equilibrium configurations that are electrically neutral, and to perturbations that are purely hydrodynamical and gravitational—i.e., that are free of electromagnetic fields.

There are two very different methods by which one can derive the equations governing small adiabatic perturbations of holes and stars. One method is to write down the Einstein field equations for a perturbed hole or star, and linearize them about an equilibrium configuration. [See, e.g., Regge and Wheeler (1957) for the case of a nonrotating hole; Teukolsky (1973), using the Newman-Penrose formalism, for a rotating hole; Taub (1962) and Chandrasekhar (1964) for radial perturbations of a nonrotating star; Thorne and Campolattaro (1967) for nonradial perturbations of a nonrotating star.] The second, and more elegant, method is to (i) take an exact variational principle for the vacuum Einstein field equations in the case of a hole, and for the coupled Einstein equations and hydrodynamical equations in the case of a star; (ii) apply the Jacobi method of second variation to obtain another variational principle whose Euler equations govern the linearized perturbations of an equilibrium star or hole; (iii) apply Hamiltonian or Lagrangian techniques to manipulate this variational principle and its Euler equations into usable form. This method was formulated in general relativity theory by Taub (1969); and it has been used by Moncrief (1974*a, b*) to study small perturbations of nonrotating holes and nonrotating, isentropic stars, and by Schutz (1972) to study perturbations of rotating, non-isentropic stars.

The first method (perturbation of the Einstein equations) is usually easier than the second; but it does not lead in any automatic way to an identification of the functions describing the dynamical degrees of freedom of the system. The second method (Jacobi-Taub) does do this rather automatically and it also produces a conserved energy-momentum-like tensor for the perturbation variables. For an elucidation of the relationship between the two methods see § II of Friedman and Schutz (1975*b*).

An ultimate goal of either method is to find a *complete set of gauge-invariant dynamical variables*—i.e., an array of functions Ψ defined on the background spacetime of the equilibrium configuration such that (i) Ψ and $\partial\Psi/\partial t$ can be specified freely at some initial moment of time t; (ii) the subsequent evolution of Ψ is determined completely by a set of hyperbolic differential equations; (iii) once Ψ is known throughout spacetime, all features of the perturbation (including the metric in some specific gauge) can be calculated by applying appropriate differential operators to Ψ and performing algebraic manipulations; (iv) if all features of the perturbation are known in some arbitrary gauge, then there exists a prescription for computing Ψ from them. Properties (i) and (ii) guarantee that the Ψ are dynamical variables; property (iii) guarantees their completeness; and property (iv), their gauge invariance.

Once a complete set of dynamical variables Ψ is known, and once one has in hand the wave equations governing the evolution of Ψ and prescriptions for passing back and forth between Ψ and all other features of the perturbation, then one is ready to study the stability of the equilibrium configuration and the radiation emitted by its pulsations. Actually, even without the full mathematical apparatus, one can learn much about stability and radiation.

The dynamical degrees of freedom in the gravitational field (radiation) endow general-relativistic pulsations with a character very different from Newtonian pulsations. For example, whereas the normal modes of oscillation of a Newtonian star form a complete set in the sense that any arbitrary pulsation can be expanded in terms of them, gravitational radiation causes the normal modes of a relativistic star or black hole to not form a complete set. As a result, the way in which one manipulates Ψ and its equations in analyses of stability bears little resemblance to the way one studies Newtonian pulsations of Newtonian stars. Instead, one's manipulations of Ψ closely resemble the mathematics of potential scattering in quantum mechanics since there, as here, one deals with a center of activity (scattering potential; star or hole) which couples to an external wave that can bring in or carry off energy.

I turn now from these general remarks to some details of perturbation theory in the specific cases of nonrotating holes (§ i), rotating holes (§ ii), nonrotating stars (§ iii), and rotating stars (§ iv).

i) *Nonrotating Holes*

Theorems by Israel (1968) and by Hawking (1972) show that the general, static, nonrotating black hole is described by the Schwarzschild (1916) metric

$$ds^2 = (1 - 2M/r)(-dt^2 + dr_*{}^2) + r^2(d\theta^2 + \sin^2\theta d\varphi^2) , \qquad (27a)$$

$$r_* = r + 2M \ln (r/2M - 1) . \qquad (27b)$$

Here M is the mass of the hole; t is the time coordinate; r_* is Wheeler's (1955) tortoise coordinate; θ and φ are the angular coordinates; $r \equiv$ (Schwarzschild radial coordinate) is defined in terms of r_* by the inverse of equation (27b); and here and henceforth I use units in which $c = G = 1$. The horizon of the hole is located at $r_* = -\infty (r = 2M)$; the external universe is the entire region $-\infty < r_* < +\infty (2M < r < \infty)$. For a detailed discussion of this equilibrium structure of a nonrotating hole see, e.g., chapters 31 and 32 of MTW.

It is now known (after much work by many people; see below) that the general dynamical perturbation of a Schwarzschild hole can be analyzed as follows. (i) One can resolve the perturbation into a sum over tensor spherical harmonics characterized by the usual quantum numbers l, m, and parity $\pi = (-1)^l$ ["even-parity perturbations"] or $\pi = (-1)^{l+1}$ ["odd-parity perturbations"]. Perturbations with $l = 0$ and 1 are trivial: they describe a nondynamical addition of mass or angular momentum to the hole, and a translation of the hole from one location to another. (ii) For each set of values of $l \geq 2$, m, π there exists a single dynamical variable $\psi_{lm\pi}$ which is a *real* function of radius r_* and time t in the background Schwarzschild geometry of equation (27). In the absence of sources, this variable satisfies the Regge-Wheeler (1957) wave equation

$$\partial^2\psi/\partial r_*^2 - \partial^2\psi/\partial t^2 - V\psi = 0 \qquad (28a)$$

$$V \equiv \left[\frac{l(l+1)}{r^2} - \frac{6M}{r^3}\right]\left(1 - \frac{2M}{r}\right) \approx \frac{l(l+1)}{r_*^2} \qquad \text{as } r_* \to \infty$$

$$\approx \frac{l(l+1)}{(2M)^2}\exp\left(\frac{r_*}{2M}\right) \qquad \text{as } r_* \to -\infty .$$

$$(28b)$$

Note that this wave equation depends only on l, not on m or π; and its effective potential V is everywhere positive and has short range—i.e., dies out faster than $1/r_*$ as $r_* \to \pm\infty$. (iii) The components of the metric or Riemann tensor for an arbitrary dynamical perturbation of the hole can be constructed by taking arbitrary solutions $\psi_{lm\pi}$ of this wave equation, applying certain differential operators to them, multiplying them by certain tensor spherical harmonics, and then summing over all integers $l \geq 2$, m between $-l$ and l, and $\pi = (-1)^l$ or $(-1)^{l+1}$.

Thus, the quantities $\psi_{lm\pi}$ form a complete set of dynamical variables for perturbations of a Schwarzschild black hole; and the analysis of such perturbations reduces, in large measure, to a study of the Regge-Wheeler wave equation (28).

It has not been an easy task to bring the theory into this simple form. Major efforts by a number of workers over an 18-year period were required. The principal steps along the way were: (i) The original formulation of the problem in a very special gauge (very special coordinate system) by Regge and Wheeler (1957), and their derivation of the wave equation (28) for odd-parity perturbations; (ii) a cleaning up of algebraic errors in the Regge-Wheeler analysis by

Edelstein and Vishveshwara (1970); (iii) the derivation by Zerilli (1970a, b) of a radial wave equation with the form

$$\partial^2\psi/\partial r_*{}^2 - \partial^2\psi/\partial t^2 - V_Z\psi = 0,$$

$$V_Z = \begin{pmatrix} \text{a complicated function of } l, M, \text{ and } r \text{ that is} \\ \text{different from the Regge-Wheeler potential (28b)} \end{pmatrix} \quad (29)$$

for the even-parity perturbations; (iv) a conversion of the Regge-Wheeler-Zerilli theory from gauge-dependent to gauge-independent form by Moncrief (1974a); (v) a derivation by Bardeen and Press (1973), using the Newman-Penrose formalism, of an alternative equation

$$\frac{\partial^2\psi}{\partial r_*{}^2} - \frac{\partial^2\psi}{\partial t^2} - 4\frac{r - 3M}{r^2}\frac{\partial\psi}{\partial t} - \frac{l(l+1)(r-2M)+2M}{r^3}\psi = 0 \quad (30)$$

for the even-parity perturbations; (vi) a proof by Chandrasekhar (1975) that the Bardeen-Press equation (30), the Zerilli equation (29), and the Regge-Wheeler equation (28) can all be obtained from each other by changes of variables, and that therefore even-parity perturbations can be constructed by applying suitable differential operators and algebraic manipulations to solutions of the Regge-Wheeler odd-parity equation (28); (vii) an explicit formulation, by Chandrasekhar and Detweiler (1975a), of the transformation back and forth between the Zerilli and Regge-Wheeler equations.

Because one did not know, until the recent work of Chandrasekhar, that the Regge-Wheeler equation describes even-parity perturbations as well as odd, the literature is filled with complicated calculations using Zerilli, Bardeen-Press, and other equations, which can now be recast in far simpler form using the Regge-Wheeler equation.

Let us focus attention, for the moment, on perturbations with angular dependence described by a single spherical harmonic—i.e., fixed l, m, π. Of particular interest is the further specialization to a sinusoidal time dependence

$$\psi = \text{Real } [\psi_\omega(r_*)e^{-i\omega t}] . \quad (31)$$

The angular frequency ω can be real or complex. For any given value of ω there exist four particularly useful sinusoidal solutions (Chrzanowski and Misner 1974; see Fig. 8): An "in" solution, which describes radiation that comes in from $r_* = \infty$, partially scatters off and partially passes through the effective potential V, and then goes partially down the hole and partially back out to infinity (Fig. 8a); an "up" mode, which describes radiation that comes up from the past horizon of the Schwarzschild metric (or, equivalently, up from the surface of the star that collapsed long ago to form the hole), interacts with the potential, then goes partially back down the hole and partially out to infinity (Fig. 8b); an "out" solution, for which radiation comes partially from infinity

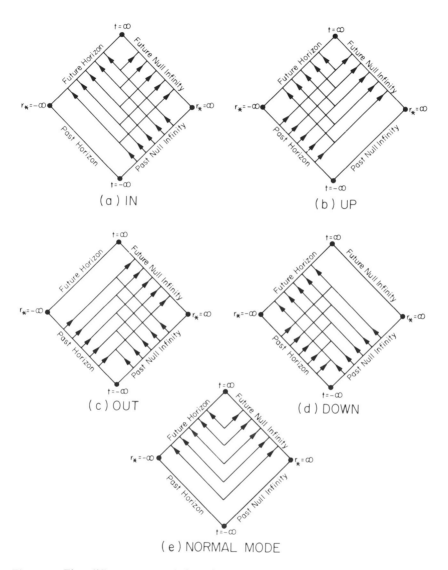

Figure 8. Five different types of sinusoidal perturbations of Schwarzschild or Kerr spacetime. Each figure is a "Penrose diagram" (cf. § 34.2 of MTW) of the spacetime region between the horizons and infinity. The internal lines show the asymptotic directions of propagation of the waves near the horizons and near infinity. Only the normal modes (Fig. 8e) are free of waves coming in from infinity and up from the horizon. When the spacetime represents the external field of a black hole, the "past horizon" is replaced by the world line of the collapsing stellar surface that produced the hole.

and partially from the past horizon, and after interaction with the potential goes entirely out to infinity and not at all down the hole (Fig. 8c); and a "down" mode, for which radiation coming from both the horizon and infinity goes entirely down the hole and not at all out to infinity (Fig. 8d). Although these four solutions are very different from each other physically, they are intimately connected mathematically:

$$\psi_\omega^{\text{out}} = (\psi_\omega^{\text{in}\dagger})^\dagger , \qquad \psi_\omega^{\text{up}} = (\psi_\omega^{\text{down}\dagger})^\dagger , \qquad (32)$$

as one can readily see from equations (28) and (31). Here a dagger denotes complex conjugation.

For very special, discrete values of the frequency ω there exist solutions in which no radiation comes up from the past horizon and none comes in from infinity (Fig. 8e). Such solutions describe *"normal-mode oscillations of the black*

Table 1 The Quadrupole, Octupole, and Hexadecapole Eigenfrequencies of Pulsation of a Nonrotating Black Hole, as Computed by Chandrasekhar and Detweiler (1975a)

$l = 2$	$l = 3$	$l = 4$
$0.37367 - 0.08896i$	$0.59945 - 0.09271i$	$0.80918 - 0.09416i$
$0.34844 - 0.27469i$	$0.58201 - 0.28116i$	$0.79657 - 0.28439i$
	$0.42629 - 0.37273i$	$0.56010 - 0.42329i$

Note: The numbers listed are angular eigenfrequencies ω ($\psi \propto e^{-i\omega t}$ at fixed radius r_*), measured in units of $M^{-1} \equiv (GM/c^3)^{-1} = 2\pi \times (32312 \text{ Hz})(M/M_\odot)^{-1}$, where M is the mass of the hole. The eigenfrequencies shown here are the only ones found for $l = 2, 3, 4$ in the numerical work of Chandrasekhar and Detweiler. However, it is conceivable that others exist.

hole."[4] That black holes should possess normal modes was originally suggested by Press (1971); but the actual proof that normal modes exist and the numerical calculation of the lowest few eigenfrequencies and eigenfunctions were accomplished only recently, by Chandrasekhar and Detweiler (1975a). It appears probable, from the work of Chandrasekhar and Detweiler, that there exist only a finite number of normal modes for each value of l. For $l = 2$ they found two eigenfrequencies, and corresponding to each eigenfrequency there are 10 degenerate modes—each with different values of azimuthal quantum number $m = -2$, $-1, 0, 1, 2$ and parity $\pi = -1, +1$. For $l = 3$ they found three eigenfrequencies, each $2 \times (2l + 1) = 14$-fold degenerate. For $l = 4$ they also found three eigenfrequencies, each now 18-fold degenerate. (The spin-2 nature of the gravitational field prevents the existence of monopole, $l = 0$, or dipole, $l = 1$, dynamical perturbations.) The eigenfrequencies for $l = 2, 3, 4$, as computed by Chandrasekhar and Detweiler (1975a), are listed in Table 1.

[4] Sometimes the alternative term "quasi-normal mode" is used. The "quasi" is appended to emphasize the fact that the eigenfrequency is (usually) not real—i.e., the oscillations damp or grow exponentially with time.

Press (1971) has suggested, on the basis of numerical experiments with scalar-wave propagation, that for large l there should exist slowly damped normal modes with

$$\text{Real } (\omega) = (27)^{-1/2}l/M \,, \tag{33}$$

where M is the mass of the hole. Goebel (1972) has given a simple analytic justification for Press's asymptotic formula: One can try to construct, from normal modes with $m \approx l \approx l_0 \gg 2$, a beam of high-frequency gravitational waves that circle the black hole's equator on the (unstable), circular, geodesic, null orbit at Schwarzschild radius $r = 3M$. If such a construction succeeds (something one cannot guarantee *a priori*), then the beam will contain $m \approx l_0$ cycles and will require a coordinate time $\Delta t = 2\pi r(1 - 2M/r)^{-1/2} = 2\pi(27)^{1/2}M$ to circle the hole once. Hence, at fixed r it will oscillate with the eigenfrequency $2\pi l_0/\Delta t = (27)^{-1/2}l_0/M$.

The Press-Goebel large-l formula (33) gives a surprisingly good approximation to the actual eigenfrequencies of the most slowly damped mode for $l = 2$, 3, 4: Compare

$$(27)^{-1/2}l = 0.38490 \quad \text{and} \quad \text{Real } (\omega M) = 0.37367 \quad \text{for} \quad l = 2 \,;$$

$$(27)^{-1/2}l = 0.57735 \quad \text{and} \quad \text{Real } (\omega M) = 0.59945 \quad \text{for} \quad l = 3 \,;$$

$$(27)^{-1/2}l = 0.76980 \quad \text{and} \quad \text{Real } (\omega M) = 0.80918 \quad \text{for} \quad l = 4 \,.$$

As I discussed in § IIa (especially Fig. 1), the normal modes of a black hole are excited strongly by a radially infalling particle; and presumably they will also be excited strongly in the gravitational collapse by which a new hole is born, and in any collision between two holes or between a hole and a compact object. Computations by Vishveshwara (1970*b; Fig.* 3) also show the normal modes to be excited when incoming gravitational waves are scattered by a black hole—though in 1970 the concept of black-hole vibrations had not yet been formulated, so Vishveshwara did not understand what he was seeing in his computer calculations.

The entire subject of black holes is of astrophysical relevance only if a hole is stable against small perturbations. For a normal mode with $\omega = \sigma + i\alpha$, the time dependence at fixed r is $e^{-i\omega t} = e^{-i\sigma t}e^{\alpha t}$. For $\alpha < 0$ (for ω in the lower-half complex plane) the oscillations are damped and thus stable; for $\alpha > 0$ (ω in the upper-half complex plane) they grow exponentially, and are thus unstable. All the normal modes of Table 1 are stable—and in fact they damp rather rapidly:

$$\frac{(\text{Amplitude after one cycle of oscillation})}{(\text{Amplitude before that cycle})} = \exp\left(\frac{2\pi\alpha}{\sigma}\right)$$

$$= 0.22406 \text{ for the most slowly damped } l = 2 \text{ mode} \,. \tag{34}$$

An analytic proof that all normal modes of a Schwarzschild hole are stable was given by Vishveshwara (1970a) at a time when the full mathematical formalism was not available. Now, with the full formalism in hand, one can prove stability of the normal modes trivially (Press and Teukolsky 1973): By virtue of the Regge-Wheeler equation (28a), the normal-mode eigenfunctions $\psi(r_*)$ satisfy the eigenequation

$$d^2\psi/dr_*{}^2 + (\omega^2 - V)\psi = 0 \qquad V \text{ positive and of short range ,} \quad (35a)$$

with boundary conditions

$$\psi = Ae^{+i\omega r_*} \text{ as } r_* \to \infty , \qquad \psi = Be^{-i\omega r_*} \text{ as } r_* \to -\infty . \qquad (35b)$$

Unstable modes have $\text{Im}(\omega) > 0$, so ψ dies out exponentially with radius as $r_* \to \pm \infty$. For such modes equations (35) form a self-adjoint, linear eigenvalue problem, which means that ω^2 must be real—and, hence [since $\text{Im}(\omega) \neq 0$], negative. The eigenequation then takes the form $d^2\psi/dr_*{}^2 = $ (positive quantity) $\times \psi$—which manifestly has no solution with the required dying-out form at $r_* \to \pm \infty$. Thus, there exist no unstable normal modes. One can readily extend the proof to show that there also exist no real-frequency normal modes.

Stability of the normal modes does *not* automatically imply stability of the black hole. This is because there are so few normal modes (e.g., only 2 for $l = 2$ and m, π fixed) that one cannot possibly expand an arbitrary perturbation in terms of them; i.e., the normal modes do not form a complete set for perturbations purely outgoing at infinity and downgoing at the horizon. Fortunately, one can prove stability of a Schwarzschild hole quite easily by a Liapunov-type method, without reference to the normal modes. (The argument was first stated explicitly by Moncrief 1974c, though it is trivial once one has the full perturbation formalism.) In the proof one focuses attention on dynamical perturbations of fixed l, m, π (but not fixed ω)—which, by virtue of the properties of spherical harmonics, form a complete set. Each such perturbation satisfies the Regge-Wheeler equation (28), and to be physically acceptable it must have no "in" radiation at infinity, and no "up" radiation at the horizon:

$$\psi \approx f_+(t - r_*) \text{ as } r_* \to \infty , \qquad \psi \approx f_-(t + r_*) \text{ as } r_* \to -\infty . \qquad (36)$$

From the Regge-Wheeler equation (28) one readily derives an "energy-like" conservation law for such a perturbation:

$$\frac{d}{dt} \int_a^b \left[\left(\frac{\partial\psi}{\partial r_*}\right)^2 + \left(\frac{\partial\psi}{\partial t}\right)^2 + V\psi^2\right]dr_* = 2\frac{\partial\psi}{\partial r_*}\frac{\partial\psi}{\partial t}\bigg|_a^b , \qquad (37)$$

where a and b are arbitrary, fixed radii. For a and b in the asymptotic regions near the horizon and near infinity, the boundary conditions (36) guarantee that the "energy fluxes" $2(\partial\psi/\partial r_*)(\partial\psi/\partial t)$ are nonpositive, and hence that the left-hand side of (37) is nonpositive—which, in turn, implies that ψ cannot grow without bound at any r_*, i.e., that whatever the initial perturbation may be, the black hole is stable.

One attempt (Tomita 1974) has been made to check whether this stability conclusion remains true when one considers second-order ($\propto \psi^2$) corrections to the first-order perturbation equations. At second order the downgoing gravitational waves induce a change in the location of the future horizon; so any adequate analysis must use a coordinate system that covers the future horizon in a nonsingular manner. Ingoing Eddington-Finkelstein coordinates are probably the best; see, e.g., Box 31.2 of MTW. Unfortunately, the analysis by Tomita used outgoing Eddington-Finkelstein coordinates, which behave badly on the future horizon; and, as a result, it produced unreliable, pathological conclusions. One's experience with Vaidya (1953)-type treatments of high-frequency perturbations suggests strongly that no pathologies will be encountered at second order; see, e.g., Lindquist *et al.* (1965); Israel (1967).

When calculating explicitly the time development of an arbitrary initial first-order perturbation, and when calculating the coupling of a perturbation to its source (e.g., to a particle falling into a hole), it is often helpful to use Green's functions. One can conveniently construct Green's functions as sums and integrals over sinusoidal modes with real frequencies. Retarded Green's functions have the form

$$G(x, x') = \sum_{l,m,\pi} \int_{-\infty}^{+\infty} \Phi_{lm\pi\omega}^{\mathrm{out}\ \dagger}(x')\Phi_{lm\pi\omega}^{\mathrm{up}}(x)d\omega \quad \text{if} \quad r' < r$$

$$= \sum_{l,m,\pi} \int_{-\infty}^{+\infty} \Phi_{lm\pi\omega}^{\mathrm{down}\ \dagger}(x')\Phi_{lm\pi\omega}^{\mathrm{in}}(x)d\omega \quad \text{if} \quad r' > r \quad (38)$$

(Chrzanowski and Misner 1974; Chrzanowski 1975). Here x and x' are events in spacetime; r and r' are the radial coordinates of x and x'; and the Φ's are appropriately chosen dynamical functions, with spherical-harmonic angular dependence and sinusoidal time dependence included, that describe the up, out, in, and down modes of Figure 8. The choice of Φ depends on whether one wants a Green's function for the Regge-Wheeler dynamical variable, or for the full metric perturbation in some given gauge, or for some other quantity. Explicit choices of Φ and their uses are discussed in the case of the Kerr metric (rotating hole) by Chrzanowski and Misner (1974), Chrzanowski (1975), and Detweiler (1976); but explicit discussions in the Schwarzschild case have not been given. The Schwarzschild case is simple enough that for computations with sources it has been adequate in the past to work directly with the Regge-Wheeler equation or the equivalent Zerilli equation. See, e.g., Zerilli (1970*a*, *b*) and Breuer *et al.* (1975) for formulations of Regge-Wheeler-type equations with sources; and Davis *et al.* (1971, 1972) for computations with such equations.

The waves produced by generic initial perturbations and by generic sources are expected to consist, roughly speaking, of three components: (i) waves that propagate rather directly out of the source region; (ii) damped, oscillatory waves from normal-mode oscillations that are excited by the initial perturbations or sources; (iii) "tails" that remain behind after the direct radiation has left and after the normal-mode oscillations have been exponentially damped, and that die out at fixed radius r_* as a power law in time t. The direct and normal-mode waves show up clearly in the particle-infall example of Figure 1; the tails do not, because they are so tiny that they were still negligible when the numerical computations stopped. However, the tails *must* exist in all generic situations, as one can see from a trivial extension of the analysis by Price (1972a, b) (see also Unt and Keres 1972 and Thorne 1972); and they play important roles in various issues of principle, such as the "Newman-Penrose conserved quantities" (Bardeen and Press 1973) and the issue of precisely how fast a perturbed black hole returns to its equilibrium state (Price 1972a, b).

ii) *Rotating Holes*

Theorems by Israel (1968), Carter (1973), Hawking (1972), and Robinson (1975) show that the general, stationary, rotating black hole is described by the Kerr (1963) metric. One can introduce coordinates in which the Kerr metric takes a form similar to equation (27)—but with added complications due to the angular momentum of the hole. The metric, and all other properties of the hole, are determined completely (modulo coordinate transformations) by two parameters: the hole's total mass-energy M, and its rotation parameter

$$a_* \equiv (\text{total angular momentum})/M^2 . \tag{39}$$

The rotation parameter can take on any value between 0 and 1—though it is generally believed that if a_* is too close to 1, e.g., $a_* \gtrsim 0.9999$, then random perturbations will drive a_* downward in an astrophysically short time; see, e.g., Bardeen *et al.* (1973); Thorne (1974); Page (1976b). In the limit $a_* \to 0$ a Kerr hole becomes a Schwarzschild hole. For detailed reviews of the theory of Kerr black holes see, e.g., chapter 33 of MTW; Carter (1973); Bardeen (1973); Stewart and Walker (1973); and Eardley and Press (1975).

Perturbations of a Kerr black hole can be analyzed in a manner analogous to perturbations of a Schwarzschild hole, though the details of the analysis are considerably more complicated. Teukolsky (1973) has identified two complex functions $\psi(t, r_*, \theta, \varphi)$, either one of which can be used as a complete, gauge-invariant, dynamical variable from which to reconstruct all features (metric, curvature tensor, . . .) of a generic, dynamical perturbation of the Kerr metric. These functions are the perturbations of the Newman-Penrose components Ψ_0 and Ψ_4 of the Weyl conformal tensor; and by being complex, they contain both the even-parity and the odd-parity parts of the perturbations. Teukolsky only conjectured that each of his ψ's contains full information about the perturbations. Wald (1973) proved the conjecture; and subsequently Chrzanowski (1975) gave a detailed prescription for constructing the metric perturbation by applying differential operators and algebraic manipulations to ψ, and Cohen and Kegeles (1975) gave a proof that Chrzanowski's prescription is correct.

Teukolsky (1973) has shown that, although the angular dependence of ψ cannot be factored out *ab initio* to give a complete set of dynamical variables depending on t and r_* alone, nevertheless if one first Fourier analyzes ψ, then the angular separation can be achieved:

$$\psi(t, r_*, \theta, \varphi) = \Sigma_{l,m} \, S_{lm}(\theta) e^{im\varphi} \int e^{-i\omega t} \psi_{\omega lm}(r_*) d\omega \, . \qquad (40)$$

Here $S_{lm}(\theta)$ are "spin-weighted spheroidal harmonics" which reduce to spin-weighted spherical harmonics in the limit as the hole becomes nonrotating. The radial eigenfunction $\psi_{\omega lm}$ satisfies the "Teukolsky equation," which is a complicated second-order differential equation with coefficients depending on the mass and rotation parameter of the hole, M and a_*, and on l, m, and ω.

All details of perturbations of Kerr holes can be learned, in principle, by appropriate analyses of the Teukolsky equation; and the recent literature is filled with detailed studies and applications of it, some of which I shall describe below. (For a review of other applications and for an extensive bibliography see Breuer 1975.) Until 1976 all studies and applications of the Teukolsky equation were technically difficult because of its complexity. Fortunately Chandrasekhar and Detweiler (1975b, 1976) and Chandrasekhar (1976) have recently found changes of variables which bring the Teukolsky equation into the form of a standard radial wave equation with a short-range potential:

$$\frac{d^2\psi}{dr_*{}^2} + (\omega^2 - V)\psi = 0 \, ,$$

$$V = V(r_*) = \begin{cases} \text{a function that depends on } M, a_*, \omega, l, m; \text{ that is real} \\ \text{for some values of these parameters but complex for} \\ \text{others; and that dies out exponentially at } r_* \to -\infty \\ \text{and as } r_*{}^{-2} \text{ at } r_* \to +\infty \end{cases} \, . \qquad (41)$$

Because the potential depends on frequency, this equation is not self-adjoint. Nevertheless, it is a considerable simplification over the Teukolsky equation. For any given set of M, a_*, l, m and real ω, there is a Chandrasekhar-Detweiler change of variables which makes V real (Detweiler 1977); therefore, in studying real-frequency modes, one need not deal with complex potentials.

One can formulate the evolution of scalar, neutrino, and electromagnetic fields in the gravitational field of a black hole in a manner mathematically analogous to that for gravitational perturbations; see, e.g., Teukolsky (1973). I shall not go into this topic here because my primary interest is gravitational radiation.

For a Kerr black hole, as for Schwarzschild, five different types of sinusoidal perturbations $\psi_{\omega lm}$ are of particular interest: "in," "up," "out," "down," and "normal-mode" perturbations (Fig. 8). From "in," "out," "up," and "down" one can construct Green's functions in the manner of equation (38) (Chrzanowski and Misner 1974; Chrzanowski 1975); and these Green's functions can be used to compute the future time evolution of any initial perturbation, as well as the response of the gravitational field to any given perturbative source.

As in the Schwarzschild case, so also for Kerr, normal modes exist only for certain discrete frequencies. Nobody has yet computed all the $l = 2$ or 3 or 4 normal-mode frequencies as functions of the hole's rotation parameter a_*; but one does know, at least, that *all of the normal modes are stable*—i.e., their frequencies all lie in the lower-half complex plane. Press and Teukolsky (1973) formulated the plan of the proof of stability and carried out half of it success-fully; Hartle and Wilkins (1974) subsequently carried out the second half. The basic plan was as follows: Consider the sinusoidal "in" perturbations for fixed l and m. At radial infinity they consist of a superposition of ingoing waves and outgoing waves. One defines a complex number[5]

$$\tilde{Z} \equiv \frac{(\text{amplitude of outgoing waves at infinity})}{(\text{amplitude of ingoing waves at infinity})}$$

$$\equiv \text{ superradiance amplitude.} \tag{42}$$

For fixed l and m, this quantity is a function of the hole's rotation parameter a_* and of the dimensionless product $M\omega$ of the hole's mass M and the sinusoidal frequency ω. Any value of ω which makes $\tilde{Z}$ infinite is a normal-mode eigen-frequency, since for that ω the ingoing waves disappear, and the ingoing mode collapses into a normal mode. Thus, there is a one-to-one correspondence be-tween normal-mode frequencies, ω_n, and poles of $\tilde{Z}$ in the complex frequency plane. For $a_* = 0$ (Schwarzschild hole) we know that all the poles are in the lower-half frequency plane—i.e., they are all stable. As a_* is gradually turned up from 0 toward 1, a pole could get into the upper-half frequency plane, where it is unstable, by one of three routes (that these are the only possible routes was proved by Hartle and Wilkins 1974): (i) The pole could come in from infinity—but Press and Teukolsky (1973) give an analytic proof that the poles are con-fined to a bounded domain and that therefore this never happens. (ii) The pole could emerge from a branch point in the upper-half frequency plane. Hartle and Wilkins (1974), by an analytic investigation of the analyticity properties of $\tilde{Z}$, show that the several branch points which lie in the upper-half plane are "im-potent" in the sense that poles cannot emerge from them. (iii) The pole could emerge from a branch point on the real axis (there exists one at $\omega = 0$), or it could smoothly move across the real axis from below. Both of these possibilities require that, for some a_*, there exist a pole of $\tilde{Z}$ on the real axis or arbitrarily close to it. Press and Teukolsky (1973) have computed $\tilde{Z}(a_*, M\omega)$ numerically

[5] Actually, my $\tilde{Z}$ is equal to the $Z_{\text{out}}/Z_{\text{in}}$ of Press and Teukolsky.

for real frequencies ω in the bounded domain where poles can lie $[0 \leq \omega \leq m\Omega_+$ where $\Omega_+ =$ (angular velocity of rotation of hole)], for $0 \leq a_* < 1$, and for a variety of low values of l and m. They see no indications whatsoever of any poles; and the values of $\tilde{Z}$ behave sufficiently smoothly that one would be very surprised if real-frequency poles had been missed or if they occur for values of l and m not investigated. Moreover, analytic analyses show an absence of poles in the regions where one might most expect them: in neighborhoods of $\omega = 0$ for all a_* and of $\omega = m\Omega_+ \equiv m \times$ (angular velocity of rotation of hole) for a_* near 1; see Starobinsky (1973), Starobinsky and Churilov (1973), Teukolsky and Press (1974). Thus, it seems safe to conclude that no poles ever get onto the real axis or into the upper-half frequency plane, and that therefore all normal modes of all Kerr black holes are exponentially damped (stable).

This conclusion has been strengthened by further numerical work of Stewart (1975). For selected values of l, m, and a_* he computes the "winding number" of a function of ω, which is essentially $1/\tilde{Z}$, as ω moves around closed curves in the upper-half frequency plane. He finds that the winding number (which equals the number of zeros of $1/\tilde{Z}$ or poles of $\tilde{Z}$ enclosed within the curve) is always zero. Thus, for the cases he computes—as for the cases Press and Teukolsky compute—there are no unstable normal modes.

As in the Schwarzschild case, the normal modes do not by any means constitute a complete set of functions in which to expand an arbitrary perturbation. Consequently, stability of the normal modes does not automatically imply complete stability of the hole. In the Schwarzschild case one proves complete stability, without reference to the normal modes, by constructing a positive definite, nonincreasing energy integral (eq. [37]). In the Kerr case no such energy integral has been found, so one must seek a proof of complete stability in some other direction.

Press and Teukolsky (1973), and independently Detweiler and Ipser (1973a), have sketched a proof of complete stability along the following lines: Let $\psi(t, r_*, \theta, \varphi)$ be one of Teukolsky's (1973) complete dynamical variables. Specify ψ and $\partial\psi/\partial t$ arbitrarily but smoothly at time $t = 0$

$$\psi(0, r_*, \theta, \varphi) = \psi_0, \quad \partial\psi/\partial t(0, r_*, \theta, \varphi) = \dot{\psi}_0 \tag{43}$$

—but demand (for simplicity) that ψ_0 and $\dot{\psi}_0$ vanish near the horizon and near infinity (at $|r_*| \gg M$). The subsequent evolution of ψ, with waves outgoing at infinity and downgoing at the horizon, is governed by the Teukolsky equation. It is reasonable to assume (but has not been *proved*) that at any fixed radius r_* ψ does not diverge, at late times, faster than $e^{\alpha_0 t}$ for some positive α_0. In this case standard results from Fourier-transform theory (e.g., Morse and Feshbach 1953, § 4.8) guarantee that the Fourier transform $\psi_\omega(r_*, \theta, \varphi)$ of $\psi(t, r_*, \theta, \varphi)$ is defined and is an analytic function of ω above the line Imaginary $(\omega) = \alpha_0$ in the complex frequency plane. Thus, one can express ψ as

$$\psi = \int_{\mathcal{C}_0} \psi_\omega e^{-i\omega t} dt \tag{44}$$

where $\mathcal{C}_0$ is a curve above that line (Fig. 9). One can then try to deform the contour of integration downward until it lies entirely below the real axis. In such a deformation, if one passes over a pole of ψ_ω, one picks up its residue as a term that must be added onto the Fourier integral. Thus, after downward deformation to a new contour $\mathcal{C}_1$, ψ has the form

$$\psi = \int_{\mathcal{C}_1} \psi_\omega e^{-i\omega t} d\omega + \Sigma_j R_j \exp\left(-i\omega_j t\right), \tag{45}$$

where j labels the poles of ψ_ω above curve $\mathcal{C}_1$, ω_j are their frequencies, and $R_j(r_*, \theta, \varphi)$ are their residues. If one succeeds in deforming the curve below the real axis, then at late times ψ will be dominated entirely by the residues of poles above the real axis—which means that for such poles $R_j \exp\left(-i\omega_j t\right)$ must by themselves satisfy the Teukolsky equation, which in turn means that $R_j \exp\left(-i\omega_j t\right)$ are normal modes. Thus, if the deformation can be achieved, then ψ can be written as

$$\psi = \text{(a Fourier integral that dies away at late times)}$$
$$+ \text{(a sum of unstable normal modes)}. \tag{46}$$

From this one concludes that ψ *dies out at large t (i.e., the black hole is completely stable) if and only if all of the normal modes are stable.*

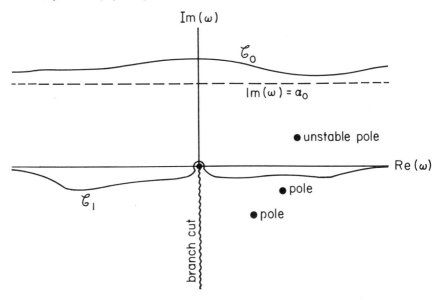

Figure 9. The contours $\mathcal{C}_0$ and $\mathcal{C}_1$ in the complex frequency plane which one uses in a proof of the complete stability of a Kerr black hole.

This conclusion relies on one's ability to deform the contour $\mathcal{C}_0$ until it falls below the real axis. Neither Press and Teukolsky (1973) nor Detweiler and Ipser (1973a) attempted to prove that such a deformation is possible. However, thanks to the analyticity studies of Hartle and Wilkins (1974) and the Green's function analyses of Chrzanowski (1975), one can now sketch such a proof (Teukolsky 1976; Chrzanowski 1976)—though the full details of the proof have not yet been worked out, and one has no firm guarantee that the proof will go through as sketched.

The only thing which might impede a downward deformation of the curve $\mathcal{C}_0$ is branch points of ψ_ω. Now, $\psi_\omega(r_*, \theta, \varphi)$ should be expressible as a spatial integral of a Green's function $G_\omega(r_*, \theta, \varphi; r_*', \theta', \varphi')$ folded into the initial data $\psi_0(r_*', \theta', \varphi')$ and $\dot{\psi}_0(r_*', \theta', \varphi')$. Presumably the branch points of ψ_ω will be the same as the branch points of G_ω. Now G_ω can be expressed as a product of sinusoidal "in" and "up" solutions (cf. eqs. [38] and [32]), whose analyticity properties have been studied by Hartle and Wilkins (1974). Presumably the "impotent" branch points that Hartle and Wilkins found in the upper-half plane will present no barrier to deformation of the contour. However, the branch point at $\omega = 0$ is certain to present a barrier: it is required, in the form

$$\psi_\omega \sim G_\omega \sim \omega^n \ln \omega \quad \text{for } n \text{ an integer}, \tag{47a}$$

to produce the power-law tail of ψ,

$$\psi \sim t^{-(n+1)} \tag{47b}$$

which must dominate the decay of ψ at late times (Price 1972a, b).

The above argument suggests that the contour $\mathcal{C}_0$ can be deformed into the contour $\mathcal{C}_1$ of Figure 9—which, although it does not lie entirely below the real axis, is sufficiently below that axis to guarantee the late-time behavior (46) for ψ. As a result, the black hole is completely stable if and only if its normal modes are all stable (which they seem to be, according to the previous discussion).

I turn now from the stability of a Kerr hole to the details of its stable normal modes. Detweiler (1977) has shown that one can learn much about the normal modes of fixed l and m by studying the function

$$Q(a_*, M\omega) \equiv \frac{|\text{amplitude of } \psi_{\omega l m}^{\text{in}} \text{ on future horizon}|^2}{|\text{amplitude of } \psi_{\omega l m}^{\text{in}} \text{ at past null infinity}|^2}, \tag{48}$$

which for real ω is related to the Press-Teukolsky "superradiance amplitude" $\tilde{Z}$ (eq. [42]) by

$$Q = \frac{\omega Z}{\omega - m\Omega_+}, \quad Z \equiv |\tilde{Z}|^2 - 1 \equiv \text{superradiance function}. \tag{49}$$

Here Ω_+ is the angular velocity of rotation of the hole. Detweiler shows that any normal modes ω_n lying near the real-frequency axis

$$|\text{Im}\,(\omega_n)| \ll |\text{Re}\,(\omega_n)|$$

will show up on the real axis as resonances in $Q(\omega)$; and that

$$\text{Re}\,(\omega_n) = \text{peak frequency of resonance},$$

$$\text{Im}\,(\omega_n) = \text{half-width of resonance at half-maximum}. \qquad (50)$$

He also shows that generic sources near the horizon, without much frequency structure of their own, will generate gravitational waves with an energy spectrum of the form

$$\frac{dE}{d\omega} = \begin{pmatrix}\text{slowly varying function of} \\ \omega \text{ characteristic of source}\end{pmatrix} \times Q(\omega), \qquad \omega \text{ real}. \qquad (51)$$

Thus, any resonances in $Q(\omega)$ will show up as peaks in the radiated energy spectrum. The physics behind this, of course, is that generic sources excite normal-mode vibrations of the hole which then radiate waves with a spectrum proportional to the resonance (50) in $Q(\omega)$.

Detweiler (1977) has gone on to compute Q as a function of ωM and a_* for several values of l and m. He finds that for each $l = m = 2, 3, 4, \ldots$ and for $a_* \gtrsim 0.9$, Q has one very sharp resonance; and that as $a_* \to 1$, the half-width of the resonance goes to zero and its frequency approaches from above the Press-Teukolsky (1973) "critical frequency for superradiance"

$$\omega_{\text{crit}} = m\Omega_+ = \frac{ma_*}{2M[1 + (1 - a_*{}^2)^{1/2}]}. \qquad (52)$$

[This resonant behavior cannot be seen directly in the Teukolsky-Press 1974 plots of $Z(\omega)$ because the factor $\omega - m\Omega_+$ linking Q and Z in equation (49) suppresses the resonance in Z.] Figure 10 shows several of Detweiler's resonances and shows the dependence of the $l = m = 2$ resonance on the rotation parameter of the hole.

Detweiler's analysis shows that the more rapidly a hole rotates, the easier it is to excite the normal modes with $l = m$ and $\omega \approx \omega_{\text{crit}}$. Once excited, the normal modes damp only very slowly—presumably because they are quite efficient in extracting rotational energy from the hole and radiating it away to infinity (close analog of the "Penrose [1969] process" and of the Zel'dovich [1971, 1972]-Misner [1972] "superradiance phenomenon"). If rapidly rotating holes are common remnants of gravitational collapse (which is not at all obvious), and if the near-horizon sources of perturbations of such holes have nonnegligible amplitude at the rather high normal-mode frequencies $\omega \approx \omega_{\text{crit}} = 2\Omega_+, 3\Omega_+, \ldots$ (which also is not at all obvious), then this phenomenon could have a strong influence on the spectrum of gravitational waves from black-hole events. See § IIa(i).

iii) *Nonrotating Stars*

Turn attention now from black holes to fully relativistic stellar models. The theory of pulsations of relativistic stars has been formulated only for the adiabatic case (no nuclear energy generation or heat transfer coupled to the pulsations). For this reason, there is no need to consider the energy-generation or heat-transfer properties of the equilibrium models that one perturbs. One can confine attention to models in which the entropy and chemical abundances are distributed according to some predetermined rule. The modern theory of such

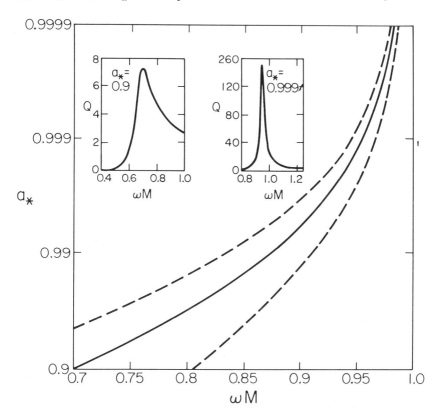

Figure 10. The Detweiler (1977) spectral function $Q(\omega)$ which quantifies the way in which a generic source near a Kerr black hole excites the hole's normal modes of pulsation (eqs. [48]–[51]). The insets show $Q(\omega)$ for holes with rotation parameter $a_* = 0.9$ and $a_* = 0.999$, and for $l = m = 2$. The large graph shows the frequency ω of the central peak of the resonance (*solid curve*), and the frequencies of the resonance half-maximum points (*dashed curves*) as functions of a_*, for $l = m = 2$. The curves for $l = m = 3$ and $l = m = 4$ are quite similar, except for a displacement upwards in frequency from these by factors of $\sim 3/2$ and ~ 2. [Based on calculations by Detweiler (1977).]

models, for the nonrotating case, was formulated by Oppenheimer and Volkoff (1939) and is reviewed in chapter 23 of MTW. The literature is full of specific examples of such models, especially models of neutron stars (see Arnett and Bowers 1974 and references cited therein).

Perturbation theory for relativistic stars is more complicated than for holes because of the additional degrees of freedom in the matter (usually assumed to be a perfect fluid) of which the stars are made. The fluid degrees of freedom are intimately coupled to the gravitational-wave degrees of freedom by the perturbed Einstein field equations. The result is a dynamical evolution problem (eq. [54] below) with two families of characteristics—i.e., two signal propagation speeds: the speed of sound and the speed of gravity waves ($\equiv$ speed of light). The coupling of gravity to fluid is so intimate that for general perturbations of nonrotating, spherical stars nobody has managed to construct coupled, second-order wave equations involving only the dynamical degrees of freedom.

Instead, what one must deal with are the dynamical degrees of freedom (an array of gravitational and fluid functions which I shall symbolize by Ψ) plus several other gravitational-field variables (an array symbolized by Φ). These arrays can be resolved into spherical-harmonic components so that there are separate dynamical and auxiliary arrays $\Psi_{lm\pi}$, $\Phi_{lm\pi}$ associated with each set of values of the spherical-harmonic indices l and m, and parity π. For given l, m, π at some initial moment of time one can specify Ψ and $\partial\Psi/\partial t$ arbitrarily throughout the star's interior and exterior. (This fact identifies Ψ as dynamical degrees of freedom.) One can then solve a set of initial-value equations of the form

$$\mathfrak{D}\Phi = \mathfrak{M}\Psi ,$$

$$\mathfrak{D}, \mathfrak{M} \equiv \begin{pmatrix} \text{matrix operators involving radius } r \text{ and first- and} \\ \text{second-order radial derivatives } \partial/\partial r_*, \partial^2/\partial r_*{}^2 \end{pmatrix} \quad (53)$$

to determine the initial values of the auxiliary variables Φ. The subsequent evolution is then given by a wave equation of the form

$$\partial^2\Psi/\partial t^2 + \mathfrak{L}\Psi + \mathcal{P}\Phi = 0 ,$$

$$\mathfrak{L}, \mathcal{P} \equiv (\text{operators of the same form as } \mathfrak{D} \text{ and } \mathfrak{M}) , \quad (54)$$

where at each moment in the evolution Φ must be determined from the initial-value equations (53). Ideally one would like to remove the term $\mathcal{P}\Phi$ from the dynamical equations; but this has proved impossible, except in special cases. Once the time evolution of the dynamical variables Ψ has been computed and the Φ have been derived from the Ψ by solving the initial-value equations (53), one can apply differential operators and algebraic manipulations to (Ψ, Φ) and thereby obtain all details of the pulsating star's structure, metric, and Riemann curvature.

A formalism of the above type has been derived for radial pulsations ($l = 0$), by Taub (1962) in the special case of isentropic stars, and by Chandrasekhar (1964) in the general case; for a review see chapter 26 of MTW. For radial pulsations the fluid has one dynamical degree of freedom, its radial displacement $\xi(r)$; and gravity has none (monopole gravitational waves cannot exist). Thus, Ψ reduces to the single function $\xi(r)$; and there is only one set of characteristics, that associated with sound waves. For these radial perturbations Chandrasekhar and Taub managed to decouple all the auxiliary variables Φ from the dynamical equation—i.e., they obtained a dynamical equation of the form (54) with $\mathcal{P} = 0$. Chandrasekhar (1964) showed that, when specialized to normal modes, his dynamical equation was self-adjoint; and he derived a variational principle for it. Subsequently Taub (1969) showed how to obtain the Chandrasekhar variational principle by the Jacobi method of second variation.

Chandrasekhar (1964) showed, using his variational principle, that general relativity induces an instability against spherical collapse in stars which, according to Newtonian theory, would be stable. This instability can be quantified by the critical adiabatic index, Γ_{crit}, such that stars with radius-independent adiabatic indices $\Gamma_1 < \Gamma_{crit}$ are unstable, and stars with $\Gamma_1 > \Gamma_{crit}$ are stable. Chandrasekhar (1964) derives the relation

$$\Gamma_{crit} = 4/3 + K \cdot 2M/R \qquad (55)$$

in the nearly Newtonian limit, where K is a constant of order unity that depends on the star's Newtonian structure, and $2M$ and R are the star's Schwarzschild radius and actual radius. Ipser (1970), using Chandrasekhar's formalism, derives values of Γ_{crit} substantially larger than $4/3$ for fully relativistic stellar models. This "relativity-induced instability" is important for the physics of white dwarfs, neutron stars, and supermassive stars; cf. chapter 24 of MTW and chapter 11 of Zel'dovich and Novikov (1971).

Dipole ($l = 1$) pulsations of a nonrotating star are analogous to radial pulsations in their lack of gravitational dynamical degrees of freedom. (Dipole gravitational waves cannot exist.) Campolattaro and Thorne (1970) have derived a formalism of the above type (eqs. [53] and [54]) for dipole pulsations; and Detweiler (1975b) has constructed a variational principle for the normal-mode solutions, $\Psi \propto e^{-i\omega t}$, and has proved that the normal-mode eigenvalue problem is self-adjoint.

Quadrupole and higher-order pulsations ($l \geq 2$) possess fluid *and* gravitational degrees of freedom. For such pulsations a formalism of the type (53), (54) has been constructed by Thorne and Campolattaro (1967); and a variational principle for solutions with sinusoidal time dependences, $\Psi \propto e^{-i\omega t}$, has been constructed by Detweiler and Ipser (1973b). For the special case of nonradial perturbations of *isentropic* stars [in which case the pressure and density perturbations Δp and $\Delta\rho$ are related to the unperturbed pressure and density distributions by $\Delta p / \Delta \rho = (dp/dr)(d\rho/dr)^{-1}$] Moncrief (1974b) has simplified the theory (see also Cazzola and Lucaroni 1972). Using a Hamiltonian formalism which begins with the Jacobi method of second variation, Moncrief has (i) put the formalism into gauge-invariant language, and (ii) decoupled the auxiliary variables Φ from the dynamical equation (54). It is not clear today whether such a decoupling can be achieved in the nonisentropic case.

The theory of sinusoidal pulsations, $\Psi \propto e^{-i\omega t}$ and $l \geq 2$, is conceptually more simple for stars than for holes because a star has no horizons. The horizons endow a hole with four distinct types of sinusoidal modes for each choice of ω: "up," "down," "in," and "out" (Fig. 8). By contrast, for each (real or complex) ω—and for each l, m, π—a star has precisely one sinusoidal mode. Typically that mode involves waves which come in from past null infinity, excite the star's pulsations, and get scattered and reemitted off toward future null infinity:

$$\Psi = \Psi_\omega(r_*)e^{-i\omega t} \; ; \quad \Psi_\omega \approx Z_{\text{in}}e^{-i\omega r_*} + Z_{\text{out}}e^{+i\omega r_*} \text{ as } r_* \to \infty \; . \quad (56)$$

Here Z_{in} is the amplitude of the incoming waves; and Z_{out}, the amplitude of the outgoing waves. For very special, discrete choices of frequency ω, Z_{in} vanishes; i.e., the pulsations couple to purely outgoing waves. These are the *normal-mode* pulsations of the star.[6]

For a star made of perfect fluid all of the normal modes have "even parity," $\pi = (-1)^l$. This is because odd-parity perturbations $[\pi = (-1)^{l+1}]$ consist only of a steady, differential rotation of the star plus freely propagating gravitational waves that couple not at all to the star (Thorne and Campolattaro 1967). (Torsional oscillations of a crystalline neutron star have odd parity and couple to odd-parity gravitational waves [Thorne and Żytkow 1977]; but I shall confine myself here to perfect-fluid stars.)

For each l and m there is an infinite sequence of even-parity normal modes—a situation analogous to Newtonian stellar theory, but different from black holes where the number of normal modes is finite. However, the normal modes of a relativistic star do not form a complete set of functions in terms of which to expand the future time development of spatially bounded, arbitrary initial perturbations. As with black holes (and unlike Newtonian stars), the gravitational radiation prevents completeness of the normal modes. An example of a perturbation that cannot be expanded in normal modes is this: The initial state is a quiescent neutron star, plus a wave packet of gravitational radiation at radius $r = 37$ light-years. The radiation subsequently propagates toward the star. Most of it passes by at large radii, but some of it interacts with the star, producing small pulsations that subsequently radiate away their energy.

The normal modes simply do not have enough functional freedom in themselves to handle a situation like the above. One can convince oneself of this by

---

[6] The terms "quasi-normal mode" and "outgoing normal mode" are sometimes used rather than simply "normal mode." The "quasi" emphasizes that the oscillation frequency is complex; cf. n. 4. The "outgoing" emphasizes that the radiation is purely outgoing.

focusing attention on nearly Newtonian stars where the normal modes are adequate to describe all fluid motions with nearly conserved energy, but where their external waves are far from a complete set of functions for the entire vacuum, nearly flat-space, external universe.

Although the normal modes are not complete, Ipser (1975) has managed to prove a remarkable theorem: *A relativistic, nonrotating, perfect-fluid stellar model is completely stable against small, adiabatic, nonradial perturbations with no incoming radiation if and only if all of its $l \geq 1$ normal modes are stable.* (The same theorem, for radial pulsations ($l = 0$), follows from the completeness of the radial normal modes; Chandrasekhar 1964.) Ipser's proof is of the Liapunov type—i.e., it involves showing that a certain positive-definite functional of the dynamical variables cannot increase with time if the normal modes are all stable.

There exist a number of important lemmas and theorems about the normal modes of a fully relativistic star: (i) Unstable normal modes of any l $(0, 1, 2, \ldots)$ have purely imaginary eigenfrequencies—i.e., an unstable mode grows exponentially in time, without any sinusoidal oscillations (Chandrasekhar 1964 for $l = 0$; Detweiler 1975*b* for $l = 1$; Detweiler and Ipser 1973*b* for $l \geq 2$). (ii) For $l = 0$ and 1, stable normal modes have purely real frequencies (Chandrasekhar 1964 for $l = 0$; Detweiler 1975*b* for $l = 1$). (iii) For $l \geq 2$, stable normal modes possess eigenfrequencies with negative, nonzero imaginary parts—i.e., they always damp out exponentially with time due to radiation reaction (Detweiler and Ipser 1973*b*). (iv) A star with density that everywhere decreases outward, $d\rho/dr < 0$, possesses a zero-frequency normal mode with $l \geq 1$ if and only if the relativistic Schwarzschild discriminant

$$S(r) \equiv \frac{dp}{dr} - \Gamma_1 \frac{p}{\rho + p} \frac{d\rho}{dr} \tag{57}$$

vanishes over some finite region inside the star (Islam 1970 for $l = 1$; Detweiler and Ipser 1973*b* for $l \geq 2$). (v) Consider a sequence of stellar models parametrized by a real variable λ, such that the structure of the models is a smooth function of λ. Then the normal-mode frequencies will be continuous functions of λ; i.e., as λ increases, each normal-mode frequency ω_n moves smoothly along a trajectory in the complex plane (Detweiler and Ipser 1973*b*).

By combining the above results with the well-known fact that Newtonian models with $d\rho/dr < 0$ and $S > 0$ are stable for $l \geq 1$, Detweiler and Ipser (1973*b*) prove the following stability criterion for $l \geq 2$; and Detweiler (1975*b*) proves it for $l = 1$: *Consider a relativistic, nonrotating star with $d\rho/dr < 0$ everywhere. If its Schwarzschild discriminant is positive everywhere, $S > 0$, then all of its nonradial normal modes are stable (and therefore [Ipser 1975] the star is completely stable against nonradial perturbations). Moreover, the star is neutrally stable against nonradial perturbations if and only if $S = 0$ over some finite region*

—and in that case (Thorne 1969; Islam 1970) there exist an infinite number of zero-frequency normal modes for each $l \geq 1$.

These stability criteria complement the following result which is almost certainly true, but has not yet been proved with full rigor: *A nonrotating star with $d\rho/dr > 0$ and/or $S < 0$ in some finite region possesses an infinite number of unstable normal modes for each $l \geq 1$* ("Rayleigh-Taylor" and "convective" instabilities; Kovetz 1967; Thorne 1966 and 1969; Schutz 1970; Islam 1970; Detweiler and Ipser 1973b; Detweiler 1975b).

The mathematical formalism for pulsations with $l \geq 2$ has been applied to specific neutron-star models by Thorne (1969) and by Detweiler (1975a).

Table 2 Quadrupole Normal-Mode Eigenfrequencies of Neutron-Star Models

Equation of State	ρ_c (g cm^{-3})	$M/M_\odot$	R (km)	$2M/R$	n	ω (s^{-1})	Reference
P...........	3×10^{14}	0.209	17.4	0.0354	0	$5800-0.04i$	D
HW.........	3×10^{14}	0.405	20.8	0.0574	0	$5250-0.08i$	D, T
P...........	5.15×10^{14}	0.367	12.7	0.0854	0	$7930-0.3i$	D
Vγ.........	5.15×10^{14}	0.678	12.6	0.159	0	$8980-0.06i$	D, T
P...........	1×10^{15}	0.854	10.8	0.233	0	$12130-3i$	D
P...........	3×10^{15}	1.63	9.06	0.530	0	$18070-12i$	D
Vγ.........	3×10^{15}	1.95	9.93	0.580	0	$20210-5i$	D, T
					1	$40400-0.6i$	T
					2	$61200-0.4i$	T

Note: The equations of state are: P, Pandharipande 1971; HW, Harrison and Wheeler (see Harrison *et al.* 1965); Vγ, Tsuruta and Cameron 1966b. Each equilibrium configuration is characterized by its central density ρ_c, its mass M, and its radius R. Angular eigenfrequencies ω ($\psi \propto e^{-i\omega t}$) are shown for the fundamental quadrupole mode ($n = 0$) and, in one model, for the first two "harmonics" ($n = 1, 2$). The references are: D, Detweiler 1975a; T, Thorne 1969.

Thorne computes the lowest few quadrupole ($l = 2$) eigenfrequencies and eigenfunctions by studying the resonances and phase shifts they produce, on the real-frequency axis, in the quantity

$$\frac{(\text{amplitude of pulsation at center of star})}{(\text{amplitude of gravitational waves at } r_* \to \infty)}.$$

(This resonance method is analogous to the one depicted in Fig. 10 for Kerr black holes.) Detweiler computes the eigenfrequencies using the Detweiler-Ipser (1973b) variational principle. Table 2 shows some of the Thorne and Detweiler eigenfrequencies. Notice that the eigenperiods

$$P = 2\pi/\text{Real}\,(\omega) \tag{58a}$$

of the fundamental quadrupole modes typically lie between 0.3 and 1.0 milliseconds; and the amplitude damping times

$$\tau_d = -1/\text{Im}\,(\omega) \tag{58b}$$

are typically of order 0.1 to 20 seconds.

Detweiler (1975a) has also computed the quadrupole eigenfrequency of an incompressible fluid sphere as a function of

$$\text{(gravitational radius)}/\text{(actual radius)} = 2M/R.$$

He finds that the ratio of damping time to period

$$\tau_d/P = \text{Re}\,(\omega)/2\pi\,\text{Im}\,(\omega)$$

is large for $2M/R \ll 1$ (because a nearly Newtonian star cannot radiate very effectively); it decreases to a minimum of about 280 when $2M/R \approx 0.5$; and thereafter, as $2M/R$ increases toward $8/9$ (where the central density becomes infinite), τ_d/P rises toward infinity. Presumably this rise of τ_d/P in the ultra-relativistic regime occurs because the strong curvature of spacetime prevents radiation from escaping to infinity. Detweiler's remarkable conclusion is that, whereas nonrotating holes have $\tau_d/P \approx 0.67$ (Table 2), nonrotating relativistic stars cannot have τ_d/P smaller than ~ 100.

iv) *Rotating Stars*

When discussing rotating stars, I shall always deal with fully relativistic, rapidly and (possibly) differentially rotating stellar models, unless I state otherwise. In Newtonian theory such models can be stationary without being axially symmetric. For example, the Dedekind triaxial ellipsoids are stationary in an inertial frame, but have strong internal circulation and much angular momentum; and the Jacobi triaxial ellipsoids are stationary in a rotating frame, but have rapid and rigid rotation in an inertial frame; see, e.g., Chandrasekhar (1969). In general relativity, gravitational-radiation reaction prevents the existence of stationary Jacobi-type configurations; but presumably stationary Dedekind-type configurations still exist—though nobody has yet devised a method for constructing fully relativistic examples.

In the case of axial symmetry—but with full relativity and rapid rotation—there has been much work on the theory of stationary, equilibrium configurations of perfect fluid. Hartle and Sharp (1967), and Bardeen (1970) have given variational principles which govern the structures of such models, and have extracted from them theorems about the structures. Numerical relaxation techniques for constructing such models have been devised and implemented by Wilson (1972, 1973), Bonazzola and Schneider (1974), Butterworth and Ipser (1975, 1976), and Butterworth (1976). Unfortunately, as discussed by Butterworth and Ipser, Wilson's method and models are unreliable because of an inadequate matching to the external gravitational field; and the Bonazzola-Schneider method breaks down when the star is so relativistic and the rotation so rapid that "ergotoroids" form inside the star. Thus, the rotating, axially symmetric stars I shall consider will be of the Butterworth-Ipser type, to which the Hartle-Sharp and Bardeen variational principles are applicable.

Special insight can be gained in some situations by dealing with limiting cases of the generic, rotating, axially symmetric, equilibrium models. These include: (i) *Fully relativistic, rapidly rotating disks* (Bardeen and Wagoner 1971). (ii) *Slowly rotating, fully relativistic models* (theory developed by Hartle 1967 and applied to neutron stars by Hartle and Thorne 1968; homogeneous models, which give particular insight, constructed by Chandrasekhar and Miller 1974). (iii) *Rapidly rotating, post-Newtonian models* (method for constructing general models developed by Seguin 1973 using the Chandrasekhar 1965 post-Newtonian formalism; homogeneous models studied in a series of papers by Chandrasekhar, culminating with Chandrasekhar and Elbert 1974, and also by Bardeen 1971). (iv) *Rapidly rotating, Newtonian models* (methods of constructing general models developed by Stoeckly 1965 and by Ostriker and Mark 1968; homogeneous models—the classical "ellipsoidal figures of equilibrium"—reviewed by Chandrasekhar 1969).

Perturbation theory for rotating, relativistic stars is in a highly incomplete state. Much effort is still being expended on it; and many new, important results can be expected in the next few years. However, by now the basic outline is beginning to emerge:

The most important generic results are those derived recently by Friedman and Schutz (1975*a, b*). Friedman and Schutz allow the equilibrium configuration to be completely generic (full relativity; rapid, differential rotation; non–axially symmetric or axially symmetric, as one wishes; perfect fluid or shear-stressed, as one wishes)—but, of course, they insist that it be stationary in an asymptotically inertial frame. The perturbations are restricted to normal-mode oscillations (sinusoidal time dependence, $\Psi \sim e^{-i\omega t}$; no incoming radiation at infinity). Friedman and Schutz derive a Lagrangian and a variational principle for the normal modes; and from their variational principle they extract the following important results: (i) A theorem that, so long as a normal mode radiates (which Friedman and Schutz *assume* is always the case for $\omega \neq 0$ and for non-zero stellar rotation), it cannot have a real frequency. (ii) A theorem that, along a smoothly varying sequence of equilibrium configurations, normal modes can go unstable only via a zero-frequency mode (ω can pass from the lower-half complex plane to the upper-half plane only by going through $\omega = 0$). (iii) The construction of a quadratic functional C (integral, over any spacelike or null hypersurface, of an expression quadratic in the metric and matter perturbations) with the properties that (*a*) if C is positive definite with respect to all perturbation functions satisfying the initial-value equations, then all normal modes are stable [$\text{Im}(\omega) < 0$]; and (*b*) along a smoothly varying sequence of models which have C positive definite at one end, a normal-mode instability sets in at the configuration where C ceases to be positive definite. In applying these C-criteria for stability one can use any hypersurface of integration and any gauge that one wishes; the result will be independent of one's choice.

Because the Friedman-Schutz analysis is so general, it does not provide one with an explicit prescription for selecting out a complete set of dynamical variables Ψ to vary, or for solving the initial-value equations to obtain the auxiliary variables Φ explicitly in terms of the Ψ. However, for axially symmetric perturbations of axially symmetric, perfect-fluid configurations an explicit choice of Ψ and Φ, an explicit prescription for solving the initial-value equations, and the explicit form of the normal-mode variational principle in terms of those Ψ and Φ have been given by Chandrasekhar and Friedman (1972, 1973a; uniform rotation) and by Will (1974; differential rotation). See Chandrasekhar (1974) for a review.

It is likely—but nobody has tried to prove it—that a rotating equilibrium configuration is completely stable against arbitrary perturbations with no incoming waves at infinity, if and only if its normal modes are stable. In the absence of a proof, all we know about complete stability are the results derived by Schutz (1972) using the method of second variation: For general perturbations (no sinusoidal assumption) Schutz (i) constructs a variational principle; (ii) constructs gauge-dependent, local conservation laws for energy and angular momentum; (iii) shows that the integral E of his gauge-dependent energy density over a hypersurface of constant time t (a) is gauge invariant, (b) equals the total "Keplerian-measured" mass-energy associated with the perturbations, and (c) decreases with time as radiation flows away from the star in accord with the Isaacson (1968) analysis of energy loss by gravitational waves. Most importantly, Schutz shows that his energy E is a Liapunov functional for the analysis of stability: If E is positive definite with respect to all perturbations satisfying the initial-value equations, then the star is completely stable. It seems likely that the converse holds (E not positive definite implies instability); but no proof exists.

The above formalisms, when they are ultimately applied to the Butterworth-Ipser equilibrium configurations, should provide us with an important test of the following conjecture (Chandrasekhar 1974; Friedman and Schutz 1975b): *An axially symmetric, relativistic configuration of perfect fluid can experience dynamical[7] instabilities under two circumstances:* (i) *when its adiabatic index Γ_1 is too low; and* (ii) *when its rotation is too rapid. In the latter case the instability is produced by a coupling of gravitational-radiation reaction to the rotation; and, depending on the structure of the equilibrium configuration, the dominant instability may be of either of two types:* (a) *a "bar-mode instability" (angular dependence*

---

[7] The word "dynamical" is intended to divert attention away from "convective-type" instabilities such as those that occur, in the case of zero rotation, when the Schwarzschild discriminant is negative; see equation (57) and associated discussion. However, one should keep in mind that such instabilities in some realistic circumstances (e.g., some red supergiants) can be as catastrophic as "dynamical instabilities" (e.g., can completely disrupt the star; cf. Keeley 1970). For a discussion of some convective-type instabilities in general relativistic, rapidly rotating stars see Seguin (1975).

$e^{im\varphi}$ with $m = 2$), or (b) an "ergotoroid instability" ($e^{im\varphi}$ with the largest m's being the most unstable).

This conjecture is based primarily on perturbation calculations for special types of equilibrium configurations:

Low-Γ_1 instability. The "low-Γ_1 instability" is the generalization, to rotating configurations, of the standard "$\Gamma_1 < 4/3$" instability for radial perturbations of nonrotating stars. In the Newtonian, rotating case it is an instability of axially symmetric modes ($e^{im\varphi}$ with $m = 0$; cf. Lebovitz 1970); and presumably this remains true in the relativistic, rotating case. In the nonrotating case, general-relativistic effects strengthen the instability (eq. [55] above, and associated discussion). When relativity is weak, its destabilizing effects can be counterbalanced by slow rotation, producing a critical adiabatic index (generalization of eq. [55])

$$\Gamma_{\text{crit}} = \frac{4}{3} + K\frac{2M}{R} - \frac{2}{9}\frac{\Omega^2 I}{|W|} \tag{59}$$

(Fowler 1966; Chandrasekhar and Lebovitz 1968). Here Ω is the slow and uniform angular velocity, I is the trace of the second moment of the star's mass distribution, and W is the star's gravitational potential energy. In the case of full relativity, the slow rotation induces small changes in the normal-mode frequencies of radial oscillations. These changes have been analyzed by Hartle and colleagues, and by Chandrasekhar and Friedman; for references see Hartle and Friedman (1975). The rotation not only produces a change in the real part of the eigenfrequency, $\delta\text{Real}(\omega^2) \propto \Omega^2$; it also produces a distortion of the star, which causes the (formerly spherical) pulsations to radiate gravitational waves—and, by radiation reaction, to acquire an imaginary part in their eigenfrequency: $\text{Im}(\omega^2) \propto \Omega^4$ (cf. Hartle and Thorne 1977). Insight into the combined effects of rapid rotation and full relativity on the "low-Γ_1 instability" must await future applications of the Friedman-Schutz (1975a, b) formalism to the Butterworth-Ipser (1975, 1976) configurations.

Bar-mode instability. The bar-mode instability is best understood by considering, initially, *Newtonian configurations of incompressible fluid.* (Such configurations have given useful insight, for nearly three centuries, into the behavior of rapidly rotating, self-gravitating systems; and they are no less useful today than they ever have been. Chandrasekhar's 1969 book is now the standard treatise on the subject.) The axially symmetric, rigidly rotating, incompressible configurations are called "Maclaurin spheroids." They make up a one-parameter family running from a spherical, nonrotating configuration at one end to a completely flattened, rotating disk at the other end. It is conventional to parametrize the family by the eccentricity e of the spheroid's meridional section; e runs from 0 at the spherical end to 1 at the disk end (Fig. 11). The configuration of given e can be chosen to have any density ρ and mass M that one wishes; ρ and M are trivial "scale factors" which we can ignore. Figure 11

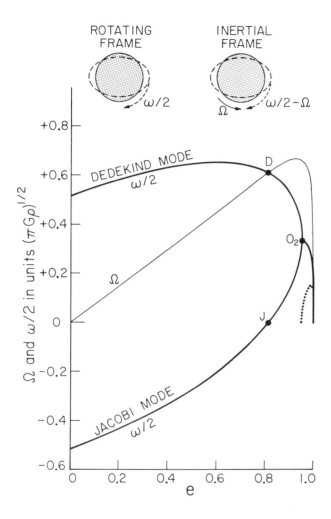

Figure 11. The bar-mode eigenfrequencies of Maclaurin spheroids, as computed in chapter 5 of Chandrasekhar (1969). Plotted horizontally is the eccentricity e of the meridional section of a Maclaurin spheroid. Plotted vertically (*thin curve*) is the angular velocity Ω with which the spheroid must rotate to have eccentricity e. Also plotted vertically (*thick curve*) is one-half the eigenfrequency ω of oscillation of the bar-mode perturbation, as measured in a frame rotating with the unperturbed spheroid. Beyond the point O_2 the eigenfrequencies of the two bar modes are complex $\omega = \sigma \pm i\alpha$. The solid line is σ; the dotted line is α. All ω, σ, α, and Ω are measured, in this figure, in units of $(\pi G\rho)^{1/2}$, where ρ is the density of the incompressible fluid making up the spheroid.

shows the angular velocity Ω of rigid rotation for a Maclaurin spheroid, as a function of its eccentricity e.

The least stable normal modes of a Maclaurin spheroid are its bar modes. (Chandrasekhar 1969 uses the phrase "toroidal modes.") They involve a deformation of the equatorial plane into a bar shape ("$m = 2$" deformation)

$$\delta r = \delta(x^2 + y^2)^{1/2} = r \cdot \text{Real} \{\exp[-i(\omega t + 2\varphi)]\} . \qquad (60)$$

Here φ is angle and ω is the eigenfrequency, both measured in a frame that rotates with the equilibrium spheroids ("rotating frame"). For positive frequency, the bar rotates clockwise with angular velocity $\omega/2$; for negative frequency it rotates counterclockwise with angul arvelocity $-\omega/2$ (see upper left insert in Fig. 11). As seen in an inertial frame (upper right insert in Fig. 11), where the unperturbed spheroid rotates counterclockwise with angular velocity Ω, the bar-shaped perturbation rotates clockwise with angular velocity $\omega/2 - \Omega$.

The nonrotating spheroid ($e = \Omega = 0$) has two bar modes, one rotating counterclockwise ($\omega < 0$; "Jacobi mode"); the other rotating clockwise ($\omega > 0$; "Dedekind mode"); both with $|\omega/2| = 0.51640(\pi G\rho)^{-1/2}$. As the spheroid gets spun up (increasing e and Ω), the eigenfrequencies of its two bar modes change as indicated in Figure 11. At

$$e = 0.81267 , \quad \Omega = 0.61774(\pi G\rho)^{1/2} , \qquad (61)$$

the Jacobi mode has $\omega/2 = 0$ (point J in Fig. 11); and the Dedekind mode has $\omega/2 = \Omega$ (point D). This means that the Jacobi mode has zero frequency (the bar is at rest) in the rotating frame; and the Dedekind mode has zero frequency (the bar is at rest) in the inertial frame. These zero-frequency modes deform the spheroid into an equilibrium, triaxial ellipsoid—a "Jacobi ellipsoid" in the case of the Jacobi mode; a "Dedekind ellipsoid" in the case of the Dedekind mode. One says that the Jacobi and Dedekind equilibrium configurations "bifurcate" from the Maclaurin equilibrium configurations at $e = 0.81267$. Beyond the bifurcation point the Jacobi and Dedekind modes are dynamical once again ($\omega/2$ and $\omega/2 - \Omega$ nonzero). At the point O_2 the Jacobi and Dedekind modes become identical, and beyond that point their eigenfrequencies become complex, $\omega = \sigma \pm i\alpha$, so that one mode is stable and the other unstable.

Relativistic corrections to Newtonian gravity produce a major change in the bar modes (Chandrasekhar 1970): Gravitational radiation reaction damps both bar modes (gives ω a negative imaginary part) below the Jacobi-Dedekind bifurcation point ($e < 0.81267$). However, above the Jacobi-Dedekind point it makes the Dedekind mode unstable ($\text{Im}[\omega] > 0$)! Note that, as a result, the instability of the Maclaurin spheroids sets in, in relativity theory (point D), through a zero-frequency mode

$$\omega - 2\Omega = (\text{eigenfrequency in inertial frame}) = 0$$

in accordance with the Friedman-Schutz (1975*b*) theorem; whereas in Newtonian theory it sets in through a mode with $\omega - 2\Omega = -\Omega \neq 0$ (point O_2).

As Chandrasekhar (1970) has emphasized, viscosity and gravitational-radiation reaction have opposite effects on the bar modes: Beyond the bifurcation point radiation reaction destabilizes the Dedekind mode while leaving the Jacobi mode stable; but viscosity destabilizes the Jacobi mode while leaving the Dedekind mode stable.

A similar behavior holds for the Jacobi and Dedekind equilibrium configurations (Chandrasekhar 1970): Because Jacobi configurations have rigid, non–axially symmetric rotation, they radiate gravitational waves; and the resulting radiation reaction causes them to evolve. However, the rigidity of their rotation saves them from viscous forces. Dedekind configurations, by contrast, have non-rotating bar shapes as viewed in an inertial frame, so they radiate no gravitational waves and are free from radiation-reaction forces. However, they possess large internal circulation (differential rotation), which couples to viscosity and thereby produces evolution.

Miller (1974) has computed the evolution of Newtonian, perfect-fluid ellipsoids when subjected to the perturbative driving force of gravitational radiation reaction. She shows quantitatively the manner in which an unstable Maclaurin spheroid beyond the bifurcation point is driven, via its Dedekind mode of oscillation, away from the Maclaurin family and through a sequence of weakly pulsating "Riemann S ellipsoids" until it ends up as a nonrotating Dedekind ellipsoid. She also shows how a Jacobi ellipsoid is driven, through a sequence of weakly pulsating Riemann S ellipsoids, until it winds up either as a stable Maclaurin spheroid below the bifurcation point or as a stable Dedekind ellipsoid above the bifurcation point. Very recently Detweiler and Lindblom (1977) have examined evolution of the ellipsoids under the combined driving force of radiation reaction and viscosity. They find some tendency for these two driving forces to cancel each other out. In all these evolutions substantial amounts of (nearly) monochromatic gravitational radiation are emitted, with the radiation frequency $\omega - 2\Omega$ evolving as the ellipsoid evolves.

By extrapolating the weak-relativity results of Chandrasekhar and Miller to neutron stars (where relativity is not weak), Friedman and Schutz (1975*a*, corrected for an error in the numerical coefficient) conclude that radiation reaction will spin down a too-rapidly-rotating neutron star of mass $M \approx 1\,M_\odot$ and $R \approx 10$ km in a time

$$\tau \sim (0.3 \text{ s})(t_0/t_c)^{-3}(t_0/t_c - 1)^{-5}, \qquad t_0 > t_c. \tag{62}$$

Here t_0 is the ratio of kinetic energy of rotation to gravitational potential energy when the neutron star first forms; $t_c \approx 0.14$ is the critical value of t at which the bar-mode instability sets in for Maclaurin spheroids; and τ is the e-folding time for $t/t_c - 1$ to approach zero. For large t_0 the gravitational radiation from this spin-down can be stronger than that from compressional oscillations of a new-born neutron star. However, reliable estimates must await further numerical work using the Friedman-Schutz formalism and Butterworth-Ipser configurations, and using a Chandrasekhar-Friedman (1973b)–Will (1974) criterion for locating "bifurcation points" along sequences of fully relativistic, axially symmetric, rotating, equilibrium configurations.

Ergotoroid instability. The ergotoroid instability is a phenomenon confined to highly relativistic, very rapidly rotating equilibrium configurations. Such configurations were shown by Wilson (1972, 1973) to possess toroidally shaped "ergoregions"—i.e., regions, like the ergosphere of a Kerr black hole, in which the Killing vector representing time translations at infinity, $\partial/\partial t$, is locally spacelike. Friedman (1977) has recently demonstrated that, although small azimuthal perturbations (circulating density waves; $\delta\rho \propto e^{-i(\omega t - m\varphi)}$) in a star's ergoregion have positive energy as measured locally, they have negative energy as measured by a distant observer. As a result, when the perturbations radiate gravitational waves, the radiation reaction drives them into larger-amplitude motion. They are unstable. The instability is strongest for modes of high azimuthal quantum number m—i.e., for "fine-grained" density ripples—and, consequently, the gravitational radiation emitted will have a complicated high-frequency structure.

It is not yet clear which rotation-induced instability—bar-mode or ergotoroid —will dominate in rapidly rotating, newborn neutron stars or in rapidly rotating supermassive stars. Further insight requires much new numerical work.

d) Conclusion

When gravitational waves are ultimately detected—as I expect they will be —experimenters will begin to give us detailed information about the waveforms, $h_{ij}^{TT}(t - r)$, emitted by various sources. Theorists will then be faced with the task of interpreting the waveforms—i.e., of extracting astrophysical information from them. We are far from ready for that task. However, the perturbation-theory analyses described above give us an initial foundation; for example: We are beginning to understand how the frequencies and damping times of waves from a black hole depend on the angular momentum and mass of the hole. We are aware that highly monochromatic radiation can come from normal modes of extremely rapidly rotating holes, from compressional and torsional oscillations of stars, and from bar-mode instabilities in rapidly rotating stars; but we do not yet understand how one might distinguish these sources from each other. We know that radiation with high-frequency structure may come from ergo-toroid instabilities in rapidly rotating stars; but we know nothing about what frequencies should really dominate for a given source, or how the star and its radiation will evolve in response to the ergotoroid instability.

Note added in proof: Recently Friedman and Schutz (1977) have discovered a new type of instability, which is driven by gravitational radiation reaction acting on azimuthally propagating sound waves in *any* rotating star. Presumably the time scale for the instability is exceedingly long unless the star is strongly relativistic and rapidly rotating.

Relativistic perturbation theory for holes and stars will be 20 years old, but still quite incomplete, when this book gets published; and computer analyses of large-amplitude, fully relativistic motions (black-hole collisions; highly nonspherical stellar collapse) will only be giving their very first results. In view of the effort expended to reach our present state of knowledge, and in view of the knowledge needed in an era of incoming observational data, I think it not unreasonable to pursue these theoretical studies with increased effort during the next few years.

For helpful discussions during preparation of the written version of this lecture I thank Paul Chrzanowski, Steven Detweiler, David Douglass, James Hartle, Bill Hamilton, James Ipser, Richard Price, Larry Smarr, Saul Teukolsky, and Joseph Weber. I also appreciate the hospitality of hideout providers in Logan, Utah; Patagonia, Arizona; and Wimberly, Texas.

References

Anderson, A. J. 1971, *Nature*, **229**, 547.
Anderson, J. D. 1974, in Bertotti (1974), p. 163.
Anderson, P. W., and Palmer, R. G. 1971, *Nature Phys. Sci.*, **231**, 145.
Arnett, W., and Bowers, R. 1977, *Ap. J. Suppl. Ser.*, **33**, 415.
Bahcall, J., and Ostriker, J. P. 1975, *Nature*, **256**, 23.
Bardeen, J. M. 1970, *Ap. J.*, **162**, 71.
———. 1971, *Ap. J.*, **167**, 425.
———. 1973, in *Black Holes*, ed. C. and B. S. DeWitt (New York: Gordon & Breach).
Bardeen, J. M., Carter, B., and Hawking, S. W. 1973, *Commun. Math. Phys.*, **31**, 161.
Bardeen, J. M., and Petterson, J. A. 1975, *Ap. J. (Letters)*, **195**, L65.
Bardeen, J. M., and Press, W. H. 1973, *J. Math. Phys.*, **14**, 7.
Bardeen, J. M., and Wagoner, R. V. 1971, *Ap. J.*, **167**, 359.
Baym, G., and Pines, D. 1971, *Ann. Phys. (USA)*, **66**, 816.
Bertotti, B. ed. 1974, *Proceedings of the International School of Physics "Enrico Fermi", Course 56: Experimental Gravitation* (New York: Academic Press).
Blandford, R. D. 1976, *M.N.R.A.S.*, **176**, 465.
Bludman, S. A., and Ruderman, M. A. 1975, *Ap. J. (Letters)*, **195**, L19.
Bolton, C. T. 1975, *Ap. J.*, **200**, 269.
Bonazzola, S., and Schneider, J. 1974, *Ap. J.*, **191**, 273.
Boughn, S. P., Fairbank, W. M., McAshan, M. S., Paik, H. J., Taber, R. C.,

Bernat, T. P., Blair, D. G., and Hamilton, W. O. 1974, in DeWitt-Morette, ed. (1974), p. 40.

Braginsky, V. B. 1965, *Usp. Fiz. Nauk*, **86**, 433; English transl. in *Soviet Phys. —Uspekhi*, **8**, 513.

———. 1970, *Physics Experiments with Test Bodies* (Moscow: Nauka); English transl. published as NASA-TT F762, available from National Technical Information Service, Springfield, VA.

———. 1974a, lectures in Bertotti (1974), p. 235.

———. 1974b, lecture in DeWitt-Morette (1974), p. 28.

———. 1975, lectures at 1975 Erice meeting; proceedings to be published in Weber (1977).

Braginsky, V. B., and Manukin, A. B. 1974, *Measurements of Small Forces in Physical Experiments: Techniques and Fundamental Limitations* (Moscow: Nauka); English transl. in preparation by University of Chicago Press.

Braginsky, V. B., and Vorontsov, Yu. I. 1974, *Usp. Fiz. Nauk*, **114**, 41; English transl. in *Soviet Phys.—Uspekhi*, **17**, 644 (1975).

Braginsky, V. B., Vorontsov, Yu. I., and Krivchenkov, V. D. 1975, *Zhur. Eksp. Teor. Fiz.*, **68**, 55; English transl. in *Soviet Phys.—JETP*, **41**, 28 (1975).

Breuer, R. A. 1975, *Gravitational Perturbation Theory and Synchrotron Radiation*, Vol. 44 of Lecture Notes in Physics (Berlin: Springer-Verlag).

Breuer, R. A., Tiomno, J., and Vishveshwara, C. V. 1975, *Nuovo Cimento*, **25B**, 851.

Butterworth, E. M. 1976, *Ap. J.*, **204**, 561.

Butterworth, E. M., and Ipser, J. R. 1975, *Ap. J. (Letters)*, **200**, L103.

———. 1976, *Ap. J.*, **204**, 200.

Campolattaro, A., and Thorne, K. S. 1970, *Ap. J.*, **159**, 847.

Canuto, V., and Chitre, S. M. 1974, in *IAU Symposium No. 53, Physics of Dense Matter*, ed. C. Hansen (Dordrecht: Reidel), p. 133.

Carr, B. J. 1975, *Ap. J.*, **201**, 1.

———. 1976, *Ap. J.*, **206**, 8.

Carr, B. J., and Hawking, S. W. 1974, *M.N.R.A.S.*, **168**, 399.

Carter, B. 1973, in *Black Holes*, ed. C. and B. S. DeWitt (New York: Gordon & Breach).

Cazzola, P., and Lucaroni, L. 1972, *Phys. Rev. D*, **6**, 950.

Chandrasekhar, S. 1931, *Ap. J.*, **74**, 81.

———. 1964, *Ap. J.*, **140**, 417.

———. 1965, *Ap. J.*, **142**, 1488.

———. 1969, *Ellipsoidal Figures of Equilibrium* (New Haven, Conn.: Yale University Press).

———. 1970, *Phys. Rev. Letters*, **24**, 611.

———. 1974, in DeWitt-Morette ed. (1974), p. 63.

———. 1975, *Proc. Roy. Soc. London A*, **343**, 289.

———. 1976, *Proc. Roy. Soc.*, **348**, 39.

Chandrasekhar, S., and Detweiler, S. 1975a, *Proc. Roy. Soc. A*, **344**, 441.

———. 1975b, *Proc. Roy. Soc. A*, **345**, 145.

———. 1976, *Proc. Roy. Soc. A*, **350**, 165.

Chandrasekhar, S., and Elbert, D. 1974, *Ap. J.*, **192**, 731.
Chandrasekhar, S., and Friedman, J. L. 1972, *Ap. J.*, **176**, 745.
———. 1973a, *Ap. J.*, **181**, 481.
———. 1973b, *Ap. J.*, **185**, 1.
Chandrasekhar, S., and Lebovitz, N. R. 1968, *Ap. J.*, **152**, 267.
Chandrasekhar, S., and Miller, J. C. 1974, *M.N.R.A.S.*, **167**, 63.
Chrzanowski, P. L. 1975, *Phys. Rev. D*, **11**, 2042.
———. 1976, private communication.
Chrzanowski, P. L., and Misner, C. W. 1974, *Phys. Rev. D*, **10**, 1701.
Chung, K. P. 1973, *Nuovo Cimento*, **14B**, 293.
Clark, J. W., and Chao, N.-C. 1972, *Nature Phys. Sci.*, **236**, 37.
Cohen, J. M. 1970, *Ap. and Space Sci.*, **6**, 263.
Cohen, J. M., and Kegeles, L. S. 1975, *Phys. Letters*, **54A**, 5.
Cunningham, C. T. 1976, *Ap. J.*, **208**, 534.
Davidsen, A., and Ostriker, J. P. 1973, *Ap. J.*, **179**, 585.
Davies, R. W. 1974, in *Colloques Internationaux C.N.R.S. No. 220—Ondes et radiations gravitationelles*, ed. Y. Choquet-Bruhat (Paris: Editions du C.N.R.S.), p. 33.
Davis, M., Ruffini, R., Press, W. H., and Price, R. H. 1971, *Phys. Rev. Letters*, **27**, 1466.
Davis, M., Ruffini, R., and Tiomno, J. 1972, *Phys. Rev. D*, **5**, 2932.
Detweiler, S. L. 1975a, *Ap. J.*, **197**, 203.
———. 1975b, *Ap. J.*, **201**, 440.
———. 1977, *Proc. Roy. Soc. A*, **352**, 381.
———. 1977, paper in preparation.
Detweiler, S. L., and Ipser, J. R. 1973a, *Ap. J.*, **185**, 675.
———. 1973b, *Ap. J.*, **185**, 685.
Detweiler, S. L., and Lindblom, L. 1977, *Ap. J.*, **213**, 193.
DeWitt-Morette, C., ed. 1974, *IAU Symposium No. 64, Gravitational Radiation and Gravitational Collapse* (Dordrecht: Reidel).
Doroshkevich, A. G., Sunyaev, R. A., and Zel'dovich, Ya. B. 1974, in *IAU Symposium No. 63, Confrontation of Cosmological Theories with Observational Data*, ed. M. S. Longair (Dordrecht: Reidel), p. 213.
Doroshkevich, A. G., Zel'dovich, Ya. B., and Novikov, I. D. 1967, *Astr. Zhur.*, **44**, 295 (1967); English transl. in *Soviet Astr.—AJ*, **11**, 233.
Douglass, D. H. 1976, work in progress at University of Rochester.
Douglass, D. H., Gram, R. Q., Tyson, J. A., and Lee, R. W. 1975, *Phys. Rev. Letters*, **35**, 480.
Drever, R. W. P., Hough, J., Bland, R., and Lessnoff, G. W. 1973, *Nature*, **246**, 340.
Dyson, F. 1972, unpublished remarks at the Sixth Texas Symposium on Relativistic Astrophysics, New York City, December.
Eardley, D. M. 1974, *Phys. Rev. Letters*, **33**, 442.
———. 1975, *Phys. Rev. D*, **12**, 3072.
Eardley, D. M., and Press, W. H. 1975, *Ann. Rev. Astr. Ap.*, **13**, 381.
Edelstein, L. A., and Vishveshwara, C. V. 1970, *Phys. Rev. D*, **1**, 3514.

Estabrook, F. B., and Wahlquist, H. D. 1975, *Gen. Rel. Grav.*, **6**, 439.
Fairbank, W. M., Boughn, S. P., Paik, H. J., McAshan, M. S., Opfer, J. E., Taber, R. C., Hamilton, W. O., Pipes, B., Bernat, T., and Reynolds, J. M. 1974, in Bertotti ed. (1974), p. 294.
Fowler, W. A. 1966, *Ap. J.*, **144**, 180.
Frank, J., and Rees, M. J. 1976, *M.N.R.A.S.*, **176**, 633.
Fricke, K. J. 1973, *Ap. J.*, **183**, 941.
———. 1974, *Ap. J.*, **189**, 535.
Friedman, J. L. 1977, paper to be submitted to *Phys. Rev. D.*
Friedman, J. L., and Schutz, B. F. 1975a, *Ap. J. (Letters)*, **199**, L157.
———. 1975b, *Ap. J.*, **200**, 204.
———. 1977, *Ap. J.*, in press.
Giacconi, R., and Gursky, H. 1974, *X-Ray Astronomy* (Dordrecht: Reidel).
Gibbons, G. W. 1975, *Commun. Math. Phys.*, **44**, 245.
Gibbons, G. W., and Schutz, B. F. 1972, *M.N.R.A.S.*, **159**, 41P.
Giffard, R. 1977, *Phys. Rev. D*, **10**, 2478.
Ginzburg, V. L., and Zheleznyakov, V. V. 1975, *Ann. Rev. Astr. Ap.*, **13**, 511.
Goebel, C. J. 1972, *Ap. J. (Letters)*, **172**, L95.
Hamilton, W. O. 1975, lectures at 1975 Erice meeting; proceedings to be published in Weber (1977).
Harrison, B. K., Thorne, K. S., Wakano, M., and Wheeler, J. A. 1965. *Gravitation Theory and Gravitational Collapse* (Chicago: University of Chicago Press).
Hartle, J. B. 1967, *Ap. J.*, **150**, 1005.
———. 1970, *Ap. J.*, **161**, 111.
———. 1973, *Ap. and Space Sci.*, **24**, 385.
Hartle, J. B., and Friedman, J. L. 1975, *Ap. J.*, **196**, 653.
Hartle, J. B., and Sharp, D. H. 1967, *Ap. J.*, **147**, 317.
Hartle, J. B., and Thorne, K. S. 1968, *Ap. J.*, **153**, 807.
———. 1977, paper in preparation.
Hartle, J. B., and Wilkins, D. C. 1974, *Commun. Math. Phys.*, **38**, 47.
Hartquist, T. W., and Cameron, A. G. W. 1977, *Astrophys. and Space Sci.*, in press.
Hawking, S. W. 1971a, *M.N.R.A.S.*, **152**, 75.
———. 1971b, *Phys. Rev. Letters*, **26**, 1344.
———. 1972, *Commun. Math. Phys.*, **33**, 323.
———. 1974, *Nature*, **248**, 30.
———. 1975, *Commun. Math. Phys.*, **43**, 199.
Hills, J. G. 1975, *Nature*, **254**, 295.
Ipser, J. R. 1970, *Astrophys. and Space Sci.*, **7**, 361.
———. 1975, *Ap. J.*, **199**, 220.
Ipser, J. R., and Price, R. H. 1977, *Ap. J.*, in press.
Isaacson, R. A. 1968, *Phys. Rev.*, **166**, 1272.
Islam, J. 1970, *M.N.R.A.S.*, **150**, 237.
Israel, W. 1967, *Phys. Letters*, **24A**, 184.
———. 1968, *Commun. Math. Phys.*, **8**, 245.
Jones, B. 1976, *Rev. Mod. Phys.*, **48**, 107.

Kafka, P. 1975, lectures at 1975 Erice meeting; proceedings to be published in Weber (1977).

Keeley, D. A. 1970, *Ap. J.*, **161**, 643.

Kerr, R. P. 1963, *Phys. Rev. Letters*, **11**, 237.

Kovetz, A. 1967, *Zs. f. Ap.*, **66**, 446.

Lamb, F. K. 1975, in Proceedings of Seventh Texas Symposium on Relativistic Astrophysics, *Annals New York Acad. Sci.*, **262**, 331.

Lamb, F. K., Pethick, C. J., and Pines, D. 1973, *Ap. J.*, **184**, 271.

Lebovitz, N. 1970, *Ap. J.*, **160**, 701.

Lee, M., Gretz, D., Steppel, S., and Weber, J. 1976, *Phys. Rev. D*, **14**, 893.

Lifshitz, E. M. 1946, *Zh. Eksp. Teor. Fiz.*, **16**, 587.

Lightman, A. P., and Shapiro, S. L. 1976, *Ap. J.*, **203**, 701.

Lindquist, R. W., Schwartz, R. A., and Misner, C. W. 1965, *Phys. Rev.*, **137**, B1364.

Lovelace, R. V. E. 1976, *Nature*, **262**, 649.

Lynden-Bell, D. 1969, *Nature*, **223**, 690.

Lynden-Bell, D., and Rees, M. J. 1971, *M.N.R.A.S.*, **152**, 461.

Middleditch, J., and Nelson, J. 1976, *Ap. J.*, **208**, 567.

Miller, B. D. 1974, *Ap. J.*, **187**, 609.

Mironovskii, V. N. 1965, *Astr. Zhur.*, **42**, 977; English transl. in *Soviet Astronomy—AJ*, **9**, 752.

Misner, C. W. 1972, *Phys. Rev. Letters*, **28**, 994.

———. 1974, in DeWitt-Morette, ed. (1974), p. 3.

Misner, C. W., Thorne, K. S., and Wheeler, J. A. 1973, *Gravitation* (San Francisco: W. H. Freeman); cited in text as MTW.

Moncrief, V. 1974a, *Ann. Phys.*, **88**, 323.

———. 1974b, *Ann. Phys.*, **88**, 343.

———. 1974c, *Phys. Rev. D*, **9**, 2707.

Morse, P. M., and Feshbach, H. 1953, *Methods of Theoretical Physics* (New York: McGraw-Hill).

Moss, G. E., Miller, L. R., and Forward, R. L. 1971, *Appl. Optics*, **10**, 2495.

Norman, C. A., and ter Haar, D. 1973, *Astr. and Ap.*, **24**, 121.

Novikov, I. D., and Thorne, K. S. 1973, in *Black Holes*, ed. C. DeWitt and B. S. DeWitt (New York: Gordon & Breach), p. 343.

Oppenheimer, J. R., and Volkoff, G. 1939, *Phys. Rev.*, **55**, 374.

Ostriker, J. P., and Mark, J. W.-K. 1968, *Ap. J.*, **151**, 1075.

Ostriker, J. P., Peebles, P. J. E., and Yahil, A. 1974, *Ap. J.* (*Letters*), **193**, L1.

Paczyński, B. 1975, private communication.

Page, D. N. 1976a, *Phys. Rev. D*, **13**, 198.

———. 1976b, *Phys. Rev. D*, **14**, 3260.

Page, D. N., and Hawking, S. W. 1976, *Ap. J.*, **206**, 1.

Page, D. N., and Thorne, K. S. 1974, *Ap. J.*, **191**, 499.

Pandharipande, V. R. 1971, *Nucl. Phys.*, **A178**, 123.

Peebles, P. J. E. 1966, *Ap. J.*, **146**, 542.

———. 1974, in *IAU Symposium No. 58, The Formation and Dynamics of Galaxies.* ed. J. R. Shakeshaft (Dordrecht: Reidel), p. 55.

Peebles, P. J. E., and Dicke, R. H. 1968, *Ap. J.*, **154**, 891.

Penrose, R. 1969, *Riv. Nuovo Cimento Ser. I*, **1**, Numero Speciale, 252.

Pines, D., Shaham, J., and Ruderman, M. 1972, *Nature Phys. Sci.*, **237**, 83.

———. 1974, in *IAU Symposium No. 53, Physics of Dense Matter*, ed. C. Hansen (Dordrecht: Reidel), p. 189.

Press, W. H. 1971, *Ap. J. (Letters)*, **170**, L105.

Press, W. H., and Teukolsky, S. A. 1973, *Ap. J.*, **185**, 649.

Press, W. H., and Thorne, K. S. 1972, *Ann. Rev. Astr. Ap.*, **10**, 335.

Price, R. H. 1972*a*, *Phys. Rev. D*, **5**, 2419.

———. 1972*b*, *Phys. Rev. D*, **5**, 2439.

Pringle, J. E., and Rees, M. J. 1972, *Astr. and Ap.*, **21**, 1.

Pringle, J. E., Rees, M. J., and Pacholczyk, A. G. 1973, *Astr. and Ap.*, **29**, 179.

Rappaport, S., Joss, P. C., and McClintock, J. E. 1976, *Ap. J. (Letters)*, **206**, L103.

Rees, M. J. 1974, in *Colloques Internationaux C.N.R.S. No. 220—Ondes et radiations gravitationnelles*, ed. Y. Choquet-Bruhat (Paris: Editions du CNRS), p. 203.

Regge, T., and Wheeler, J. A. 1957, *Phys. Rev.*, **108**, 1063.

Robinson, D. C. 1975, *Phys. Rev. Letters*, **34**, 905.

Rosen, N., and Shaviv, G., eds. 1975, *General Relativity and Gravitation: Proceedings of GR7* (Jerusalem: Keter Publishing House).

Rosi, L. A., and Zimmerman, R. L. 1976, *Ap. and Space Sci.*, **45**, 447.

Ruderman, M. 1969, *Nature*, **223**, 597.

Ruderman, M. 1976, *Ap. J.*, **203**, 213.

Ruffini, R., and Wheeler, J. A. 1971, in *Proceedings of the Conference on Space Physics* (Paris: European Space Research Organization), p. 45.

Saslaw, W. C. 1975, in *Dynamics of Stellar Systems, Proceedings of IAU Symposium No. 69*, ed. A. Hayli (Dordrecht: Reidel), p. 379.

Schutz, B. F., Jr. 1970, *Ap. J.*, **161**, 1173.

———. 1972, *Ap. J. Suppl.*, **24**, 343.

Schwarzschild, K. 1916, *Sitzber. Deut. Akad. Wiss. Berlin, Kl. Math.-Phys. Tech.*, p. 424.

Sciama, D., Weber, J., Kafka, P., Drever, R., and Tyson, A. 1975, "Gravitational Waves: Panel Discussion," in *General Relativity and Gravitation: Proceedings of GR7*, ed. G. Shaviv and J. Rosen (Jerusalem: Keter Publishing House), p. 243.

Seguin, F. H. 1973, *Ap. J.*, **179**, 289.

———. 1975, *Ap. J.*, **197**, 745.

Shakura, N. I., and Sunyaev, R. A. 1973, *Astr. and Ap.*, **24**, 337.

Shapiro, S. L. 1973*a*, *Ap. J.*, **180**, 531.

———. 1973*b*, *Ap. J.*, **185**, 69.

Shapiro, S. L., Lightman, A. P., and Eardley, D. M. 1976, *Ap. J.*, **204**, 187.

Shwartzman, V. F. 1971, *Astr. Zhur.*, **48**, 479; English transl. in *Soviet Astr.—AJ*, **15**, 377.

Silk, J., and Arons, J. 1975, *Ap. J. (Letters)*, **200**, L131.

Smarr, L., Cadez, A., DeWitt, B., and Eppley, K. 1976, papers in preparation.

Spitzer, L., Jr. 1971, in *Semaine d'étude sur les noyeaux des galaxies* (Pontificae Academiae Scientarum, Script Varia no. 35), p. 443.

Spitzer, L., Jr., and Shull, J. M. 1975, *Ap. J.*, **201**, 773.

Starobinsky, A. A. 1973, *Zh. Eksp. Teor. Fiz.*, **64**, 48; English transl. in *Soviet Phys.—JETP*, **37**, 28.

Starobinsky, A. A., and Churilov, S. M. 1973, *Zh. Eksp. Teor. Fiz.*, **65**, 3; English transl. in *Soviet Phys.—JETP*, **38**, 1.

Stewart, J. M. 1975, *Proc. Roy. Soc. London A*, **344**, 65.

Stewart, J., and Walker, M. 1973, *Black Holes—The Outside Story; Springer Tracts in Modern Physics*, No. 69 (Berlin: Springer-Verlag).

Stoeckly, R. 1965, *Ap. J.*, **142**, 208.

Talbot, R. J. 1976, *Ap. J.*, **205**, 535.

Tammann, G. A. 1974, in *Supernovae and Supernova Remnants*, ed. C. B. Cosmovici (Dordrecht: Reidel).

Taub, A. 1962, in *Les Théories Relativistes de la Gravitation* (Paris: Editions du CNRS), p. 173.

———. 1969, *Commun. Math. Phys.*, **15**, 235.

Taylor, J. H., Hulse, R. A., Fowler, L. A., Gullahorn, G. E., and Rankin, J. M. 1976, *Ap. J. (Letters)*, **206**, L53.

Teukolsky, S. A. 1973, *Ap. J.*, **185**, 635.

———. 1976, private communication.

Teukolsky, S. A., and Press, W. H. 1974, *Ap. J.*, **193**, 443.

Thorne, K. S. 1966, *Ap. J.*, **144**, 201.

———. 1969, *Ap. J.*, **158**, 1.

———. 1972, in *Magic without Magic: John Archibald Wheeler*, ed. J. Klauder (San Francisco: W. H. Freeman and Co.), p. 231.

———. 1974, *Ap. J.*, **191**, 507.

Thorne, K. S., and Braginsky, V. B. 1976, *Ap. J. (Letters)*, **204**, L1.

Thorne, K. S., and Campolattaro, A. 1967, *Ap. J.*, **149**, 591; and **152**, 673.

Thorne, K. S., and Żytkow, A. 1977, paper in preparation.

Tipler, F. J. 1975, *Ap. J.*, **197**, 199.

Tomita, K. 1974, *Prog. Theor. Phys.*, **52**, 1188.

Tremaine, S. D. 1976, *Ap. J.*, **203**, 345.

Tremaine, S. D., Ostriker, J. P., and Spitzer, L., Jr. 1975, *Ap. J.*, **196**, 407.

Tsuruta, S., and Cameron, A. G. W. 1966a, *Canadian J. Phys.*, **44**, 1863.

———. 1966b, *Canadian J. Phys.*, **44**, 1895.

Tsuruta, S., Canuto, V., Lodenquai, J., and Ruderman, M. 1972, *Ap. J.*, **176**, 739.

Unruh, W. 1977, *Phys. Rev.*, in press.

Unt, V. and Keres, P. 1972, *Proceedings of the Estonian Academy of Sciences: Physics and Mathematics*, **21**, 17.

Vaidya, P. C. 1953, *Nature*, **171**, 260.

Vishveshwara, C. V. 1970a, *Phys. Rev. D*, **1**, 2870.

———. 1970b, *Nature*, **227**, 936.

Wagoner, R. V. 1973, *Ap. J.*, **179**, 343.

Wagoner, R. V., Fowler, W. A., and Hoyle, F. 1967, *Ap. J.*, **148**, 3.

Wald, R. M. 1973, *J. Math. Phys.*, **14**, 1453.

Weber, J. 1960, *Phys. Rev.*, **117**, 306.

———. 1961, *General Relativity and Gravitational Waves* (New York: Interscience), esp. chapter 8.

———. 1969, *Phys. Rev. Letters*, **22**, 1302.

———. 1970, *Phys. Rev. Letters*, **24**, 6.

———. ed. 1977, Proceedings of the *International School of Cosmology and Gravitation, Course 4: Gravitational Waves* (held in Erice, Sicily, in March 1975) (New York: Plenum Press), in press.

Wheeler, J. A. 1955, *Phys. Rev.*, **97**, 511.

Will, C. M. 1974, *Ap. J.*, **190**, 403.

Wilson, J. R. 1972, *Ap. J.*, **176**, 195.

———. 1973, *Phys. Rev. Letters*, **30**, 1082.

Wolfe, A. M., and Burbidge, G. R. 1970, *Ap. J.*, **161**, 419.

Zel'dovich, Ya. B. 1971, *Pis'ma Zh. Eksp. Teor. Fiz.*, **14**, 270; English transl. in *Soviet Phys.—JETP Letters*, **14**, 180.

———. 1972, *Zh. Eksp. Teor. Fiz.*, **62**, 2076; English transl. in *Soviet Phys.— JETP*, **35**, 1085.

Zel'dovich, Ya. B., and Novikov, I. D. 1966, *Astr. Zhur.*, **43**, 758; English transl. in *Soviet Astr.—AJ*, **10**, 602 (1967).

———. 1967, *Relyativistskaya Astrofizika* (Moscow: Nauka); English transl.: Zel'dovich and Novikov (1971).

———. 1971, *Relativistic Astrophysics*, Vol. 1 (Chicago: University of Chicago Press); expanded translation of Zel'dovich and Novikov (1967).

Zel'dovich, Ya. B., Novikov, I. D., and Starobinsky, A. A. 1974, *Zh. Eksp. Teor. Fiz.*, **66**, 1897; English transl. in *Soviet Phys.—JETP*, **39**, 933 (1975).

Zerilli, F. J. 1970*a*, *Phys. Rev. Letters*, **24**, 737.

———. 1970*b*, *Phys. Rev. D*, **2**, 2141.

9 Singularities of Spacetime

Roger Penrose

1. Introduction

Combined evidence from observation and theory points to the existence of singularities in the spacetime continuum of our universe: in the first place, there is the big-bang singularity representing its initial state; second, there are the singularities inside black holes, representing the final state of collapse of massive stars. Other types of spacetime singularities might also exist. I shall review here our present understanding of the status of these singularities without, however, attempting to be comprehensive. This account will, to some extent, represent my own personal view. I am glad to be able to present this account as an article in Chandra's honor—particularly in view of the fact that it was his work of 40 years ago which led us, apparently inescapably, into the perplexities and singularities of the black-hole phenomenon.

Let us first ask what is meant by the phrase "spacetime singularity." In fact it has rather a spectrum of meanings, depending upon the context in which it is used. When speaking descriptively or in loose physical terms, we refer to singularities as regions of infinite spacetime curvature. Physically, spacetime curvature finds its expression in the gravitational tidal forces exerted on freely moving bodies. Thus, if a singularity is approached, we expect these tidal forces to mount to infinity. Ordinary physical objects could not survive too close to such a singularity. The tidal forces there would be expected to become so strong that molecules, atoms, and nuclei would in their turn be ripped to pieces; and there is no reason to suppose that elementary particles themselves would not be subject to the same fate. Tidal forces may be envisaged so large that they correspond to radii of spacetime curvature as little as the characteristic strong-interaction size: 10^{-13} cm. And it would seem that precious little of existing elementary-particle physics could survive for a background spacetime as curved as the elementary particles themselves. Furthermore, without particles with which to probe the geometry, it is hard to see how the description of a continuous background spacetime could then be meaningfully maintained.

This notwithstanding, it appears that the majority of physicists who actively concern themselves with quantization of (and in) gravitational fields believe that a drastic revision of physical theory is not required at this stage. Instead,

Roger Penrose is at the Mathematical Institute, Oxford, England.

they argue, the classical notions of spacetime geometry must hold good until radii of spacetime curvature are encountered of the order of the Planck length: 10^{-33} cm. But certainly at such more extreme situations there seems no question that new physics would be required. For this is the scale at which quantum fluctuations of the spacetime metric would be large enough to render the normal classical picture of a (pseudo-) Riemannian manifold completely inadequate as a picture of reality. Even if the required theory took merely the form of a successful application of standard quantization techniques to Einstein's equations (which, in my view, is most unlikely), this would nevertheless still have to qualify as "new" physics in view of the impasse to which such attempts seem so far to have led (see Isham *et al.* 1975).

Thus, the *physical* interpretation of the term "spacetime singularity" is a region in which curvatures have become so enormously large that our normal physical theories break down. (Whether one expects this to occur for radii of 10^{-13} cm, 10^{-33} cm, or some other value is a question to which I shall return later.) But from the point of view of a precise *mathematical* treatment, such an interpretation is rather negative and unmanageable. In the absence of a good physical theory of such high-curvature regions, we must concentrate, instead, on those regions where the normal pseudo-Riemannian picture of spacetime still holds good. We need assume not that the accuracy of our mathematical model is perfect, but that it holds good to a high degree down to the kind of dimension (say 10^{-13} cm or 10^{-33} cm) that we are concerned with. For radii of curvature smaller than that dimension, our model would be meaningless as a picture of reality. If desired, those portions of high curvature can be deleted from the model, leaving a pseudo-Riemannian manifold which is mathematically extendable to a somewhat larger one (namely, to the one obtained by reinstating the deleted portions). However, normally it is mathematically more convenient to include the highly curved portions and to work instead with a maximally extended manifold. In any case, the singularities themselves are not to be part of the pseudo-Riemannian manifold (which is viewed as consisting of mathematically "nonsingular" points only). Instead, these singularities constitute a kind of boundary (or portion of a boundary) for the spacetime manifold. (Considered in this way, the singularities for specific models need not always actually correspond to regions of infinite curvature.) The boundary can be constructed according to any of several precise (though unfortunately inequivalent) mathematical prescriptions. I shall describe only one of these prescriptions here (in § 2), namely, that of ideal points (or causal boundary), since it is easy to apply and seems closest to the physics; and a number of familiar examples will be discussed.

In § 3 I shall briefly review the reasons for believing that gravitational collapse, with or without symmetries, can lead inevitably to the existence of spacetime singularities; and I shall describe the standard picture of gravitational collapse to a black hole. When large deviations from spherical symmetry are present, the validity of the standard picture requires the hypothesis of cosmic censorship, according to which naked singularities are forbidden from arising in a collapse. I shall mention various versions of this hypothesis in § 4 and relate it to such questions as global hyperbolicity (the existence of a Cauchy hypersurface) and the principle of surface-area increase for black holes' event horizons.

A recent development in the theory of black holes is Hawking's quantum-mechanical evaporation process. This can have observational significance only if very small black holes (of radius down to 10^{-13} cm or less) can have been produced in a sufficiently chaotic initial big bang. But the process has in any case a considerable theoretical interest. I shall discuss (in § 5) some aspects of this and present some views of my own (which differ in certain essential respects from those of Hawking) as to the implications for thermodynamic time-asymmetry and its relation to the detailed structure of spacetime singularities.

2. TIPs and TIFs

There are various different ways that boundary points, representing singularities, may be adjoined to an (inextendable) spacetime manifold (Geroch 1968; Seifert 1971; Schmidt 1971, 1972; Hawking and Ellis 1973). The construction that has been regarded as the most mathematically satisfactory is the b-boundary of Schmidt (and its conformal extension the c-boundary; see Schmidt 1974). However, his procedure is rather nonintuitive and is difficult to apply in practice. Furthermore, according to a recent result of Bosshard (1975) it can sometimes produce anomalous results—even when applied to so apparently straightforward an example as a closed Friedmann universe. For such reasons I shall adopt here the much simpler notion of causal boundary (Geroch et al. 1972; Hawking and Ellis 1973). Though it may lack the richness of structure possessed by the b-boundary, and though there is the awkwardness that the past and future boundary points cannot easily be fitted together into a single scheme, there is a basic simplicity about the construction which makes the idea physically direct and easy to apply.

I shall suppose that the spacetime $\mathfrak{M}$ is strongly causal (although past- and future-distinguishing [Geroch et al. 1972] would suffice). Then any point $p \in \mathfrak{M}$ is uniquely defined by the set $I^+(p)$ ($\subset \mathfrak{M}$) of points lying to the chronological future of p. Similarly, p is uniquely defined by its chronological past $I^-(p)$. Writing $x \ll y$ for the statement "a future-directed timelike curve from x to y exists in $\mathfrak{M}$," where $x, y \in \mathfrak{M}$, we have (Penrose 1972) $I^+(p) := \{x \mid p \ll x\}$ and $I^-(p) := \{y \mid y \ll p\}$. Also, for any subset S of $\mathfrak{M}$ we use the notation

$$I^+[S] := \bigcup_{p \in S} I^+(p) \quad \text{and} \quad I^-[S] = \bigcup_{p \in S} I^-(p) .$$

Thus each point p may be represented either by the set $I^+(p)$ or by the set $I^-(p)$.

The advantage of representing p in this way is that a slight generalization of these sets enables us to represent not just ordinary points p of $\mathfrak{M}$ but certain extra "ideal points" which may be thought of as constituting a kind of boundary to $\mathfrak{M}$. This generalization rests on the fact that there are two intrinsic properties of the sets $I^+(p)$ and $I^-(p)$ which are almost, but not quite, sufficient to characterize them as sets of this kind; and the extra sets which arise give the ideal points.

In the first place, each set $I^+(p)$ is a *future-set* and each $I^-(p)$ is a *past-set*. (A future-set is a subset F which is its own chronological future: $F = I^+[F]$; a past-set P is its own chronological past: $P = I^-[P]$.) In the second place, each of $I^+(p)$ and $I^-(p)$ is actually *irreducible* in the sense that $I^+(p)$ cannot be expressed as the union of two future-sets each distinct from $I^+(p)$, and similarly for $I^-(p)$. An irreducible future-set is called an IF; and an irreducible past-set an IP. Those IFs which are of the form $I^+(p)$ for some $p \in \mathfrak{M}$ are called PIFs (proper IFs), while those *not* of this form are called TIFs (terminal IFs). Similarly, the IPs are subdivided into PIPs, having the form $I^-(p)$, and TIPs. By remarks above, the PIFs are in a natural one-to-one correspondence with the points of $\mathfrak{M}$; and, similarly, so are the PIPs. But the TIFs give us something extra, and so also do the TIPs. These correspond to the ideal points for $\mathfrak{M}$.

Roughly speaking, we may regard each TIF as a set "$I^+(q)$," where q is not a point of $\mathfrak{M}$ but an ideal *past boundary point* of $\mathfrak{M}$; similarly, each TIP may be regarded as "$I^-(r)$," where r is an ideal *future boundary point* of $\mathfrak{M}$. (For some spacetime models it seems natural to identify certain past boundary points with certain corresponding future boundary points, but I shall not enter into this here.) Thus, we think of the TIF itself as the set of points of $\mathfrak{M}$ lying to the (chronological) future of the ideal point which the TIF serves to define. Similarly, each TIP is the chronological past of the ideal point it defines.

There is another somewhat more intuitive way of discussing IPs and IFs which arises because of the following theorem (Geroch *et al.* 1972):

Theorem. *The IFs of $\mathfrak{M}$ are precisely the sets of the form $I^+[\gamma]$; and the IPs of $\mathfrak{M}$, those of the form $I^-[\gamma]$, where in each case γ is some timelike curve in $\mathfrak{M}$. The same holds if γ is allowed to be null.*

If γ has a past endpoint p, then clearly $I^+[\gamma] = I^+(p)$ and we have a PIF. The TIFs arise when γ is *past-endless* in $\mathfrak{M}$ (i.e., it continues indefinitely, in $\mathfrak{M}$, into the past, perhaps approaching a singularity). Correspondingly, $I^-[\gamma]$ is a PIP or TIP according as γ has a future endpoint or is *future-endless* in $\mathfrak{M}$. Thus, we may think of the past boundary of $\mathfrak{M}$ as supplying ideal past endpoints to timelike (or null) curves which are past-endless in $\mathfrak{M}$ and the future boundary as correspondingly supplying ideal future endpoints to future-endless timelike (or null) curves. For two such curves γ and γ', the condition for their past endpoints, whether ideal or not, to coincide is $I^+[\gamma] = I^+[\gamma']$; similarly the condition for their future endpoints, ideal or not, to coincide is $I^-[\gamma] = I^-[\gamma']$.

The ideal points thus constructed need not be singular points for $\mathfrak{M}$, but may be points at infinity. As an example, let $\mathfrak{M}$ be Minkowski space. If γ and γ' are (endless) timelike straight lines in $\mathfrak{M}$, then clearly $I^+[\gamma] = I^+[\gamma'] = \mathfrak{M}$ and $I^-[\gamma] = I^-[\gamma'] = \mathfrak{M}$, so γ and γ' share the same future ideal boundary point (called i^+) and also the same past ideal boundary point (called i^-). But there

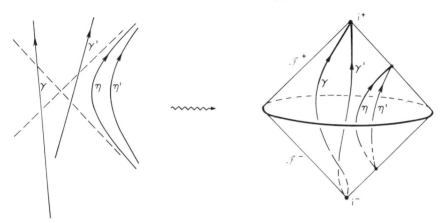

Figure 1. Timelike straight lines γ, γ' in Minkowski space acquire the ideal endpoints i^+ and i^-. The curves η, η' of uniform acceleration acquire ideal endpoints at null infinity $\mathscr{I}^+$, $\mathscr{I}^-$.

are also other boundary points for $\mathfrak{M}$. Consider the curve η of uniform acceleration, given in the usual coordinates by

$$x = A, \quad y = B, \quad z = (C + t^2)^{1/2}, \quad t \in (-\infty, \infty),$$

where A, B, and C are constants, with $C > 0$. Then the set $I^+[\eta]$ consists of the points of $\mathfrak{M}$ for which $t > -z$; and the set $I^-[\eta]$, the points for which $t < z$. Thus, whatever the values of A, B, and C, the curve η acquires the same pair of ideal points at both past and future infinity. If the z-equation is replaced by $z = t$, then we get a null straight line with the same future ideal point as before, but a different past ideal point (corresponding to the TIF given by $t > z$); if it is replaced by the null line $z = -t$, then the past ideal endpoint is the same as before, but not the future endpoint. If the z-equation is replaced by $z - D = [C + (t - D)^2]^{1/2}$, then the future ideal point is the same as before, but the past ideal point depends on D (with TIF given by $t + z > 2D$); if by $z + D = [C + (t - D)^2]^{1/2}$, then it is the future endpoint that depends on D. We can obtain other ideal points by simply performing a rotation in x, y, z to the above construction, the general TIF (other than $\mathfrak{M}$ itself) being given by

$$t > lx + my + nz + p,$$

and the general TIP (other than $\mathfrak{M}$) by

$$ t < lx + my + nz + p \, , $$

where $l^2 + m^2 + n^2 = 1$. The corresponding past and future ideal boundary points for $\mathfrak{M}$ constitute two three-dimensional systems (parametrized by l, m, n, p), each with topology $S^2 \times R$, called $\mathscr{I}^-$ (past null infinity) and $\mathscr{I}^+$ (future null infinity), respectively (see Figure 1).

As a second example, consider a Friedmann cosmological model with an initial "big bang" singularity. It is well known that such models possess *particle horizons*. Thus, if γ and γ' are the world-lines of two different substratum particles, then no point of γ' is "visible" from points of γ within a certain distance of the initial singularity; in other words, $I^+[\gamma']$ does not contain the whole of γ. It follows that $I^+[\gamma'] \neq I^+[\gamma]$, so γ' and γ have different ideal past endpoints. In fact, there is precisely one past boundary point corresponding to each sub-

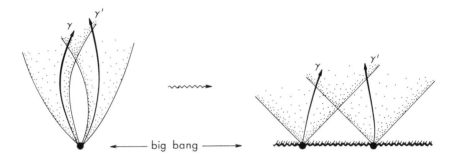

Figure 2. The TIFs representing the big bang define a three-dimensional system of singular ideal points.

stratum world-line. The initial singularity may be identified with this entire past boundary. It is spacelike (in a well defined sense to be described shortly) and has topology S^3 (if $k = +1$) or R^3 (if $k = 0$ or -1). In a similar way, we may consider future boundary points and associate them with *event horizons*. (The boundary of $I^-[\gamma]$ is the event horizon of the observer whose world-line is γ.) In a Friedmann universe model which recollapses to a final singularity, the structure of the future boundary is similar to that of the past boundary and both boundaries refer to singular points. But in a Friedmann model which expands indefinitely to infinity, the future boundary may differ considerably in structure from the past boundary. This is over and above the fact that the past boundary represents a singularity whereas the future boundary points are points at infinity.

In depicting the spacetimes with their boundaries attached, it is often convenient to use a two-dimensional representation in which null directions are oriented at $\pm 45°$. This entails that the big-bang singularity is depicted not as a single point but as a spacelike line (cf. Fig. 2). In Figure 3 a number of examples of familiar spacetimes are depicted. Each point of each diagram, with the exception of those points on the symmetry axis (and possibly certain odd points on the boundary), represents a sphere S^2. The symmetry axis (or axes) is depicted by a broken line (or lines), each point of which represents a single point of the spacetime. Past ideal boundary points or future ideal boundary points (normally thought of as distinct from each other—only in the case of anti–de Sitter space, or for the odd point in the extended Reissner-Nordstrom solution, would it be reasonable to "identify" certain past with future boundary points) are depicted by heavy straight lines if they are points at infinity and by jagged lines if they are singular points.

We need a criterion for distinguishing points at infinity from singular points. The simplest is the following (Penrose 1974a):

Definition. A TIP is called an ∞-TIP if it is of the form $I^-[\gamma]$ for some timelike curve γ of infinite proper length into the future; otherwise it is called a *singular TIP*. The ∞-$TIFs$ and singular TIFs are defined correspondingly.

Note that all the TIPs of Minkowski space are ∞-TIPs since they can be generated either by curves of uniform acceleration (for the points of $\mathscr{I}^+$) or timelike straight lines (for i^+), which are all of infinite proper length into the future. It may be remarked that any TIP in any spacetime can always be generated by some curve of *finite* proper length (for example, the future-finite curve $x = 0$, $y = 0, z + t = (z - t)^{-1/2} > 0$ in Minkowski space generates the TIP $t < z$) or, indeed, by a curve of zero proper length. However, singular TIPs can be generated *only* by such future-finite curves. So, roughly speaking, *all* observers reach the singular (future) ideal points in finite proper time, whereas the points at infinity can be "reached" by *some* observers only after infinite proper time.

There are certain alternative definitions for distinguishing points at infinity from singular points. Though apparently equivalent to the above in most "normal" circumstances, this need not be the case in every spacetime. One such alternative is the following:

Definition. A TIP (resp. TIF) is called *null-finite* if it is of the form $I^-[\gamma]$ (resp. $I^+[\gamma]$), where γ is a null geodesic of finite affine length; otherwise it is called *null-infinite*.

Each null-finite TIP and TIF may be thought of as defining a sort of singular ideal point. In fact, a null-finite TIP can be an ∞-TIP only if the curvature, in a certain well defined sense, is unbounded along the relevant null geodesic γ. So it would not be unreasonable to think of these TIPs as defining singular points, though perhaps they are best thought of as "singular points at infinity." One

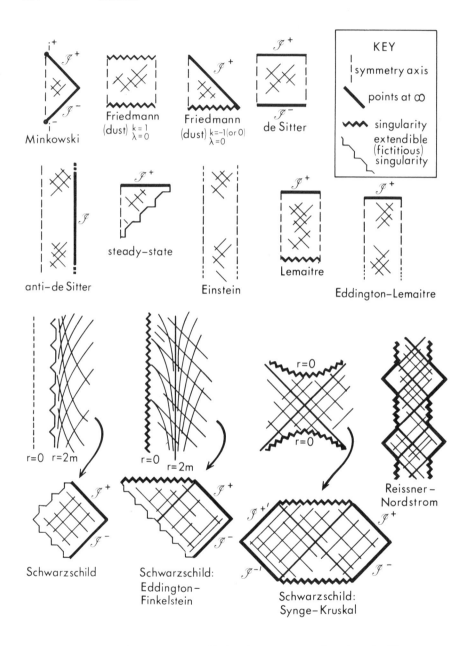

Figure 3. Diagrams illustrating the ideal point structure of various familiar space-times.

could also envisage certain null-infinite ∞-TIPs representing "singular" points at infinity (because of unbounded curvature along an appropriate generating curve γ). I shall not enter into a general discussion of this matter here, preferring simply to stick to the terminology given in the above definitions.

It is possible to define causal and chronological relations between TIPs and between TIFs in a simple way (Geroch *et al.* 1972). This enables one to discuss the spacelike, null, or timelike nature of the ideal point sets:

> **Definition.** Let P and Q be TIPs, and let F and G be TIFs. Then we say that *P causally precedes Q* if $P \subset Q$ and *F causally precedes G* if $G \subset F$. We say that *P chronologically precedes Q* if $P \subset I^-(x)$ for some $x \in Q$ and *F chronologically precedes G* if $G \subset I^+(x)$ for some $x \in F$.

Causal and chronological relations between TIPs and TIFs can also be defined but I shall not bother with them.

The above definitions can also be applied directly to IPs and IFs generally. In particular, when applied to PIPs and PIFs, it gives causal and chronological relations between points of $\mathfrak{M}$. However, in some spacetimes these are not quite the same as the usual ones (namely, x chronologically precedes y if there is a timelike curve in $\mathfrak{M}$ from x to y; x causally precedes y if there is timelike or null curve in $\mathfrak{M}$ from x to y) but differ from them in that ideal points may now be present in the chronological or causal path connecting the points.

The past boundary of $\mathfrak{M}$ is *acausal* (i.e., completely spacelike) if no two TIFs of $\mathfrak{M}$ are causally related to one another, that is, *no TIF contains another TIF*. It is *achronal* (i.e., completely spacelike or null) if no two TIFs are chronologically related to one another, that is, if *no TIF is contained in a PIF*. (Any PIF $I^+(x)$ is contained in some TIF, namely, $I^+[\gamma]$ where γ extends endlessly into the past from x. Thus if the TIF G is contained in the PIF $I^+(x)$, then $I^+[\gamma]$ chronologically precedes G; that is, we have two TIFs chronologically related.)

Thus, roughly speaking, the system of TIFs would be characterized as *timelike* by the fact that each TIF is contained in a PIF, as *null* by the fact that no TIF is contained in a PIF but each TIF is contained in another TIF, and as *spacelike* by the fact that no TIF is contained in any other IF. The situation for TIPs is similar.

3. Gravitational Collapse

Let us consider the standard picture (Penrose 1969, 1974a) of collapse to a black hole and see how the preceding ideas can be applied. First, there is the spherically symmetrical situation where a spherical body collapses inward, its external field being described by the Schwarzschild solution. The body collapses within its Schwarzschild radius $r = 2m$ and continues right down until the singularity at $r = 0$ is reached. The external field continues to within $r = 2m$, but the original static Schwarzschild coordinates cannot be retained there. Instead, a nonstatic coordinate must be used, such as those of Eddington-Finkelstein for which the metric becomes

$$ds^2 = \left(1 - \frac{2m}{r}\right) dv^2 - 2dvdr - r^2(d\theta^2 + \sin^2\theta d\phi^2)$$

where v is an advanced time parameter.

The situation is depicted in Figure 4, where the representation with null lines

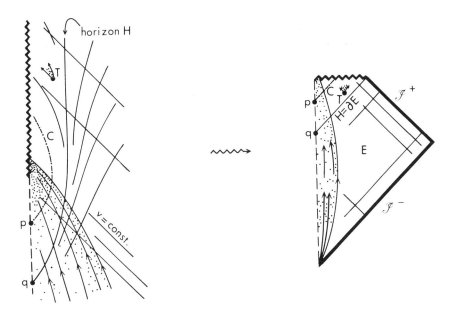

Figure 4. The standard picture of spherically symmetric collapse to a black hole.

at 45° is also given. The TIFs of the spacetime form a system essentially identical with that for Minkowski space. Thus, there is an ideal point i^- (which supplies a past endpoint to any timelike geodesic that is past-endless in $\mathfrak{M}$) corresponding to the TIF $\mathfrak{M}$; and there is a three-dimensional system $\mathscr{I}^-$ of ideal points (supplying past endpoints to the null geodesics past-endless in $\mathfrak{M}$) which is null in the sense described in the last paragraph of § 2. All the TIFs are ∞-TIFs (and null-infinite).

As to the TIPs, we have, similarly to the case of Minkowski space, a null three-dimensional system of ∞-TIPs defining the set $\mathscr{I}^+$ of ideal points of future endpoints of, for example, the various outgoing radial null geodesics. There is also a single ∞-TIP $E = I^-(i^+)$ which is actually the union of all the other ∞-TIPs. The ideal point i^+ is the future endpoint of each timelike geodesic which has infinite proper length into the future. It is also the future endpoint of, for example, the null geodesics which orbit endlessly about the final black hole at $r = 3m$, or of the null geodesic generators of the black hole's

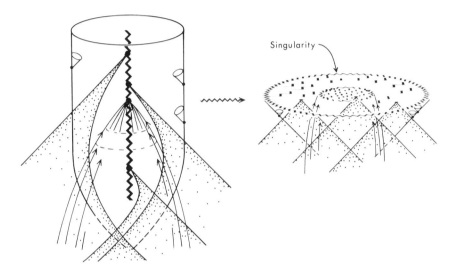

Figure 5. The singular TIPs in spherical collapse define a three-dimensional spacelike set.

absolute event horizon. The ideal point i^+ lies causally to the future of all the points of $\mathscr{I}^+$.

There are also some singular TIPs. These form a spacelike three-dimensional system. This may be ascertained by examining in detail the behavior of the null cones of the Eddington-Finkelstein metric in the neighborhood of $r = 0$ (see Fig. 5). The spacelike nature of the singular part of the future boundary implies that no observer falling into the black hole can actually "see" the singularity before he hits it. In fact any two such observers, unless they share the same future ideal endpoint, must fail to observe the final portions of each other's world-lines. Thus we have the situation of (observer dependent) event horizons (boundaries of TIPs)—a situation familiar in, for example, the de Sitter cosmology. These event horizons do not, of course, coincide with the *absolute event horizon* (Penrose 1969) H, which is the boundary of E:

$$H = \partial E = \partial I^-(i^+) = \partial I^-[\mathscr{I}^+] \,.$$

In the vacuum region, ∂E consists of the Schwarzschild hypersurface $r = 2m$, while in the matter region it is a null hypersurface—the light cone of some point q at the center of the collapsing body—which fits smoothly on $r = 2m$ at the surface of the body.

This is the standard collapse picture when exact spherical symmetry is assumed. If deviations from sphericity are small, then, since the perturbation analysis of the Schwarzschild solution (Regge and Wheeler 1957; Vishveshwara 1970; Chandrasekhar 1975) indicates stability for spherical black holes, we do not expect any significant deviations from the spherically symmetrical picture—at least in the region from the neighborhood of $r = 2m$ outward. On the other hand, if deviations from sphericity are moderately large, then we cannot rely on perturbation analysis. We can, however, appeal to the singularity theorems to derive the fact that a spacetime singularity, analogous to $r = 0$ in the symmetrical case, will still exist, provided that a suitable criterion of "irreversible collapse" has been met. For such a criterion we recall one of the singularity theorems:

Theorem (Hawking and Penrose 1970). *If a spacetime $\mathfrak{M}$ is such that:*

a) *the timelike convergence condition*[1]

$$R_{ab}t^a t^b \leq 0 \text{ for all timelike } t^a$$

holds at every point of $\mathfrak{M}$,

b) *there are no closed timelike curves in $\mathfrak{M}$,*

c) *the generality condition that*[2]

$$l_{[a}R_{b]cd[e}l_{f]}l^c l^d \neq 0$$

holds somewhere along every timelike or null geodesic, its tangent vector being l^a, is valid in $\mathfrak{M}$,

d) *either* (i) *$\mathfrak{M}$ contains a closed spacelike hypersurface S*

or (ii) *$\mathfrak{M}$ contains a trapped surface T (a closed spacelike two-surface whose null normals are everywhere converging into the future)*

or (iii) *$\mathfrak{M}$ contains a point p such that the divergence of the system C of null geodesics through p changes sign somewhere to the future of p;*

then $\mathfrak{M}$ is singular in the sense of containing a future- or past-endless timelike or null geodesic of finite affine length.

There is another version of this theorem of more relevance to the definitions given here, which can be obtained by suitably modifying the proof of the original version. This is:

---

[1] If Einstein's equations without cosmological term $R_{ab} - \frac{1}{2}Rg_{ab} = -8\pi T_{ab}$ hold, then this is called the *strong energy condition* (Hawking and Ellis 1973) (or simply the energy condition), namely, $T_{ab}t^a t^b \geq \frac{1}{2}Tt_a t^a$ for all timelike t^a. The Ricci tensor sign is chosen consistent with this.

[2] Square brackets denote skew-symmetrization.

Theorem (Penrose 1975). *If a spacetime* $\mathfrak{M}$ *is such that conditions* (*a*), (*b*), (*c*), *and* (*d*) *of the previous theorem hold, then* $\mathfrak{M}$ *is singular in the sense of containing a singular TIP or a singular TIF.*

In these results, condition (i) of (*d*) is relevant only to closed cosmological models, while either (ii) or (iii) can be used as a criterion characterizing irreversible collapse. Referring to Figure 4 for the spherically symmetrical case, we find that a two-surface T lying in the vacuum region and defined by v = const., r = const. $< 2m$, is an example of a trapped surface. Its null normals are the pairs of null geodesics through T lying in the various (θ = const., ϕ = const.)-planes. Since both these systems of null geodesics proceed into the future to decreasing values of r, it follows that their areas of cross-section must be decreasing in future directions (since each sphere v = const., r = const., has area $4\pi r^2$), that is, these null normals are converging, in accordance with the definition of a trapped surface. Similarly the future light cone C of the point p indicated in Figure 4 moves in toward decreasing values of r after initially diverging away from p. Thus C satisfies the condition required of it for (iii).

Now, the point about the properties (ii) and (iii) is that they are inequalities of the kind whose validity we would expect to be unaffected if a fairly small, but finite, change is made in the spacetime geometry. Thus we can envisage a collapse which deviates, to a certain degree, from exact spherical symmetry; and we would still expect (ii) or (iii) to hold. The geometry near $r = 0$, where the singularity occurs in the symmetrical case, might well be drastically changed. But our theorem tells us that while the detailed structure of the singularity may now be quite different, a spacetime singularity of some sort must still be present. This, of course, is supposing that the remaining requirements (*a*), (*b*), and (*c*) still hold. In fact, they are reasonable requirements for a classical spacetime described according to general relativity, where (*c*) would be expected to hold in all but a very few special cases, of "measure zero" among the totality of possible models. [Curiously, though, (*c*) tends to fail for the known exact solutions. In particular, it fails for the spherically symmetrical collapse just considered!]

The use of (iii) as a criterion of irreversible collapse is perhaps the easiest to visualize. We imagine matter (perhaps a cloud of gas or stellar material, perhaps a collection of stars, or even of galaxies) falling inward in the general direction of some central point p. A flash of (idealized) light is emitted from p, which travels outward describing the spacetime cone C. The infalling matter crosses C and in doing so will cause a certain amount of focusing of the null rays in C. (Of course, this focusing effect on light rays passing near or "through" a material body is, in the case of the Sun, one of the *observed* effects of general relativity.) If sufficient matter crosses C in a sufficiently small region (a condition easily met with quite small densities, provided the total mass is large enough), then the divergence of the null rays in C will change to a convergence in each direction and condition (iii) will be satisfied.

Once this happens, we know from the theorem that we have a singularity. But we do not yet know whether we have an absolute event horizon enclosing the singular region and preventing, as in the spherically symmetrical case first considered, any information escaping from the neighborhood of the singularity to external infinity. If a singularity is visible[3] from infinity, it is called a *naked singularity*. If there is a region near the singularity which is not visible from infinity, then we have a (nonvacuous) *absolute event horizon*, namely, the boundary $H = \partial E$ of the set E of events from which timelike curves to infinity exist. There is also the possibility that there might be both a naked singularity *and* an absolute event horizon. But if there is only the horizon and no naked singularity, this is the situation of a *black hole* (or holes). Thus, we may conclude, roughly speaking, that if irreversible collapse takes place (and general situations can be set up so that it does), then it must result in a black hole (or holes) *unless* a naked singularity occurs—an alternative possibility which would normally be viewed with considerably more alarm than that of a black hole! For whatever unknown physics actually takes place at a spacetime singularity, its effects would be relevant *observationally* if and only if the singularity is a visible one.

4. Cosmic Censorship

The hypothesis that a physically realistic collapse will not result in naked singularities is called *cosmic censorship* (Penrose 1969). It is not known at present whether some form of cosmic censorship principle is actually a consequence of general relativity. In fact, we may regard this as possibly the most important unsolved problem of classical general relativity theory. The question is not simply whether solutions of Einstein's equations exist which possess naked singularities, or even whether solutions exist which represent the collapse of a perfectly regular initial configuration to a naked singularity. Various such solutions are already known, but they represent special situations that we have no reason to believe would arise in a realistic collapse. It is, rather, whether such collapse configurations lead *stably* to the production of naked singularities, the equations of state for the collapsing matter also not being, in some appropriate sense, too special or physically unrealistic.

Indeed, it may well be that the framing of an appropriate mathematical statement of the cosmic censorship principle could represent, in itself, a major step toward a solution of the problem. The examination of particular exact models would seem to be of rather limited value. What is required is an under-

---

[3] The term "visible" is here being used somewhat loosely, but meaning, roughly, that a timelike or null curve exists from the singularity to external infinity.

standing of the generic case. It would not, it seems, even be greatly helpful to have a full appreciation of the generic static or stationary solution of Einstein's equations. We know from the Israel-Carter-Hawking-Robinson theorem (Israel 1967; Carter 1971, Hawking 1972; Robinson 1975) (essentially) that the stationary asymptotically flat vacuum spacetimes which are not described by Kerr metrics (with $a \leq m$) cannot be black holes; so presumably these all have naked singularities. From this point of view it is not surprising that the various (non-Kerr) stationary asymptotically flat vacuum solutions which have been proposed at various times (Weyl - Levi Civita, cf. Curzon 1924; Tomimatsu-Sato 1972; Witten 1974; etc.) all possess naked singularities. They tell us little, if anything, about the validity of cosmic censorship. In those circumstances in which a collapse does result in a black hole, the process of settling down to a stationary state leads to all multipole moments of quadrupole or higher character becoming adjusted to those values given by the Kerr field (so it is to be expected) with the emission of gravitational radiation. So *if* cosmic censorship is true, these non-Kerr stationary solutions have no direct relevance.

As regards the nonstationary asymptotically flat collapse models with naked singularities (e.g., Yodzis *et al.* 1973; Steinmüller *et al.* 1975; Szekeres 1975; cf. also Grischuk 1967), two main queries refer to the equations of state and to the generality of the solutions. The use of a pressureless fluid (dust) in some of these models raises some questions since we expect caustics, and hence infinite densities, in the fluid world-lines. While it might be that a very soft equation of state (e.g., that of Hagedorn 1968) has relevance when the situation is extreme enough, the question must be raised as to why the situation should become that extreme in the first place unless the initial setup involved a considerable amount of delicate (and intrinsically improbable) careful alignment. This brings us back to the question of generality. It would seem that this is really the key issue. Exact models might technically possess naked singularities which could be argued to be of a "harmless" kind (like those arising in other fields of physics where delta-functions are employed as a mathematical convenience, not being supposed to imply that the physical quantities that they describe are actually infinite). The issue at hand is whether such singularities would persist when the initial conditions—*and* equations of state—are replaced by something appropriately generic. It is not yet clear how this could be satisfactorily formulated, particularly as regards the equations of state. However, we might hope to express the general question in the form: "Is the production of naked singularities in a collapse a stable property?" If it never is, then we could say that the cosmic censorship hypothesis is valid.

There is a possible line of attack for obtaining a contradiction with the cosmic censorship hypothesis which depends on the fact that one of its implications, appropriately stated, is the *principle of area increase* for the absolute event horizon. The first published precise version of this principle was given by Hawking (1972). Some alternative versions (which are rather more general) will be stated shortly. By use of the area principle we can infer that certain inequalities must hold which relate the geometry of the initial setup to the surface area of the final black hole. This area, in turn, relates to the mass of the final configuration; and since this mass cannot exceed the initial mass, we arrive at a consistency condition on the initial geometry. In appropriate circumstances (e.g., collapsing shells of radiation [Penrose 1973], time-symmetric initial state for black holes [Gibbons 1972], colliding extreme relativistic particles or black holes [Penrose 1976]) this can lead to precise mathematical inequalities whose validity is a consequence of cosmic censorship. In no case has a situation been yet found in which these inequalities are violated, whereas in some fairly general circumstances proofs of the inequalities have actually been obtained (Gibbons 1972; Penrose 1976; Jang 1976). This may be regarded as providing some indirect support for cosmic censorship.

However, it could also be argued that the cases which have so far been handled have simply not been concerned with situations which are aspherical enough for cosmic censorship to be violated. As the stability analysis of black holes (Teukolsky 1973; Press and Teukolsky 1973; Hartle and Wilkins 1974; Chandrasekhar and Detweiler 1976; Stewart 1975) already appears to show (and as do arguments involving test fields on black hole backgrounds [Penrose 1973]), one might in any case expect that any violation of cosmic censorship would have to involve a collapse configuration rather far from the original spherically symmetrical exact model. One possibility, for example, is that sufficient angular momentum would be required, corresponding to a value given by a Kerr metric with $a > m$, that is, with angular momentum (in gravitational units) larger than the mass squared. If such a configuration were able, without loss of angular momentum, to collapse down to a Kerr $(a > m)$ state (with central ring singularity spanned, say, by a disk of matter), then we should have a naked singularity. This is an entirely speculative possibility. But it serves to illustrate that catastrophic gravitational collapse to a naked singularity cannot, at the moment, be ruled out. And there seems no reason why even conditions (d)(ii) and (d)(iii) of the singularity theorems of § 3 need ever be satisfied for such a collapse.

As mentioned earlier, even a clear mathematical statement of the cosmic censorship hypothesis is at present lacking. On the other hand, if we ask merely for a criterion which can be applied to a *given* spacetime that it be free of naked singularities, then a number of different definitions can be given. Three of these are Hawking's conditions (Hawking 1972; Hawking and Ellis 1973) of *future asymptotic predictability*, *strong future asymptotic predictability*, and *regular predictability* (in order of increasing restrictiveness, other properties than just absence of naked singularities being also incorporated). Hawking shows that in a regular predictable spacetime the principle of area increase must hold. The conditions are intended to apply only to spacetimes which are asymptotically flat, and they become a little complicated to state precisely. I shall not bother to give them in detail here (cf. Hawking and Ellis 1973), but suggest some other criteria which are somewhat simpler to state and of considerably wider applicability.

The first point to make is that if a hypothesis for excluding naked singularities is to be adopted, this hypothesis should not exclude singularities of the big-bang type—for otherwise one would presumably be ruling out the actual universe![4] Thus, although we expect the big-bang singularity to be "visible," it should not be classed as a proper naked singularity. We may make a distinction by requiring that a true naked singularity q be "naked" also in the time-reversed sense (Penrose 1974a, b). Thus, not only shall we require some point r lying to the future of the singularity q, but also some other point p lying to the past of q. Whether these past and future relations are to be taken in the causal or chronological sense makes no essential difference. (The exact locations of p and r can be redefined if necessary to accommodate this.) Since there are no points to the past of the big bang (according to normal ideas), it is *not* a naked singularity. The singularity q may be described as a singular TIP or else a singular TIF. In the former case (taking q to precede r causally) we have this TIP contained in the PIP $I^-(r)$. The TIP (by virtue of the fact that it *is* a TIP) will also have points lying to its (chronological) past. So the condition that this TIP represent a naked singularity is that it be a *singular TIP contained in a PIP*. Similarly, the condition that a TIF represent a naked singularity is that it be a *singular TIF contained in a PIF*.

Two possible statements which state absence of naked singularities for the spacetime $\mathfrak{M}$ are, therefore,

CC1: no PIP contains a singular TIP

and

CC2: no PIF contains a singular TIF.

We might also choose to exclude "naked points at infinity" to obtain the stronger statement (Penrose 1974a)

CC3: no PIP contains a TIP,

which is actually *equivalent* to its time-reverse

CC3: no PIF contains a TIF

because each can be shown to be equivalent to the time-symmetric statement

CC3: $\mathfrak{M}$ is globally hyperbolic.

---

[4] Condition (d)(iii) of the singularity theorems can be used in time-reversed form, where it states that the past null cone of some point p is refocused, i.e., that small objects of fixed intrinsic size reach a minimum angular diameter, as viewed from p, when at a certain distance from p. This is actually a condition whose observational feasibility is seriously considered by cosmologists. If satisfied, it implies that the big-bang singularity will persist even after generic perturbation.

A spacetime $\mathfrak{M}$ is said to be globally hyperbolic (Hawking and Ellis 1973; Penrose 1972; Geroch 1970) if it is strongly causal and if $I^+(x) \cap I^-(y)$ has compact closure for each x, $y \in \mathfrak{M}$. By a theorem of Geroch (1970), this is equivalent again to

CC3: $\mathfrak{M}$ admits a global Cauchy hypersurface.

Using the definitions at the end of § 2, we can reinterpret CC1 as the statement that the future singular boundary of $\mathfrak{M}$ is nowhere timelike, and CC2 as the statement that the past singular boundary is nowhere timelike. Thus, we may envisage cosmic censorship as stating something like:

"timelike singularities are unstable,"

so that in a generic collapse we would not expect timelike singularities to occur at all.[5]

In this form, the cosmic censorship hypothesis is rather stronger and more "local" than versions which depend on observations of singularities from infinity. But I feel that it offers more scope for possibilities of finding a general proof. One would expect, after all, that the formation of a singularity should be governed by reasonably local matters and not by how it is related to infinity.

The singularity formed at the center of the spherically symmetrical collapse described in § 3 is spacelike, as we have seen, so CC1, CC2, and, indeed, CC3 are all satisfied. When spherical symmetry is not assumed, the matter needs more careful consideration, however. In fact, for the exact Kerr metric with $a < m$ we have a spacetime rather similar to that of Reissner-Nordstrom, as depicted in Figure 3, and CC1 and CC2 are both violated. Thus, a suitable observer who falls into such a black hole can see the singularity; to *him* the singularity is actually naked (see Fig. 6). But there are reasons for believing that the spacetime and its singularity structure inside the event horizon are unstable (Simpson and Penrose 1973) (despite the probable stability of the spacetime from the absolute event horizon outward). It seems likely that a generically perturbed Kerr or Reissner-Nordstrom solution develops a spacelike (or null) singularity in the neighborhood of its "inner" (or Cauchy) horizon (Hawking and Ellis 1973; Penrose 1972) (see Fig. 6), so that CC3 becomes valid. (If this is the case, then the possibility, sometimes envisaged, of escaping to a "new universe" by entering a rotating black hole must be ruled out.)

If a cosmic censorship principle along this sort of line could be established, then even that would not be a completely satisfactory result. For suppose a gravitational collapse takes place which results in a singularity whose effects are visible from infinity, but where, unlike the comparatively mild type of naked singularity considered earlier, this singularity emits a "pulse" of gravitational

---

[5] The naked ring singularity of the $a > m$ Kerr metric may repay examination. Since rather few null rays escape the singularity, there is the chance that it might not decay as rapidly as "normal" timelike singularities.

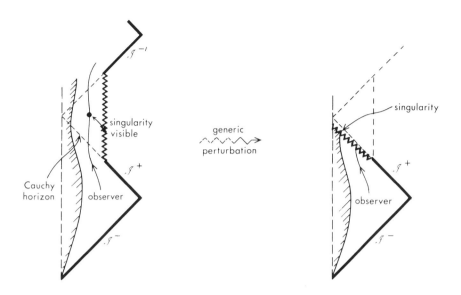

Figure 6. An observer crossing the Cauchy horizon in charged spherical collapse sees an apparently naked singularity. But this situation seems unstable against generic perturbation.

radiation of infinite amplitude—a null shell of singular curvature which destroys the universe as it expands! This should be the "naked singularity to end all naked singularities"—but nevertheless CC3 could be satisfied for the spacetime (see Fig. 7). A hypothesis to exclude such a possibility would be:

    CC4: no ∞-TIP contains a singular TIP;
and we might also postulate its time-reverse

    CC5: no ∞-TIF contains a singular TIF.
A more local version of these would be to exclude *null* singular boundary points by requiring (in addition)

    CC6: no singular TIP contains another singular TIP
and

    CC7: no singular TIF contains another singular TIF.

As a test of the utility of such postulates, we may ask whether they imply a principle of area increase for the absolute event horizon. In fact there is a variety of such results. I shall only state two of these here. First, a general definition of absolute event horizon H is required:

$$H = \partial E ,$$

where E is the "external region." We have some choice for the precise definition

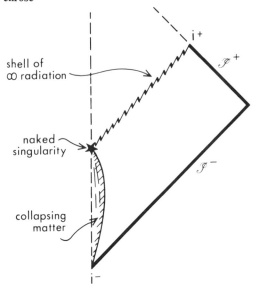

shell of
∞ radiation

naked
singularity

collapsing
matter

Figure 7. A hypothetical situation in which a "naked" singularity is formed emitting a shell of radiation of infinite intensity—which destroys the universe!

of E. The simplest would seem to be

E = set of all past endpoints of timelike curves of infinite length into the future

= union of set of all ∞-TIPs.

One might prefer, on the other hand, simply to single out some suitable subset of the ∞-TIPs, or perhaps of the null-infinite TIPs, and designate them as "external", calling them E-$TIPs$,[6] their union being, by definition, the set E. For the present purposes it makes no difference, although the characterization "absolute" for the event horizon would, in the latter case, need justification.

If S_1 and S_2 are two spacelike two-surfaces on H (possibly disconnected, but normally cross-sections of H), we can say that S_2 lies to the future of S_1 on H if S_1 is in the causal past of S_2. Since H is an achronal set (being the boundary of a future-set), this means that each point of S_1 is the past end-point of a null geodesic on H meeting S_2. The following result has been established (Penrose 1974a, 1975):

Theorem. *If $\mathfrak{M}$ satisfies CC3 and the timelike convergence condition (a), and if S_2 lies to the future of S_1 on H, then*

$$area(S_2) \geq area(S_1) .$$

---

[6] For example, those corresponding to the points of $\mathscr{J}^+$, for a weakly asymptotically simple spacetime. Then (Penrose 1969) $E = I^-[\mathscr{J}^+]$.

Another version of the area principle[7] which is considerably easier to establish uses a modified form of CC4, namely,

CC8: no E-TIP contains a null-finite TIP.

Then one can show:

> **Theorem.** *If $\mathfrak{M}$ satisfies CC8 and the null convergence condition,[8] and if S_2 lies to the future of S_1 on H, then*
>
> $$area(S_2) \geq area(S_1)$$

5. Quantum Effects and Time-Asymmetry

The discussion to this point has been entirely classical. In the normal way, one does not expect quantum-mechanical effects to have significance in gravitational theory owing to the enormous disparity in the scales involved—gravitation being normally relevant only for masses much larger than 10^{-5} g and quantum mechanics for masses much smaller than 10^{-5} g. However, one expects the presence of large curvatures near singularities—and quantum effects relating to gravity are likely to be important there. This could well be the case even in situations where the spacetime can still be treated as a classical background, the matter fields being described quantum-mechanically. For then there can be local violation of energy conditions, in addition to particle production effects. One might expect that when averaged out on a larger scale these would not *normally* lead to violation of the timelike convergence condition. But it is in any case apparent that the various classical results like the singularity theorems and area-increase principle can be trusted to have detailed validity only to the extent that such small-scale violations can be ignored.

These remarks have relevance to the question of the behavior of mini–black holes, which Hawking suggested several years ago (Hawking 1971) could be produced in a sufficiently chaotic initial big bang. They might have masses down to 10^{15} g (radius 10^{-13} cm) or even down to 10^{-5} g (radius 10^{-33} cm). More recently, Hawking has shown (Isham *et al.* 1975; Hawking 1975) that a remarkable phenomenon occurs, of significance for such mini–black holes, namely, that they would emit blackbody radiation at a temperature inversely proportional to the black-hole mass. This effect would also be present for large black holes, but utterly insignificant for ordinary astrophysics, the temperature associated with a solar-mass black hole being less than 10^{-7} K.

But a hole initially small enough that its temperature exceeds the tempera-

---

[7] Yet a different version has just recently been given by Gibbons and Hawking (1975).

[8] This is a weakened version of the timelike convergence condition (a) for which the timelike vector t^a is replaced by a null vector (Hawking and Ellis 1973). It has the advantage that Einstein's equations can be used with cosmological term, and it then becomes the weak energy condition—which is a consequence of the positive-definiteness of local energy.

ture of the background radiation in which it is immersed would emit more radiation than it absorbs. As it does so, it loses energy and therefore mass. Thus, its surface area likewise decreases, in violation of the area-increase principle. This can be attributed to a violation of the weak energy condition, in accordance with the fact that this is a situation where the matter fields must be treated quantum-mechanically.

As the mass goes down, the temperature goes up, and the effect is accelerated. For a mass smaller than about 10^{17} g, electrons and positrons would be emitted. If the mass were as small as about 10^{14} g, the temperature would reach 10^{12} K, which, according to some strong interaction models such as that of Hagedorn (1968), should be a maximum temperature. Instead of the temperature rising further, more and more species of elementary particles would be produced, and the black hole would presumably disappear in an almost instantaneous explosion (10^{35} ergs in about 10^{-23} s). Alternatively, if there were no such maximum temperature, the black hole might continue to reduce in size and mass until it reached a size of 10^{-33} cm—the Planck scale, at which quantum fluctuations in the metric would be large enough to preclude any classically meaningful spacetime geometry. The lifetime for a hole of mass 10^9 g would be about a tenth of a second, and 10^{30} ergs would be emitted. For 10^{14} g the lifetime would be 10^7 years and for 10^{15} g, a Hubble time (being inversely proportional to the cube of the mass).

For those, such as myself, who would like to believe that radii of spacetime curvature smaller than the strong-interaction scale of about 10^{-13} cm (or so) do not make consistent physical sense, there is a certain appeal in the Hagedorn model. A black hole of 10^{-14} cm would evaporate essentially instantaneously! At this final stage, the normal spacetime notions would be violated and we have the situation of a naked singularity.[9] Thus, with the Hagedorn strong-interaction model, this singularity already arises at the regime where radii of spacetime curvature are of the general order of the strong-interaction length.

In Figure 8, spacetime diagrams (analogous to those of Fig. 4) are given for the entire process of formation and final evaporation of the black hole. It is being assumed that the final naked singularity is ephemeral, no residual singularity of any kind being left in the spacetime which persists with time. There is just one singular TIF, and it violates CC2. Of the singular TIPs, apparently an S^2's worth violates CC1. Nevertheless, this naked singularity[10] cannot really be considered to be timelike. We note that these TIPs also violate CC4. However,

---

[9] This is not considered to be in violation of the cosmic censorship hypothesis, which is concerned with *classical* general relativity only.

[10] Although represented by a single point in Fig. 8, this naked singularity appears to have a different structure as a TIF than as a set of TIPs (cf. Geroch *et al.* 1972).

CC6 and CC7 are both satisfied for the whole spacetime, showing that by themselves they are insufficient to characterize absence of naked singularities.

Whether or not mini–black holes would be expected to exist in the actual universe depends on how chaotic the initial big bang was. With a soft equation of state (effectively zero pressure at high densities) such as that implied by the Hagedorn model, mini–black holes would be produced with a comparatively uniform (i.e., unchaotic) big bang. If the Hawking explosions are not observed (and some strong observational limits on this are already at hand [Carr 1975]), then this may be construed as implying a high degree of small-scale uniformity for the big bang. In fact, I shall attempt to present a completely different theoretical reason for believing in a uniform big-bang singularity. The main objection to this point of view, on the other hand, comes from the problem of galaxy formation. No plausible mechanisms have yet been put forward which do not trace back the origins of galaxies to some initial irregularities of the big bang. My own view on this is that we should wait and see if some alternative proposal for galaxy formation may not eventually be forthcoming.

The main interest in the Hawking evaporation process is, in my view, a purely theoretical one. The idea had already been put forward by Bekenstein

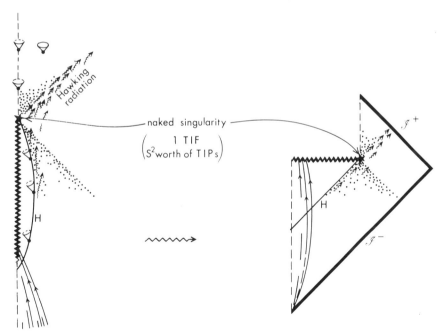

Figure 8. Spherical collapse to a black hole which subsequently evaporates via the Hawking process to a naked singularity. The ideal point structure is illustrated.

(1973, 1974) that an entropy

$$S_{\text{bh}} = \frac{1}{4} \frac{k}{\hbar} A$$

(in gravitational units, k being Boltzmann's constant, the precise value of the coefficient $\frac{1}{4}$ having been supplied later by Hawking [Isham *et al.* 1975; Hawking 1975]) should be attributed to any stationary black hole of surface area A. This he makes very plausible by means of thought experiments, where one attempts to violate the second law of thermodynamics by lowering high-entropy matter or radiation into the hole. We note the presence of $\hbar$ in this formula, which is essential for dimensional reasons. Thus, such an identification of a black-hole entropy with the area of the horizon makes physical sense only by virtue of the fact that it is *quantum* physics which is involved.

This identification of entropy leads directly (Bardeen *et al.* 1973) to the association of a *temperature* with the quantity (inversely proportional to the black hole's mass in the case of a nonrotating hole) known as the black hole's *surface gravity* (Hawking and Ellis 1973). It is this very temperature which features in the Hawking evaporation process. In fact, the Hawking process is *necessary* for the Bekenstein entropy expression to be thermodynamically consistent. But when these two features are incorporated together, there is then a remarkable elegance and coherence in the whole scheme, linking, as it does, apparently disparate elements from general relativity, thermodynamics, and quantum field theory.

However, there are also some unusual features. In the first place, the black hole has a negative specific heat. This can lead to unstable situations such as that already discussed, where a black hole can get out of equilibrium with its surroundings, the final result of this runaway situation being (in the case considered) a naked singularity. Second, there is a time-asymmetry present in the whole setup. This is manifest particularly in the geometry of Figure 8. Even if we delete the entire portion of the spacetime inside the horizon, we do not get a time-symmetrical picture. Not least, there is the striking fact that an apparently completely objective feature of the spacetime geometry has been put equal to a thermodynamic quantity—which one might have thought depended upon some subjective notion of coarse-graining.

However, despite these features, Hawking has been led to consider the hypothetical state of equilibrium (maximum entropy) of a system of fixed mass, constrained in a box with perfectly reflecting walls. When the ratio of mass to box volume lies within an appropriate range, he finds this state of equilibrium to consist of a black hole in equilibrium with blackbody radiation. He then argues that general thermodynamic considerations should imply that this configuration is time-symmetric, so that a black hole should be physically indistinguishable from a white hole (time-reverse of a black hole). Though the normal accretion process whereby a black hole acquires mass is a purely classical one, and the evaporation process whereby it loses mass is quantum-mechanical, Hawking (1975) argues that they are physically time-reverses of one another!

Striking as these ideas undoubtedly are, it seems to me that they run into severe difficulties when examined in detail. For example, it appears that large fluctuations away from equilibrium do not behave time-symmetrically (Penrose 1976). More alarming is the fact, already mentioned, that the spacetime geometry of Figure 8 is not time-symmetrical, so we are led to the view, if we accept Hawking's arguments in full, that this geometry is not objectively meaningful, even at the level of macroscopic dimensions (say of a solar-mass size black hole a few kilometers across).

My own view is that this remarkable link between black holes and thermodynamics which is supplied by the Bekenstein-Hawking ideas is really trying to tell us something different which may eventually help to solve two of the long standing problems of physics. One of these concerns the objectivity of quantum phenomena and the vexed question of the "reduction of the wave function" when an "observation" is made. I do not propose to enter into a discussion of this here, but merely point out that the question of the objectivity (i.e., observer independence) of spacetime geometry has already been mentioned above in two somewhat different contexts. The other question has to do with the statistical time-asymmetry of physics.

I take the view that, at least in normal physical circumstances and where spacetime curvature radii are much larger than 10^{-13} cm, the geometry of spacetime can be treated objectively. Then black holes and white holes (of mass greater than 10^{15} g) are different. Furthermore, it seems to me reasonable that only the black holes should exist in our universe. Since each represents a valid solution of the Einstein equations in its nonsingular part, the reason for ruling out the white hole would be that its singularity represents an unacceptable boundary condition. The view would be that it is a "high entropy" singularity acceptable only as a final boundary condition and not as an initial one. Singularities of even higher entropy could arise as the result of an asymmetric collapse to a black hole, since in this case even more information is lost in the hole than when spherical symmetry is assumed. The big bang, on the other hand, must be a low-entropy singularity. The initial *matter* distribution is, after all, normally assumed to be in thermodynamic equilibrium. The reason that the universe does not start off in a state of maximum entropy is presumably that the gravitational degrees of freedom are not used (not "thermalized") initially. With gravitation, it is irregular states that have high entropy and the smooth ones that have low entropy (in contradistinction with the normal situation). We can envisage that some suitable integrated measure of the size of the Weyl (conformal) curvature might give a measure of the entropy in the geometry. This could have a high value for the final collapse singularities—where the Weyl curvature is expected not only to diverge to infinity but also to dominate over the remaining Ricci (matter) part (Belinskii *et al.* 1970). The initial big bang, on the other hand (Zel'dovich 1973), seems to have been very uniform. This corresponds to the vanishing of Weyl curvature for the big bang, which is the case for the Friedmann-Robertson-Walker models, and which more or less serves to characterize their initial singularities.

Thus, one may make the hypothesis that initial singularities (singular TIFs) are for some reason[11] constrained to have zero Weyl curvature, whereas final singularities (singular TIPs) are not so constrained. The past and future boundary conditions for the universe must indeed be very different in structure from one another in order that the observed gross statistical time-asymmetry of our world can be attributed a cause.

References

Bardeen, J. M., Carter, B., and Hawking, S. W. 1973, *Comm. Math. Phys.*, **31**, 170.
Bekenstein, J. C. 1973, *Phys. Rev. D*, **7**, 2333.
———. 1974, *Phys. Rev. D*, **9**, 3292.
Belinskii, V. A., Khalatnikov, I. M., and Lifshitz, E. M. 1970, *Adv. in Phys.*, **19**, 525.
Bosshard, B. 1975, On the *b*-Boundary of the Closed Friedmann Model, preprint.
Carr, B. J. 1975, Caltech Orange Aid preprint No. 415.
Carter, B. 1971, *Phys. Rev. Letters*, **26**, 331.
Chandrasekhar, S. 1975, *Proc. Roy. Soc. London*, **A343**, 289.
Chandrasekhar, S., and Detweiler, S. 1976, *Proc. Roy. Soc. London*, **A345**, 145.
Curzon, H. 1924, *Proc. London Math. Soc.*, Vol. **23.**
Geroch, R. P. 1968, *J. Math. Phys.*, **9**, 450.
———. 1970, *J. Math. Phys.*, **11**, 437.
Geroch, R., Kronheimer, E. H., and Penrose, R. 1972, *Proc. Roy. Soc. London*, **A327**, 545.
Gibbons, G. W. 1972, *Comm. Math. Phys.*, **27**, 87.
Gibbons, G. W., and Hawking, S. W. 1975, preprint.
Grischuk, L. P. 1967, *Soviet Phys.-JETP*, **24**, 320.
Hagedorn, R. 1968, *Nuovo Cimento*, **56A**, 1027.
Hartle, J. B., and Wilkins, D. C. 1974, *Comm. Math. Phys.*, **38**, 47.
Hawking, S. W. 1971, *M.N.R.A.S.*, **152**, 75.

[11] Since the balance of what is important in local physics is presumably greatly different near a singularity from what it is elsewhere, there is the possibility that the interaction responsible for the time-symmetry of K^0 decay may be playing a role (Penrose 1969, 1976). I should emphasize, moreover, that no explanation can be found merely in particle creation effects (Misner 1969; Novikov and Zel'dovich 1973; Lukash and Starobinskii 1974) (say) which depend on a *time-symmetric* local physics, since all arguments would apply equally to the future singularities. Furthermore, it seems that TCP violation is required in addition to T violation.

————. 1972, *Comm. Math. Phys.*, **25**, 152.

————. 1975, Fundamental Breakdown of Physics (preprint).

Hawking, S. W., and Ellis, G. F. R. 1973, *The Large Scale Structure of Space-time* (Cambridge: Cambridge University Press).

Hawking, S. W., and Penrose, R. 1970, *Proc. Roy. Soc. London*, **A314**, 529.

Isham, C. J., Penrose, R., and Sciama, D. W. 1975, ed., *Quantum Gravity* (Oxford: Oxford University Press).

Israel, W. 1967, *Phys. Rev.*, **164**, 1776.

Jang, P. S. 1976, private communication to A. Ashtekar.

Lukash, V. N., and Starobinskii, A. A. 1974, *Soviet Phys.—JETP*, **39**, 742.

Misner, C. W. 1969, *Phys. Rev.*, **186**, 1319, 1328.

Novikov, I. D., and Zel'dovich, Ya. B. 1973, *Ann. Rev. Astr. Ap.*, **11**, 387.

Penrose, R. 1969, *Rev. Nuovo Cimento*, Ser. 1. Num. Spec., **1**, 252.

————. 1972, *Techniques of Differential Topology in Relativity* (Philadelphia: SIAM).

————. 1973, *Ann. N.Y. Acad. Sci.*, **224**, 125.

————. 1974a, in *Gravitational Radiation and Gravitational Collapse*, ed. C. DeWitt-Morett (Dordrecht: Reidel).

————. 1974b, in *Confrontation of Cosmological Theories with Observational Data*, ed. M. S. Longair (Dordrecht: Reidel).

————. 1975, *Proc. 1974 Erice Meeting*, ed. de Sabbata (in press).

————. 1976a, in *Proc. 1975 Marcel Grossman Conference*, ed. R. Ruffini (in press).

————. 1976b, to appear.

Press, W. H., and Teukolsky, S. A. 1973, *Ap. J.*, **185**, 635.

Regge, T., and Wheeler, J. A. 1957, *Phys. Rev.*, **108**, 1063.

Robinson, D. 1975, *Phys. Rev. Letters*, **34**, 905.

Schmidt, B. J. 1971, *J. Gen. Rel. and Grav.*, **1**, 269.

————. 1972, *Comm. Math. Phys.*, **29**, 49.

————. 1974, *Comm. Math. Phys.*, **36**, 73.

Seifert, H.-J. 1971, *J. Gen. Rel. and Grav.*, **1**, 247.

Simpson, M., and Penrose, R. 1973, *Int. J. Theoret. Phys.*, **7**, 183.

Steinmüller, B., King, A. R., and Lasota, J. P. 1975, *Phys. Letters*, **51A**, 191.

Stewart, J. M. 1975, *Proc. Roy. Soc. London*, **A344**, 65.

Szekeres, P. 1975, preprint.

Teukolsky, S. A. 1973, *Ap. J.*, **185**, 635.

Tomimatsu, A., and Sato, H. 1972, *Phys. Rev. Letters*, **29**, 1344.

Vishveshwara, C. V. 1970, *Phys. Rev.*, **D1**, 2870.

Witten, L. 1974, in *Gravitational Radiation and Gravitational Collapse*, ed. C. DeWitt-Morette (Dordrecht: Reidel).

Yodzis, P., Seifert, H.-J., and Muller zum Hagen, H. 1973, *Comm. Math. Phys.*, **34**, 135.

————. 1974, *Comm. Math. Phys.*, **37**, 29.

Zel'dovich, Ya. B. 1973, *Soviet Phys.—JETP*, **37**, 33.

10 The Positive-Mass Conjecture — Robert Geroch

Introduction

For some 10 years there has been in existence in general relativity a conjecture (Arnowitt *et al.* 1960*a;* Brill and Deser 1968; Brill *et al.* 1968) whose physical content is the following: An isolated gravitating system having nonnegative local mass density must have nonnegative total mass, measured gravitationally at spatial infinity.

This question is of interest not only because of its direct physical implications, but also because the answer may be expected to give some insight into the structure of general relativity itself. As an example of the former, suppose that the conjecture turned out to be false, so that one could construct, from ordinary matter, a system S^- whose total mass is negative. Connect this system by a long rod to another, ordinary, system S^+ with positive total mass. Then S^- would repel S^+ gravitationally while S^+ would attract S^-, with the consequence that the combined system would tend to "run away" in the direction of S^+. One would like to know whether or not general relativity admits such possibilities. As an example of the latter we show that the conjecture deals essentially with the regime in which general relativity is important—strong gravitational fields. Consider first a version of the conjecture in Newtonian gravitation, characterized by mass density μ and Newtonian potential φ satisfying the gravitational equation $\nabla^2\varphi = 4\pi G\mu$ in Euclidean space. Take "isolated gravitating system" to mean that μ has compact support (or, at least, goes to zero sufficiently rapidly at infinity) and that φ approaches zero at infinity. Take "local mass density nonnegative" to mean that $\mu \geq 0$. Finally, take "total mass, measured gravitationally at infinity" to be the number $M = (4\pi G)^{-1}\int_S \nabla\varphi \cdot dA$, where S is any sufficiently large two-sphere surrounding the source. With these interpretations, the conjecture is of course true, for M is just the volume integral of μ. Now, in order to improve the resemblance to general relativity, let the gravitational equation be modified to read $\nabla^2\varphi = 4\pi G[\mu - (8\pi Gc^2)^{-1}\nabla\varphi\cdot\nabla\varphi]$, i.e., include, as a source for the gravitational field, the

Robert Geroch is in the Departments of Mathematics and of Physics at the University of Chicago, Chicago, Illinois.

This research was supported in part by the National Science Foundation, under contract number GP-34721X1, and by the Sloan Foundation.

245

"mass density of the field itself." Even with this modification the conjecture remains true, as one sees by noting that this equation can be rewritten as $\nabla^2 \exp(\varphi/2c^2) = 2\pi Gc^{-2}\mu \exp(\varphi/2c^2)$. Thus, if the conjecture can be violated at all, this should occur only far from the Newtonian limit.

We first review briefly the supporting definitions and the mathematical statement of the conjecture itself. We then summarize a few of the known results (all positive). Finally, we introduce an alternative—and perhaps mathematically more tractable—formulation of part of the conjecture. Among other examples of this formulation, we apply it to obtain results to the effect that "nontrivial topology" and "wormholes connecting to other Universes" both serve to increase total energy.

The Conjecture

We first recall some definitions.

An *initial-data set* consists of a three-manifold T with positive-definite metric q_{ab} and symmetric tensor field p_{ab} such that $\mu \geq |J^a J_a|^{1/2}$, where μ and J^a are defined by

$$\mu = \tfrac{1}{2}(R - p^{ab}p_{ab} + p^2),\tag{1}$$

$$J^a = \nabla_b(p^{ab} - pq^{ab}),\tag{2}$$

where $p = p^a{}_a$, and where R and ∇_a are the scalar curvature and derivative operator, respectively, of T, q_{ab}. The motivation for this definition is the following. Let T be a spacelike three-dimensional surface in a spacetime, and let q_{ab} and p_{ab} be the induced metric and extrinsic curvature of T, respectively. Then, from the constraint equations (e.g., Choquet-Bruhat 1962) of general relativity (i.e., from Einstein's equation), the μ and J^a defined above are the energy density and momentum density, respectively, of the matter, as measured by an observer whose four-velocity is normal to T. The condition $\mu \geq |J^a J_a|^{1/2}$ thus requires that the apparent energy-momentum of the matter be timelike.

An initial-data set T, q_{ab}, p_{ab} is said to be *asymptotically flat* (Arnowitt *et al.* 1960b, 1961a, b; Geroch 1972) if there exists on T a system of coordinates, x, y, z, each with range $-\infty$ to $+\infty$, in terms of which (i) the metric q takes the form

$$ds^2 = \left(1 + \frac{M}{2r}\right)^4 (dx^2 + dy^2 + dz^2) + O(r^{-2}),\tag{3}$$

where M is some constant and $r = (x^2 + y^2 + z^2)^{1/2}$, and (ii) the components of p_{ab} are of order r^{-2}. Note in particular that asymptotic flatness implies $T = \mathbb{R}^3$ and the absence of singularities in T. The number (Arnowitt *et al.* 1960b, 1961a, b; Geroch 1972) M (which is independent of the choice of coordinates) is called the *mass* of the initial-data set. These definitions are motivated by the observation that the spatial Schwarzschild metric can be written asymptotically in the form (3), and that, so written, the number M is precisely the usual

Schwarzschild mass. More generally, an examination of the behavior of test particles in the asymptotic regime leads to the interpretation of M as the total mass of the system, measured gravitationally at infinity.

A particular class of asymptotically flat initial-data sets can be constructed as follows. Let T be a spacelike three-surface in Minkowski space which approaches a three-plane sufficiently rapidly at infinity. Let q_{ab} and p_{ab} be the induced metric and extrinsic curvature, respectively, of this T. Then q_{ab} and p_{ab} satisfy

$$R_{ab} - p_a{}^m p_{bm} + p p_{ab} = 0 , \tag{4}$$

$$\nabla_{[a} p_{b]c} = 0 . \tag{5}$$

It follows immediately that we have an initial-data set: In fact, contracting (4) and (5), and comparing with (1) and (2), we see that $\mu = 0$ and $J^a = 0$. An asymptotically flat initial-data set satisfying (4) and (5) will be said to be *flat*. Every flat initial-data set arises from the construction above, and each has $M = 0$.

We now state the conjecture. Take "gravitating system with nonnegative local mass density" to mean initial-data set, "isolated" to mean asymptotically flat, and "total mass" to mean the number M. Then

Conjecture 1. (Arnowitt *et al.* 1960; Brill and Deser 1968; Brill *et al.* 1968). *An asymptotically flat initial-data set has $M \geq 0$, with equality holding only if the data set is flat.*

Conjecture 1 is open. There are, however, several special cases for which it has been settled. Brill and Deser (1968; Brill *et al.* 1968) have shown that the conjecture is true to second order in perturbations from flat data (a result which is essentially the basis for the Newtonian calculation in the Introduction). This particular calculation is part of a more extensive argument which strongly suggests that the conjecture as a whole is true. Jang (1976) has shown that the conjecture is true in what is called the "pure kinetic case," i.e., when q_{ab} is assumed flat. Brill (1959) has shown that the conjecture is true with the additional assumptions that $p = 0$ and that the data respect an axial symmetry. Finally, it is known (Leibovitz and Israel 1970; Misner 1971; Jang 1976), although the proof is more difficult than one might have expected, that the conjecture is true in the spherically symmetric case.

Alternative Formulation

We now introduce a reformulation of a portion of Conjecture 1, and discuss some applications of this formulation. We begin with a few definitions and observations.

Let T be a compact, connected, three-dimensional manifold with positive-definite metric q_{ab}. We call T, q_{ab} *regular* if, given any point x of T, there exists a positive function φ_x on T satisfying

$$(\nabla^2 - \tfrac{1}{8}R)\varphi_x = -4\pi\delta_x , \tag{6}$$

where δ_x is the delta function at x. We note first that regularity is conformally invariant, i.e., that if T, q_{ab} is regular, then, for any positive function ω on T, so is T, $\tilde{q}_{ab} = \omega^4 q_{ab}$. Indeed, this is immediate from the conformal invariance of (6), setting $\tilde{\varphi}_x = \omega^{-1}(x)\ \omega^{-1}\varphi_x$. We note next that, if T, q_{ab} is regular, then, under some conformal transformation, the scalar curvature of T becomes strictly positive. Let ρ be any positive function on T, and set $\omega(y) = \int \rho(x)\varphi_x(y)\ dV_x$, so the function ω is also positive on T. Then the scalar curvature of the metric $\tilde{q}_{ab} = \omega^4 q_{ab}$ on T is $\tilde{R} = -8\omega^{-5}(\nabla^2 - \tfrac{1}{8}R)\omega = 32\pi\omega^{-5}\rho$, by (6). We further note that the converse of this last result—if q_{ab} on T is conformal to a metric with positive scalar curvature, then T, q_{ab} is regular—is also true, because, for $R > 0$, the operator on the left in (6) is negative-definite, whence (6) can be solved. Finally, we note that, for T, q_{ab} regular, the solution φ_x of (6) is unique. Given two such solutions, φ_x and φ'_x their difference satisfies $(\nabla^2 - \tfrac{1}{8}R)(\varphi_x - \varphi'_x) = 0$. Carrying out a conformal transformation, if necessary, we may suppose that R is positive. Then, multiplying this last equation by $(\varphi_x - \varphi'_x)$ and integrating over T, the left side is nonpositive, whence $\varphi_x = \varphi'_x$.

Let T, q_{ab} be regular, and let x be a point of T. Then near x we have $\varphi_x = 1/r + \tfrac{1}{2}M_x + O(r)$, where r denotes geodesic distance from x and where M_x is some number. Indeed, expanding in spherical harmonics one sees that such an expression would certainly hold if q_{ab} were flat in a neighborhood of x, whence, since the components of q_{ab} in a geodesic coordinate system centered at x differ only to order r^2 from flat components, this expression holds in general.

Now consider

Conjecture 2. *Let T, q_{ab} be regular with T the three-sphere S^3, and let x be a point of T. Then the number M_x is nonnegative, with equality holding only if q_{ab} is conformal to the metric three-sphere.*

We show that Conjecture 2 is a special case of Conjecture 1. Under the hypothesis of Conjecture 2, set $\hat{T} = T - x$ and $\hat{q}_{ab} = (\varphi_x)^4 q_{ab}$. Then $\hat{T}$ is the manifold $\mathbb{R}^3$ and $\hat{q}_{ab}$ is a positive-definite metric on $\hat{T}$. Equation (6), at points other than x, ensures that the scalar curvature $\hat{R}$ of $\hat{q}_{ab}$ vanishes. Thus, setting[1] $\hat{p}_{ab} = 0$, we have from (1) and (2) that $\mu = 0$ and $J^a = 0$ for $\hat{T}$, $\hat{q}_{ab}$, $\hat{p}_{ab}$, and so this is an initial-data set. In fact, this initial-data set is asymptotically flat: Let $\hat{x}$, $\hat{y}$, $\hat{z}$ be coordinates on $\hat{T}$ such that, setting $\hat{r} = (\hat{x}^2 + \hat{y}^2 + \hat{z}^2)^{1/2}$, the coordi-

---

[1] Conjecture 2 is in fact considerably less restrictive than these remarks might suggest. Setting $p = 0$ in (1) and (2), the condition $\mu \ge |J^a J_a|^{1/2}$ becomes simply $R \ge 0$, rather than the $R = 0$ we obtain from (6). This in turn would be reflected in (6) by replacing the " $=$ " by " $\le$." But the mass associated with a solution of equation (6) so modified is always greater than or equal to our M_x, so its positivity would be guaranteed by Conjecture 2. Thus, in effect, the only restriction from Conjecture 1 is $p = 0$.

nate system $x = \hat{x}/\hat{r}^2$, $y = \hat{y}/\hat{r}^2$, $z = \hat{z}/\hat{r}^2$ is a geodesic system in T centered at x, in some neighborhood of x. Then, inserting the expansion $\varphi_x = 1/r + \frac{1}{2}M_x + O(r)$ into $\hat{q}_{ab} = (\varphi_x)^4 q_{ab}$, and expressing the result in terms of $\hat{x}$, $\hat{y}$, $\hat{z}$, we obtain precisely (3) with $M = M_x$. In particular, $M_x \geq 0$ corresponds to $M \geq 0$. Finally, flatness of the initial-data set $\hat{T}$, $\hat{q}_{ab}$, $\hat{p}_{ab}$ becomes, using $\hat{p}_{ab} = 0$ and (4), the condition that the metric $\hat{q}_{ab}$ on $\hat{T}$ be flat. But flatness of $\hat{q}_{ab}$ is equivalent to conformal flatness of q_{ab}, and hence to q_{ab}'s being conformal to the metric three-sphere.

We see from the remarks above that the function φ_x satisfying (6), when used as a conformal factor, serves three purposes: (i) It changes the manifold from S^3 to $\mathbb{R}^3$ by "sending the point x to infinity." (ii) It produces an initial-data set on the resulting $\mathbb{R}^3$. (iii) It recovers asymptotic flatness. Regarding for a moment $\hat{q}_{ab}$ as a metric on T which is "singular at x," and formally computing its scalar curvature $\hat{R}$, we see that this $\hat{R}$ is concentrated at x. Thus, we can interpret regularity of T, q_{ab} as meaning that "it is possible, by a singular conformal transformation, to completely sweep the scalar curvature of T into any point x."

A possible advantage of Conjecture 2 over Conjecture 1 is that, whereas in Conjecture 1 one associates a single number M with an asymptotically flat initial-data set, in Conjecture 2 one obtains instead a number M_x for each point x with a nearly arbitrary T, q_{ab}. For T, q_{ab} regular, denote by M the *mass function*, the function on T whose value at x is M_x, so the value of the mass function at point x is the mass that would result from the initial-data set obtained by "sending x to infinity by a conformal factor." It follows immediately from the conformal behavior of φ_x that this mass function is conformally invariant in the following sense: If M is the mass function for T, q_{ab}, then $\tilde{M} = \omega^{-2}M$ is the mass function for T, $\tilde{q}_{ab} = \omega^4 q_{ab}$. Conjecture 2 states then that the mass function is strictly positive except in the case of a conformal three-sphere, when it is zero everywhere. It would perhaps be of interest to find some differential equation satisfied by the mass function. Such an equation would necessarily be conformally invariant;[2] possibly, it could lead to a proof of Conjecture 2. We also remark that one can compute from the mass function the number $\int_T M^3 \, dV$ for any regular Riemannian three-manifold. This number is a conformal invariant and, according to Conjecture 2, is positive for $T = S^3$ unless q_{ab} is conformal to a metric three-sphere.

We mention two special cases in which it might be feasible to settle Conjecture 2. The first arises from Brill's (1959) theorem, which we may state in the present formulation as follows: Conjecture 2 is true provided (i) T, q_{ab} admits a conformal Killing field with closed orbits and which vanishes on a circle C, and

---

[2] There are apparently rather few conformally invariant equations on a function of the conformal weight of the mass function. An example of one such is $M \, \nabla^2 M - \frac{1}{2}(\nabla_a M)(\nabla^a M) - \frac{1}{4}RM^2 + cM^4 = 0$, where c is any number.

(ii) the point x is on C. (Under these provisions, the result of "sending x to infinity" will be an axially symmetric initial-data set.) What about the case when provision (i) is retained, but (ii) dropped? (This would correspond to an initial-data set with a certain conformal symmetry.) Since the essence of Brill's proof is the use he makes of the symmetry, it is perhaps not overly optimistic to believe that the proof could be extended to this case. The second case is the following. Let T be the Lie group SU(2), so, as a manifold, T is the three-sphere. Choose any positive-definite metric q_{ab} at the identity of T, and let q_{ab} be the unique left-invariant metric on T which coincides with q_{ab} at the identity. Then, for a large class (but not all) of these q_{ab}'s, the resulting T, q_{ab} will be regular. In this regular case, the mass function M will also be left-invariant on T, hence constant. Thus, the number M must be some algebraic function of the original choice of a metric q_{ab} at the identity and the structure-constant tensor $C^a{}_{bc}$ of the Lie group. It would be of interest to compute this function. It is by no means inconceivable that the resulting formula will admit negative M, hence yield a counterexample to Conjecture 2, and hence to Conjecture 1.

Finally, we apply our formulation to obtain certain results suggesting that "the bending of spacetime into configurations of nontrivial topology requires the input of additional energy." First recall that the definition of asymptotic flatness above requires that T, as a manifold, be just Euclidean space $\mathbb{R}^3$. In order to admit other possibilities, one might modify this definition by requiring only that T, q_{ab} satisfy (3) outside some compact[3] set C (more precisely, that T minus compact C be identical to some asymptotically flat initial-data set minus a compact set). In this way, we would admit "wormholes in the initial-data set" (these now being included in C). We in fact wish to generalize still further, to admit "wormholes which connect to other asymptotic regimes." That is, instead of asymptotic flatness, we might impose weak asymptotic flatness: For some compact C in T, each connected component of $T - C$ satisfies (3). Each connected component, then, will represent a separate "asymptotic regime," and with each such regime we may associate a total mass M, via (3). For example, the natural initial-data set for the extended Schwarzschild metric is weakly asymptotically flat, with two asymptotic regimes. Both asymptotic regimes have the same total mass, namely, the m of the Schwarzschild solution. (We shall see shortly that it is false in general that different asymptotic regimes must have the same total mass.) We may now state the generalized energy conjecture: For a weakly asymptotically flat initial-data set, the total mass associated with each asymptotic regime is nonnegative, with equality holding for any one only if $T = \mathbb{R}^3$ and the data set is flat. This generalization, then, permits study of the effects of nontrivial topology on the total mass.

[3] It is necessary to require that C be compact in order to exclude "singularities in the initial data which might make a negative contribution to the total mass" such as would occur, for example, in the Schwarzschild solution with m negative.

The generalized conjecture becomes particularly simple in the present formulation. First, we drop the condition that the manifold T be a three-sphere. Further, we allow for "several points of T to be sent to infinity simultaneously" (each then yielding an asymptotic regime) as follows. Instead of (6), we may set

$$(\nabla^2 - \tfrac{1}{8}R)\varphi = -4\pi(\delta_{x_1} + \ldots + \delta_{x_n}) , \tag{7}$$

where $x_1, \ldots, x_n$ are n points of T (noting that regularity implies existence of a positive φ satisfying [7]). For each i, we expand φ near x_i, and obtain a corresponding mass M_i. Thus we have

Conjecture 3. *Let T, q_{ab} be regular, and let $x_1, \ldots, x_n$ be n points of T. Then the numbers M_i defined above are all nonnegative, with equality holding for any one only if $T = S^3$, $n = 1$, and q is conformal to the metric three-sphere.*

The only case in which Conjecture 3 has been settled, aside from those implied by the remarks below, is the following. Let S^2 denote the metric two-sphere, and let K be the circle with metric such that its circumference is a. Set $T = K \times S^2$, and let q_{ab} be the product metric. Then the scalar curvature is positive, and so this T, q_{ab} is regular. Since this space is homogeneous, the mass function is constant. It is not difficult to compute M in this case: The result is a positive, monotonically decreasing function of a.

We first consider the question of the sign of "topological energy," a question we may formulate as follows. Let the manifold T be the three-sphere, and let T' be T with some identifications (i.e., let T' have, as its universal covering space, T). Let metric q'_{ab} on T' be regular, and let q_{ab}, the corresponding metric on T, also be regular. Finally, let x' be any point of T' and let x be a corresponding point of T. Then the point x' of T', q'_{ab} yields some mass M', while x of T, q_{ab} yields mass M. Since, in some sense, "T' differs from T only in that the former has nontrivial topology," we are led to consider the difference, $M' - M$.

Theorem 4. *Under the setup above, $M' > M$.*

We postpone for a moment the proof.

To consider the question of the sign of "the energy of several asymptotic regimes," we proceed as follows. Let T, q_{ab}, and $x_1, \ldots, x_n$ be as in Conjecture 3, with $n > 1$, and set $M' = M_1$. Alternatively, we could retain the same T, q_{ab}, but instead admit only the single asymptotic regime $x = x_1$. We then obtain some mass M. Since "M' differs from M only in that the former is the mass in the presence of additional asymptotic regimes," we again consider $M' - M$.

Theorem 5. *Under the setup above, $M' > M$.*

Theorem 5 is easy to prove. The solution of (7) is given in terms of the solutions of (6) by $\varphi = \varphi_{x_1} + \ldots + \varphi_{x_n}$. Hence, $M' = M + \varphi_{x_2}(x_1) + \ldots + \varphi_{x_n}(x_1) > M$. In particular, Conjecture 3 for $n = 1$ implies Conjecture 3. Theorem 4 is an immediate corollary.

It might be of interest to know whether any similar results are available without the restriction, apparently required by the present formalism, of[4] $p_{ab} = 0$. Of course, one would also like to know whether these various conjectures are true or false.

References

Arnowitt, R., Deser, S., and Misner, C. 1960a, *Ann. Phys.*, **11**, 116.
———. 1960b, *Phys. Rev.*, **118**, 1100.
———. 1961a, *Phys. Rev.*, **121**, 1556.
———. 1961b, *Phys. Rev.*, **122**, 997.
Brill, D. 1959, *Ann. Phys.*, **7**, 466.
Brill, D., and Deser, S. 1968, *Ann. Phys.*, **50**, 548.
Brill, D., Deser, S., and Faddev, L. 1968, *Phys. Letters*, **26A**, 538.
Choquet-Bruhat, Y. 1962, in *Gravitation*, ed. L. Witten (New York: Wiley).
Geroch, R. 1972, *J. Math. Phys.*, **13**, 956.
———. 1973, *Ann. NY Acad. Sci.*, **224**, 108.
Jang, P. S. 1976, *J. Math. Phys.*, **1**, 141.
Leibovitz, C., and Israel, W. 1970, *Phys. Rev.*, **1D**, 3226.
Misner, C. 1971, in *Astrophysics and General Relativity*, ed. M. Chretien, S. Deser, and J. Goldstein (New York: Gordon & Breach).

[4] See n. 1.

Contributors

George Contopoulos
Astronomy Department
University of Athens
Athens, Greece

Robert Geroch
Department of Physics
and Department of Mathe-
matics
University of Chicago
Chicago, Illinois 60637

Paul Ledoux
Institut d'Astrophysique
Université de Liège
Cointe-Sclessin (Liège),
Belgium

Leon Mestel
Astronomy Centre
University of Sussex
Brighton BN1 9QH, England

T. W. Mullikin
Division of Mathematical
Science
Purdue University
West Lafayette, Indiana
47907

Jeremiah P. Ostriker
Princeton University
Observatory
Peyton Hall
Princeton, New Jersey 08540

Roger Penrose
Mathematical Institute
Oxford University
Oxford OX1 3LB, England

Martin Schwarzschild
Princeton University
Observatory
Peyton Hall
Princeton, New Jersey 08540

Kip S. Thorne
W. K. Kellogg Radiation
Laboratory
California Institute of
Technology
Pasadena, California 91125

L. Woltjer
European Southern
Observatory
CERN
CH-1211 Geneva 23
Switzerland

Index